Single-particle Cryo-EM of Biological Macromolecules

About the Series

The Biophysical Society and IOP Publishing have forged a new publishing partnership in biophysics, bringing the world-leading expertise and domain knowledge of the Biophysical Society into the rapidly developing IOP ebooks program.

The program publishes textbooks, monographs, reviews, and handbooks covering all areas of biophysics research, applications, education, methods, computational tools, and techniques. Subjects of the collection will include: bioenergetics; bioengineering; biological fluorescence; biopolymers *in vivo*; cryo-electron microscopy; exocytosis and endocytosis; intrinsically disordered proteins; mechanobiology; membrane biophysics; membrane structure and assembly; molecular biophysics; motility and cytoskeleton; nanoscale biophysics; and permeation and transport.

Single-particle Cryo-EM of Biological Macromolecules

Edited by
Robert M Glaeser
Department of Molecular and Cell Biology, University of California, Berkeley, CA 94720, USA
Lawrence Berkeley National Lab, University of California, Berkeley, CA 94720, USA

Eva Nogales
QB3, UC Berkeley, CA 94720, USA
Lawrence Berkeley National Lab, University of California, Berkeley, CA 94720, USA

Wah Chiu
Department of Bioengineering, Stanford University, Stanford, CA 94305, USA
Division of Cryo-EM and Bioimaging, SSRL, SLAC National Accelerator Laboratory, Stanford University, Menlo Park, CA 94025, USA

IOP Publishing, Bristol, UK

ISBN 978-0-7503-3039-8 (ebook)
ISBN 978-0-7503-3037-4 (print)
ISBN 978-0-7503-3040-4 (myPrint)
ISBN 978-0-7503-3038-1 (mobi)

DOI 10.1088/978-0-7503-3039-8

Version: 20210501

IOP ebooks

British Library Cataloguing-in-Publication Data: A catalogue record for this book is available from the British Library.

Published by IOP Publishing, wholly owned by The Institute of Physics, London

IOP Publishing, Temple Circus, Temple Way, Bristol, BS1 6HG, UK

US Office: IOP Publishing, Inc., 190 North Independence Mall West, Suite 601, Philadelphia, PA 19106, USA

Cover image credits

Top left: EMD: 22657 Adapted from Zhang K *et al* 2020 Resolving individual atoms of protein complex by cryo-electron microscopy *Cell Res.* 30 1136–9, CC BY 4.0.

Top middle: EMD:21452 This figure was published in Walls A C et al 2020 Structure, function, and antigenicity of the SARS-CoV-2 spike glycoprotein *Cell* 183 1735, copyright Elsevier (2020).

Top right: EMD: 22586 Data from Watson Z L *et al* 2020 Structure of the bacterial ribosome at 2 Å resolution *eLife* 9 e60482.

Bottom left: EMD: 22311/22322 From Guo H *et al* 2021 Structure of mycobacterial ATP synthase bound to the tuberculosis drug bedaquiline *Nature* 589 143–7. Reprinted by permission from Springer Nature, copyright (2021).

Bottom middle: EMD: 6347 From Zhang R *et al* 2015 Mechanistic origin of microtubule dynamic instability and its modulation by EB Proteins *Cell* 162 849–59. Copyright (2015), reprinted with permission from Elsevier.

Bottom right: EMD: 20755 Adapted from Zhang K *et al* 2019 Cryo-EM structure of a 40 kDa SAM-IV riboswitch RNA at 3.7 Å resolution *Nat. Commun.* 10 5511, CC BY 4.0.

The Editors wish to dedicate this book in memory of our colleague, Ken Downing, who surely would have partnered with us in this project had he not passed away before it was started, after a 10-year battle with cancer. Much of what we all learned from him, as mentor, peer and friend, spanning a period of more than 40 years, is now perpetuated in so many pages throughout this book. With his pure love for research, Ken inspired many other scientists in the cryo-EM field and beyond.

December 23, 1945 - August 2, 2018

Photo and dates taken from https://www.legacy.com/obituaries/sfgate/obituary.aspx?n=kenneth-downing&pid=189964785

Contents

Editor biographies

Robert M Glaeser

Robert M Glaeser is an Emeritus Professor of molecular and cell biology at UC Berkeley and an Affiliate in the Molecular Biophysics and Integrated Bioimaging Division at the Lawrence Berkeley National Laboratory. His main interests in cryo-EM continue to be methods development, currently focusing on more reliable preparation of cryo-EM grids and collaborating on the development of a ponderomotive phase plate. He and his wife, Rachel, particularly enjoy taking extended visits with friends and colleagues in England, Germany, Japan, and China, and relaxing together in the Point Reyes National Seashore area north of San Francisco.

Eva Nogales

Born in Spain, Eva Nogales studied physics at the Universidad Autónoma de Madrid and carried out her PhD work at the Synchrotron Radiation Source in the UK. She moved to the USA for her postdoctoral studies with Ken Downing at the Lawrence Berkeley National Laboratory (LBNL) where she was part of the team that first solved the structure of tubulin. Eva is currently a Professor in the Department of Molecular Cell Biology at UC Berkeley, a Senior Faculty Scientist at LBNL, and a Howard Hughes Medical Institute Investigator. Her lab uses cryo-EM to visualize large macromolecular assemblies with the goal of understanding their structure, dynamics, and regulatory interactions. Eva is married to Howard Padmore, also a physicist working at LBNL, and they have two children. She loves traveling and going back to her home country, and particularly enjoys dancing, skiing, reading, and writing.

Wah Chiu

Wah Chiu is the Wallenberg-Bienenstock Professor in the Department of Bioengineering, the Department of Microbiology and Immunology and the SLAC National Accelerator Laboratory at Stanford University, CA. Formerly, he spent three decades at Baylor College of Medicine and is a founder of the Keck Center for Quantitative Biomedical Sciences with multiple academic institutions in the Greater Houston Area. His work has made multiple transformational contributions in developing single particle electron cryo-microscopy (cryo-EM) as a tool for the structural determination of molecular machines towards atomic resolution. He has collaborated broadly with researchers around the globe on the cryo-EM of viruses, chaperonins, ion channels, RNA–protein complexes, and RNAs. He currently directs the NIH-designated Stanford-SLAC Cryo-EM Center, which is accessible to the global community for high resolution data collection and conducts training on cryo-EM methodology for new investigators.

Section authors

Paul D Adams
Molecular Biosciences and Integrated Bioimaging, Lawrence Berkeley National Laboratory, Berkeley, CA 94720, USA
Department of Bioengineering, University of California at Berkeley, Berkeley, CA 94720, USA

Pavel V Afonine
Molecular Biosciences and Integrated Bioimaging, Lawrence Berkeley National Laboratory, Berkeley, CA 94720, USA

Xiao-chen Bai
Department of Biophysics, University of Texas Southwestern Medical Center, Dallas, TX, USA

Matthew L Baker
Department of Biochemistry and Molecular Biology, Structural Biology Imaging Center, McGovern Medical School at The University of Texas Health Science Center at Houston, 6431 Fannin Street, Houston, TX 77030, USA

Anchi Cheng
Simons Electron Microscopy Center, National Resource for Automated Molecular Microscopy, New York Structural Biology Center, 89 Convent Street, New York, NY 10027, USA

Yifan Cheng
HHMI/Department of Biochemistry and Biophysics, UCSF, San Francisco, CA, USA

Wah Chiu
Department of Bioengineering, James H Clark Center, Stanford University, Stanford, CA 94305, USA
Division of Cryo-EM and Bioimaging, SSRL, SLAC National Accelerator Laboratory, Stanford University, Menlo Park, CA 94025, USA

Radostin Danev
Graduate School of Medicine, The University of Tokyo, 113-0033 Tokyo, Japan

Robert M Glaeser
Department of Molecular and Cell Biology, University of California, Berkeley, CA 94720, USA
Molecular Biophysics and Integrated Bioimaging Division, Lawrence Berkeley National Lab, University of California, Berkeley, CA 94720, USA

Timothy Grant
John W and Jeanne M Rowe Center for Research in Virology, Morgridge Institute for Research, Madison, WI, USA
Department of Biochemistry, University of Wisconsin-Madison, Madison, WI 53715, USA

Basil Greber
Division of Structural Biology, The Institute of Cancer Research, London SW7 3RP, UK

Nikolaus Grigorieff
RNA Therapeutics Institute and Howard Hughes Medical Institute, University of Massachusetts Medical School, Worcester, MA, USA

Patricia Grob
Howard Hughes Medical Institute, University of California, Berkeley, CA 94720, USA
California Institute for Quantitative Biosciences, University of California, Berkeley, CA 94720, USA

J Bernhard Heymann
Laboratory of Structural Biology Research, NIAMS, NIH, Bethesda, MD 20892, USA

Benjamin A Himes
RNA Therapeutics Institute and Howard Hughes Medical Institute, University of Massachusetts Medical School, Worcester, MA, USA

Corey F Hryc
Department of Biochemistry and Molecular Biology, Structural Biology Imaging Center, McGovern Medical School at The University of Texas Health Science Center at Houston, 6431 Fannin Street, Houston, TX 77030, USA

Brian P Hudson
Institute for Quantitative Biomedicine, Research Collaboratory for Structural Bioinformatics, Rutgers, The State University of New Jersey, Piscataway, NJ 08854, USA

Agnel Praveen Joseph
Scientific Computing Department, Science and Technology Facilities Council, Research Complex at Harwell, Didcot OX11 0FA, UK

Catherine L Lawson
Institute for Quantitative Biomedicine, Research Collaboratory for Structural Bioinformatics, Rutgers, The State University of New Jersey, Piscataway, NJ 08854, USA

Dorothee Liebschner
Molecular Biosciences and Integrated Bioimaging, Lawrence Berkeley National Laboratory, Berkeley, CA 94720, USA

Steven Ludtke
Verna and Mars McLean Department of Biochemistry and Molecular Biology, Baylor College of Medicine, Houston, TX, USA

Dmitry Lyumkis
The Salk Institute for Biological Studies, 10010 North Torrey Pines Road, La Jolla, CA 92037, USA
Department of Integrative Structural and Computational Biology, The Scripps Research Institute, La Jolla, CA 92037, USA

Sony Malhotra
Institute of Structural and Molecular Biology, Department of Biological Sciences, Birkbeck College, University of London, Malet Street, London WC1E 7HX, UK

Takanori Nakane
MRC Laboratory of Molecular Biology, Cambridge CB2 0QH, UK

Eva Nogales
Molecular and Cell Biology, QB3, and Howard Hughes Medical Institute, UC Berkeley, CA 94720, USA
Molecular Biophysics and Integrated Bioimaging Division, Lawrence Berkeley National Laboratory, Berkeley, CA 94720, USA

Melanie Ohi
Department of Cell and Developmental Biology, and Life Sciences Institute, University of Michigan, 210 Washtenaw Ave., Ann Arbor, MI 48109, USA

Grigore Pintilie
Department of Bioengineering, James H Clark Center, Stanford University, Stanford, CA 94305, USA

Jane S Richardson
Department of Biochemistry, Duke University Medical Center, Durham, NC 27710, USA

Alexis Rohou
Department of Structural Biology, Genentech, 1 DNA Way, South San Francisco, CA 94080, USA

Peter B Rosenthal
Structural Biology of Cells and Viruses Laboratory, Francis Crick Institute, London NW1 1AT, UK

John L Rubinstein
The Hospital for Sick Children, Toronto, ON, Canada
The University of Toronto, Toronto, ON, Canada

Sjors H W Scheres
MRC Laboratory of Molecular Biology, Cambridge CB2 0QH, UK

Dimitry Tegunov
Max Planck Institute for Biophysical Chemistry, 37077 Göttingen, Germany

Thomas C Terwilliger
Bioscience Division, Los Alamos National Laboratory, Los Alamos, NM 87545, USA
New Mexico Consortium, Los Alamos, NM 87545, USA

Rebecca F Thompson
School of Molecular and Cellular Biology, Faculty of Biological Sciences, University of Leeds, Leeds, UK

Maya Topf
Institute of Structural and Molecular Biology, Department of Biological Sciences, Birkbeck College, University of London, Malet Street, London WC1E 7HX, UK

Jasenko Zivanov
MRC Laboratory of Molecular Biology, Cambridge CB2 0QH, UK
Current address: Laboratory of Biological Electron Microscopy, EPFL, Lausanne, Switzerland

IOP Publishing

Robert M Glaeser, Eva Nogales and Wah Chiu

Chapter 1

Introduction and overview

1.1 Visualizing biological molecules to understand life's principles

Eva Nogales
QB3, UC Berkeley, CA 94720, USA

A structural description of a biological macromolecule is broadly considered to be an important step toward a detailed mechanistic understanding of how the macromolecule works, how mutations in its sequence give rise to disease, or how to design or improve small molecules that target it for therapeutic applications. Scientists have at their disposal only a handful of biophysical methods that allow an atomic characterization of macromolecular structure. A quick look at the protein data bank will immediately identify x-ray crystallography as the technique that has generated the largest number of high-resolution structures reported to date (please note that I will only refer to scattering-based methods in this brief historical account and nuclear magnetic resonance (NMR), which can provide the structures of small macromolecules as well as unique kinetic data and information on molecular interactions, will not be discussed).

1.1.1 A brief historical perspective on scattering-based structural biology methods

The advent of constantly improving x-ray crystallography beamlines at synchrotron sources around the world has facilitated the study of ever smaller crystals as well as crystals with very large unit cells. As the scientific community gained general access to these high-quality x-ray sources, with faster and more user-friendly data collection procedures and highly matured software tools, the power of the methodology seemed unstoppable. However, while depositions to the protein data bank have kept on growing, certain targets have remained dramatically underrepresented. Membrane proteins, given the difficulty of their overexpression and the additional challenge of dealing with their solubilization in detergents, have posed a particular challenge to the crystallization procedure. Large molecular assemblies are both more difficult to overproduce due to their size and generally more difficult to crystallize, often due to their flexibility. Unstable assemblies that rely on more fleeting interactions, and thus exist as a compositional mixture, are very challenging crystallization targets, as are biological polymers. These critical outliers demonstrate that an alternative to crystallization and x-ray diffraction is badly needed. Since as early as the studies of DNA fibers using x-ray crystallography, electron microscopy has been used as such an alternative, but only recently has it become a fully vested structural technique to systematically provide atomic resolution (or nearly so) information on biological molecules.

The use of electron microscopy to visualize biological molecules is indeed as old as the use of x-ray based methods. However, 3D reconstruction methodology and the aim of an atomic description from electron microscopy (EM) images lagged behind by a couple of decades due to the known limitations of this imaging method. Compared to the application of transmission electron microscopy to the study of inorganic materials, biological samples exist and work in aqueous solutions that are incompatible with the high vacuum of the electron microscope column, and they suffer from the unavoidable problem of radiation damage. Additionally, proteins are made of light atoms and thus EM images of unstained biological samples have low contrast. An early strategy to deal with these challenges was the use of negative stain in the preparation of biological samples for EM visualization, a method that many still use during the early stages of a project (see section 2.2). However, the negative stain may compromise the integrity of the sample and is, by its nature, limited in resolution (due to the size of the stain grain).

The alternative, with the potential to describe a biological sample in its native state(s) at high resolution, is cryo-electron microscopy (cryo-EM). The demonstration that cryo-preservation of protein samples leads to high-resolution data, within the confines of low electron doses, proved the potential of EM as a structural biology technique (Taylor and Glaeser 1974). Once simple methods of vitrification were described (Dubochet *et al* 1982) (see section 2.3) and cryo-holders became commercially available, the cryo-EM field took off. However, for several decades the resolution attained was far from that allowing atomic modeling, except for a handful of samples that were either crystallized in 2D (Henderson *et al* 1990, Kühlbrandt and Wang 1991, Nogales *et al* 1998, Murata *et al* 2000), helical in nature (Miyazawa *et al* 1999, Samatey *et al* 2001), or contained icosahedral symmetry (Liu *et al* 2010, Liang *et al* 2015). In these cases, the low signal-to-noise ratio in the cryo-EM images could be overcome by the 'redundancy' provided by geometrical constraints (2D crystals had the benefit of allowing the use of electron diffraction, but required tilting of the specimen, a large practical bottleneck). The study of samples without high symmetry, referred to as 'single particles', involved more complex computation (Frank 2002) and the resulting reconstructions were rarely at subnanometer resolution (Halic *et al* 2004, Clare *et al* 2012). In such cases, the use of a hybrid methodology, in which the atomic structures of pieces from NMR or crystallography are 'docked' into the lower-resolution map of a larger ensemble, is very helpful to describe its overall architecture.

A dramatic change finally took place with the advent of improved electron detectors, which provide images with higher contrast, as well as the higher resolution that result from the beam-induced motion correction that they enable (Li *et al* 2013).

Together with new computational approaches for data analysis (e.g. RELION (Scheres 2016) and CRYOSPARC (Punjani *et al* 2017)), these improvements now result in density maps of sufficient resolution to generate *ab initio* atomic models. As a result the protein data bank is growing in those areas that had challenged crystallography for many years. What follow are examples of distinct classes of samples that are uniquely suited to study by cryo-EM and otherwise defy structural characterization. Please note that the particular examples chosen, and in particular the specific citations, are a very small, subjective, and in some cases even arbitrary sampling of the myriad of excellent cryo-EM studies that have been published recently, in particular during the last five years.

1.1.2 Unique capabilities of cryo-EM: polymers and viruses

Polymers in the form of Fi helical assemblies, while totally resistant to crystallization, have been a preferred type of sample for cryo-EM studies. From cytoskeletal polymers to the bacterial flagella, cryo-EM has contributed structural descriptions of these samples for years (Rayment *et al* 1993, Fujii *et al* 2010). These samples have also benefited from new hardware and software improvements, in particular those incorporating iterative helical real-space principles (Egelman 2007), and today their structures are being described systematically at or near-atomic resolution. These studies include filamentous actin or microtubules most often bound by one of their many interacting factors (von der Ecken *et al* 2016, Kellogg *et al* 2018, Atherton *et al* 2019, Ma *et al* 2019).

Similarly, icosahedral viruses are difficult to crystallize into well-ordered crystals and their large unit cells are very demanding on x-ray diffraction hardware. Yet their symmetry, structural resilience, and limited flexibility make them close to ideal for cryo-EM studies (figure 1.1) (Sirohi *et al* 2016, Sugita *et al* 2018) (very large viruses pose a unique challenge at high resolution, as the curvature of the Ewald sphere, described in section 4.8.5, needs to be taken into account). The special properties of

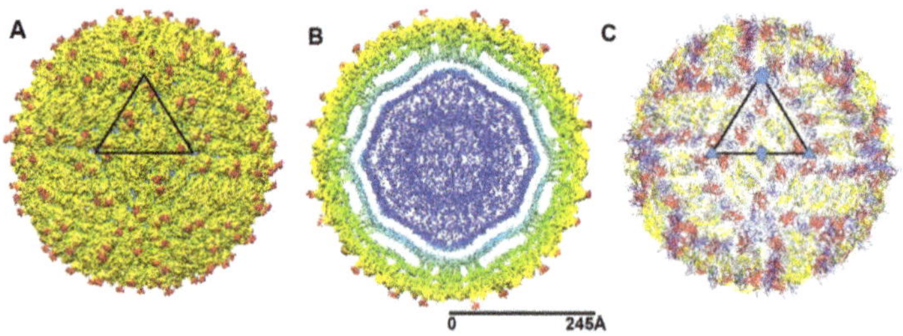

Figure 1.1. The cryo-EM structure of the Zika virus. (A) A surface-shaded, depth-cued cryo-EM density map of Zika with the asymmetric unit indicated by the black triangle. (B) Cross-section showing the radial density distribution. (C) Atomic model of the E and M proteins (color code: E protein domains I (red), domain II (yellow), and domain III (blue)). Redrawn from figure 1.1 of Sirohi *et al* (2016) with permission of AAAS. Readers may view, browse, and/or download material for temporary copying purposesonly, provided these uses are for noncommercial personal purposes. Except as provided by law, this material may not be further reproduced, distributed, transmitted, modified, adapted, performed, displayed, published, or sold in whole or in part, without priorwritten permission from the publisher.

viruses also mean that they are very high-throughput targets for cryo-EM. In general, given the unique challenges of viruses for x-ray studies, the resolution obtainable by cryo-EM is now typically higher than that obtained by crystallographic analysis. Furthermore, through symmetry expansion, cryo-EM can also be used to describe unique, non-icosahedral features within viruses of great biological significance (Liu *et al* 2019). It is safe to say that today it will be difficult to find someone trying to crystallize a virus and almost as difficult to find many pursuing fiber diffraction studies of aligned helical structures.

1.1.3 Unique capabilities of cryo-EM: integral membrane proteins

Structural biology of integral membrane proteins is an area of research that has dramatically advanced with the advent of modern cryo-EM. The publication of the first structure of the TRPV channel took the scientific community by storm (Liao *et al* 2013). Today it is difficult to find a major journal covering molecular biology that does not include the structure of an ion channel, a transporter, a membrane signal receptor, or a large respiratory complex, to cite a few (Fan *et al* 2020, Li *et al* 2020, Staus *et al* 2020). Channels with intrinsic symmetry (typically four-fold) are particularly amenable to high-resolution description by cryo-EM.

The advantages of cryo-EM over other structural approaches in the study of integral membrane proteins have to do with the requirement for a smaller amount of sample (overexpressing and purifying integral membrane proteins is particularly challenging) and with the fact that the formation of crystal contacts is hindered by the detergent surrounding the molecule. Purified channel samples that were resting in freezers after scientists had given up on their crystallization efforts are now seeing new light (electrons!) thanks to cryo-EM. Furthermore, many of the structures being solved today, and the trend will likely increase, have been reconstituted into nanodiscs or other lipid-mimicking nanostructures that better correspond to the native state of these proteins in a lipid bilayer membrane, adding to the functional value of the structural information (figure 1.2) (Liao *et al* 2013, Yin *et al* 2020).

Many of the samples in this category also have tremendous therapeutic potential. Their structural characterization, with and without small molecules affecting their

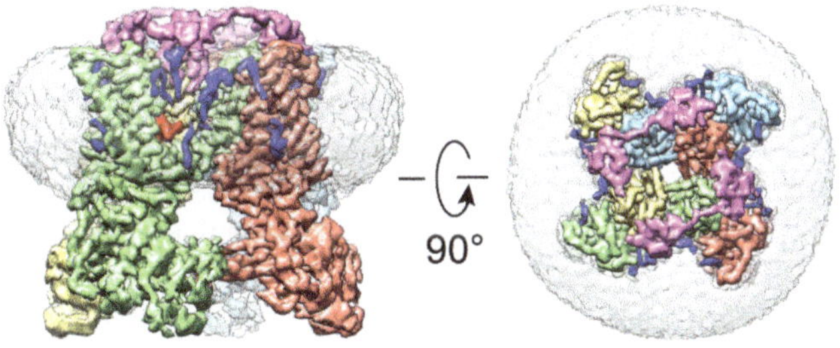

Figure 1.2. Cryo-EM density map for the TRPV1 channel in a nanodisc. Individual channel subunits are colored green, yellow, cyan, and orange, with two molecules of the double-knot toxin shown in purple, and with the resiniferatoxin agonist shown in red bound to the vanilloid-binding pocket. The densities of the nanodisc are shown in gray and ordered lipids in dark blue. Reproduced from figure 1.1 of Liao *et al* (2013) by permission from Springer Nature, copyright 2013.

function, is thus of great interest to both academic and pharmaceutical researchers (Hua *et al* 2020). Natural and man-made substrates, agonists, and antagonists are being used to describe different functional states and to learn the fundamental molecular mechanisms of whole families of challenging proteins, which previously had remained a black box.

1.1.4 Unique capabilities of cryo-EM: large assemblies

Large macromolecular assemblies often pose problems for x-ray crystallography. Even if the sample can be produced in large amounts (see below for when that is not the case) and even if it crystallizes, the crystals tend to be more poorly ordered than for small molecules, the unit cells are large, and the phasing is non-trivial. An obvious example of success for this type of sample is the ribosome. But many other 'somes' have remained impossible to solved via crystallization. Large complexes are not just amenable for cryo-EM characterization, but, for quite a while, they were almost the only samples that had enough mass (typically >1 MDa) to generate images with enough signal for alignment and classification accurate enough to lead to structures at subnanometer resolution. While the size constraints have been reduced significantly with the advent of new detectors, samples of large size continue to be major targets for cryo-EM. As a consequence, an increasing number of large macromolecular assemblies (>5 and up to 50 polypeptides) are now being characterized by this methodology, including many of the 'somes' critical to many cellular processes, such as those defining the central dogma of molecular biology. An obvious success case over the last two decades, which paved the way for the study of other large complexes, is the ribosome bound by its many partners during the translation cycle (Jobe *et al* 2019). Some large complexes, however, pose additional challenges due to their biochemical complexity, scarcity, and/or flexibility, as described below.

1.1.5 Unique capabilities of cryo-EM: scarce samples

A general type of sample that defies structural characterization by x-ray crystallography (or NMR) is that which cannot be produced in large amounts. As indicated above, some integral membrane proteins fall into that category. Most typically, this type of sample corresponds to macromolecular complexes that are difficult to overexpress in heterologous systems and/or need to be purified from endogenous sources and exist naturally in very small amounts (Nguyen *et al* 2018).

The sample amounts needed for cryo-EM are orders of magnitude lower than those needed for crystallization trials, thus opening possibilities for structural analyses that were previously impossible. These include the bacterial and eukaryotic replication (Yuan *et al* 2017, Gao *et al* 2019) and transcriptional machinery (Schilbach *et al* 2017, Boyaci *et al* 2019), or chromatin modifying complexes (Han *et al* 2020, He *et al* 2020), to cite just a few that have recently seen dramatic advances thanks to cryo-EM. Because it is even possible to concentrate the sample on the EM grid, either in a non-specific or in a specific way, to bring the number of particles per image to the realm of what is practical, samples at concentrations in the tens of nanomolars regime are now being studied (Greber *et al* 2019). There are now numerous examples of samples that

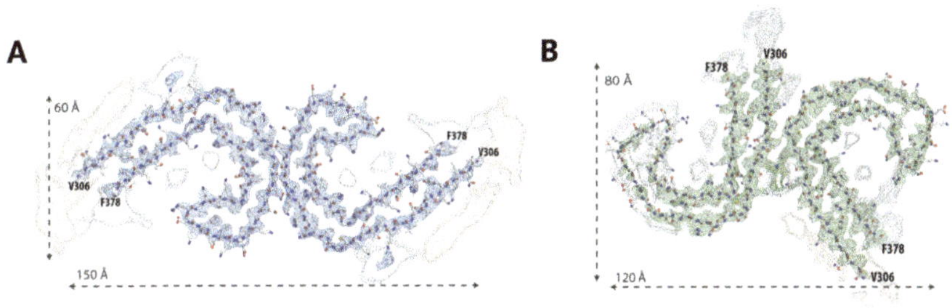

Figure 1.3. Structures of tau filaments extracted from diseased brains. Cross-section of the cryo-EM density of paired helical filaments (a) and straight filaments (b) and corresponding atomic models. Regions not modeled are shown in gray (and orange for (b)). Redrawn from figure 1.2 in Fitzpatrick *et al* (2017) by permission from Springer Nature, copyright 2017.

have been genetically tagged, typically in budding yeast via tap-tagging (but CRISPR technology can now extend application to any cell origin and type), and following a simple affinity-based purification scheme have been or are being analyzed by cryo-EM. Very excitingly, samples that are purified from human biopsies, such as those of tau fibrillary tangles from Alzheimer's patients, are now being described in atomic detail by cryo-EM (Fitzpatrick *et al* 2017, Zhang *et al* 2020) (figure 1.3).

1.1.6 Unique capabilities of cryo-EM: compositionally heterogeneous samples

A very powerful feature of cryo-EM is its potential to deal with heterogeneous samples. While heterogenous samples require more effort, larger datasets, scientific ingenuity, a non-trivial computational analysis pipeline, and may be more limited in resolution, they come with their unique perk—functional information that relates to the behavior of the molecular assembly. While some of the large assemblies we referred to above as the 'somes' tend to be stable and long-lived (the ribosome being an example and the spliceosome a notorious exception), many macromolecular assemblies rely on weak and/or reversible interactions to function and certain states in a biological process may be difficult or impossible to trap.

An early example of the power of cryo-EM for these types of studies was that of a translation-initiation state, which biochemically exists in a very small proportion as part of an equilibrium. Following the needed image-analysis pipeline, where rounds of nested classification are systematically carried out, species that were present in as low as 3% of the population were identified and described (Fernandez *et al* 2013). Indeed, cryo-EM has the potential to be able to describe this equilibrium, but some remaining experimental hurdles will first have to be solved to make such a description truly quantitative (see below). Among the more recent examples is that of different functional states of the notoriously challenging spliceosome (Zhang *et al* 2018).

A real upside of the capability of cryo-EM to deal with mixtures is that the requirement for sample purification can be relaxed. This could be of value not simply to save time/resources, but also to be able to describe several molecular species at once, perhaps obtained through a biochemical scheme that selects for a particular activity out of a complex cellular extract. In a more extreme implementation, some are exploring the possibility of studying truly complicated mixtures of mildly separated cellular fractions (Ho *et al* 2020).

1.1.7 Unique capabilities of cryo-EM: conformationally complex samples

Outside the constraints of a crystal lattice, macromolecules are free to experience the range of conformations that are energetically accessible to them for any given set of conditions. The motions connecting these states can involve small or large regions and can themselves be very limited or very large in range. From the point of view of the structural biologist, the nature of the motions can often preclude crystallization of the macromolecule altogether. Alternatively, crystallization may occur via selection of one or a few of the states. While such a lucky event may lead to high resolution information, it is not possible to know whether the state described is the state prevalent in the solution.

In principle, through the rapid vitrification of the sample while in the desired biochemical conditions, cryo-EM will provide images of each molecule trapped in the state it was sampling when it was frozen. In statistical terms, the population of particles at equilibrium should describe the population of states that one molecule will experience in time (a caveat to this statement, as for the description of the equilibrium of a compositional mixture, will be described later). This means that cryo-EM could, again in principle, be able to describe the different conformational states, if only they can be detected in images of single particles, and if the corresponding images can be grouped into separate classes for subsequent 3D reconstructions.

The problem is that, while compositional heterogeneity is, by nature, quantized, conformational heterogeneity can include (and often does) continuous motions that are much more challenging to deal with via classification schemes. Because the resolution ultimately depends on the average of many images of the molecule, if images correspond to non-identical molecules, the reconstruction will appear 'blurred', that is, it will be limited in resolution. The extent of the blurring depends on the nature of the flexibility, and goes beyond the motion of a small area, because that small area may also affect the alignment process of the full structure.

An approach often used concerns 'masking' the smaller, flexible region so that it no longer contributes to the alignment process. This process will result in better alignment and higher-resolution reconstruction for the remaining core segment. When there are multiple segments, however, all moving with respect to each other, the best strategy will be to classify apart different conformations, which has the negative effect of resulting in fewer particle images contributing to any given reconstruction. Typically, a mixture of both procedures will be needed.

Certain software implementations now allow the alignment of regions within a structure using experimental images in which the region masked out is actually subtracted from the image (rather than simply erased from the running reference used for alignment). It is even possible to carry out 'multi-body' refinement of the different regions in the structure and to gather information on the main types of motion for the different parts. Still, regions of the structure may often remain poorly

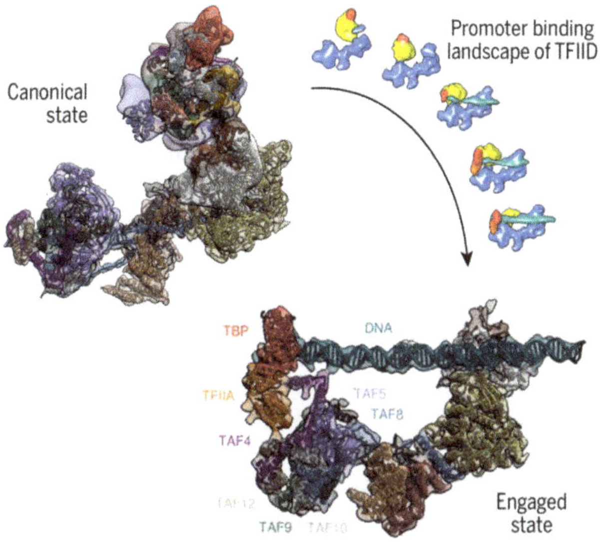

Figure 1.4. Structure and dynamics of human transcription factor II D (TFIID). The structure of apo-TFIID is shown in the so-called canonical state (top left) and that of promoter- and TFIIA-bound TFIID is depicted in the so-called engaged state (bottom right). Five states of TFIID were observed in the process of promoter binding, leading to a mechanistic model of TBP loading onto the promoter DNA (top right) (Patel *et al* 2018). Reprinted with permission from AAAS.

resolved. Such a problem is easily visualized in local-resolution maps, which can be seen as marking regions where residual motion remains, most often in the periphery. In spite of these challenges, the process will not only allow the structural description of a macromolecular assembly that would otherwise remain refractory for structural characterization, but actually often describes the range of motions experienced by the molecule that are part of its functional repertoire. Recent image-analysis algorithms now explore different principles to analyze the data assuming a continuum of states and thus hold promise to overcome the most challenging conformational heterogeneity (Punjani and Fleet 2020).

In some cases, the issues of particle size, scarcity, compositional heterogeneity, and flexibility that is of a complex and continuous nature, will all have to be faced at once (figure 1.4) (Patel *et al* 2018). Such samples, not so uncommon, are challenging even with today's hardware and software advantages, and the resolution attainable remains limited. But even in these cases it is possible to study what otherwise would have been considered quite inconceivable before and to obtain information on dynamics that provides unique insight into the biological function of the macromolecular complex.

1.1.8 Current limits of cryo-EM and things yet to come

The applicability, throughput, and resolution of cryo-EM have grown remarkably over the last decade. However, a number of bottlenecks remain that, if overcome, will release the true potential of this structural methodology.

Sample preparation is still a black art, not unlike crystallization. The preparation of thin layers of vitrified sample in which the molecules are well preserved is far from being a controlled and reproducible process. With present experimental methods, each sample requires its own optimization. Most likely a work-around rather than a true optimum is achieved in the best of cases, and then only some of the time. The work-arounds involve playing with substrates for sample deposition, surfactants, and crosslinkers, among other tricks (see section 2.5). In many cases the situation is indeed so suboptimal that up to 90% of the particle images are discarded as 'bad', probably because they did not withstand the consequences of diffusing to and touching the air–water interface before vitrification. It is the discarding of so much of the image data that limits our capacity to draw quantitative conclusions on the equilibrium distribution of the sample from the states of the sample we obtained, unless we were to assume that all states present in the solution suffer equally from these destructive forces.

The resolution limitations, which were described for the challenging samples that are dominated by continuous motions, reside in the current need to quantize states during classification. If individual segments are to be treated separately during alignment and refinement, there is a limit to the size that can be analyzed before the signal runs out to improve alignment accuracy and the final resolution of the reconstruction.

1.2 Recovery of 3D structures from images of weak-phase objects

Robert M Glaeser

Lawrence Berkeley National Lab, University of California, Berkeley, CA 94720, USA

Images that are produced by single-particle cryo-EM are usually described as representing two-dimensional (2D) projections of the three-dimensional (3D) structures of biological macromolecules. Dental x-ray images or an x-ray image of a fractured bone are also examples, familiar to most of us, of 2D projections of 3D objects.

Thin biological materials that have not been 'stained' with heavy atoms are virtually transparent to high-energy electrons. As a result, information about the projected structure is contained primarily in the spatial variations of the phase of the electron wave, and far less in variations in its intensity, immediately after it has been transmitted through the object.

The fact that the information we want is contained primarily in the phase of the transmitted electron wave is the reason why it is necessary to defocus the image before recording its intensity. This is because defocusing generates 'phase contrast', just as is true for light microscopy of unstained biological specimens (Zernike 1955). The basic ideas behind defocus-based phase contrast are initially introduced in this section, but more is added in sections 3.4, 4.3, and 4.8.

This section also introduces the basic concepts of how the 3D structure of an object can be recovered (reconstructed) from many different 2D projections, which then are elaborated more fully in chapter 4. A qualifying comment is that such 2D images must be recorded using many different orientations of the object relative to the incident electron beam. In other words, many 'independent' projections are needed in order to recover most, if not all, of the information in the third dimension.

In practice, data must actually be merged from images of tens of thousands of identical macromolecules, which is far more than would be required if the images were not as noisy as they are. The very low signal-to-noise ratio in cryo-EM images is an unavoidable consequence of the limited extent to which biological macromolecules tolerate being exposed to the electron beam. Further coverage of this topic is left to section 3.2.

1.2.1 The signal that we care about is attributed to elastic scattering of electrons

Incident electrons may randomly undergo either elastic or inelastic scattering events as they pass through a cryo-EM object. Other than the fact that the ratio of elastic to inelastic scattering does depend slightly upon the energy of the incident electron (Peet *et al* 2019), the relative cross sections for the two kinds of scattering is not something that can be varied by the user.

Inelastic scattering events are actually about three times more likely to occur than elastic scattering events for specimens made up of light atoms, e.g. carbon, nitrogen, and oxygen (Henderson 1995). Parenthetically, the mean-free-path for inelastic scattering in vitrified ice is estimated to be ~200 nm for 120 keV electrons (Grimm *et al* 1996) and ~400 nm for 300 keV electrons (Rice *et al* 2018). Even though inelastic scattering occurs three times more frequently than elastic scattering, the inelastically scattered electrons contribute nothing to the useful signal. Instead, the little signal that they do contribute is considered to be unwanted as far as cryo-EM images are concerned. It is thus beneficial to use an energy-filter lens to remove the inelastically scattered electrons (Yonekura *et al* 2006), although the benefit becomes marginal if the sample thickness is less than 50 nm and the electron energy is high, e.g. 300 keV. The contribution made by the inelastically scattered electron is thus ignored in the following.

1.2.2 The electron accumulates information as it passes through a specimen

Electrons can be described as waves and image formation in cryo-EM cannot be understood without referring to the phase of the electron wave—see, for example, chapter 3 of Glaeser *et al* (2007). Summarizing briefly, it is advantageous to use well-collimated (parallel) illumination, in which case the incident electron wave, shown in figure 1.5, is a plane wave. To explain what is meant by a 'plane wave', both the amplitude and the phase of the wave are constant over any given plane that is normal to the direction of propagation.

As the electron wave passes through the specimen, however, its phase at each point becomes modified to a different extent by interactions with the Coulomb potential of the atoms in its path. The spatially modulated wave front finally leaves the object, at which point it is referred to as the exit wave $\Psi_{\text{exit wave}}(x, y)$. Provided that the sample is thin enough, the exit wave can be represented as the product of the incident plane wave, e^{-ik_0z}, and an object-dependent transmittance function, $T(x, y)$:

$$\Psi_{\text{exit wave}}(x, y) = T(x, y)e^{-ik_0z}, \tag{1.1}$$

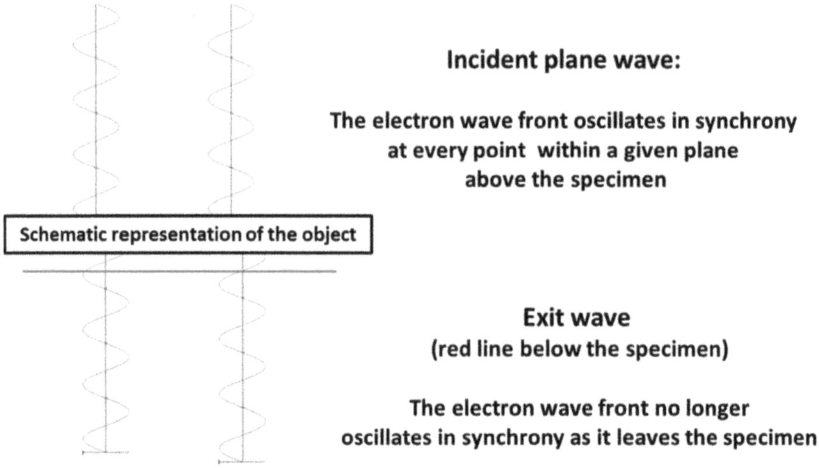

Figure 1.5. A diagram to illustrate the idea that the amplitude and the phase of an incident plane wave may be constant across the face of an object, but, after the wave has passed through the object, the phase may vary from point to point across the exit wave.

where (x, y) are the coordinates of a given point in the plane of the exit wave, $k_0 = \frac{2\pi}{\lambda}$ is the magnitude of the wave vector of the electron, λ is the electron wavelength, and z is the direction of propagation of the electron.

Small modulations also occur in the amplitude of the exit wave. These amplitude modulations result primarily from spatial variations in inelastic scattering, although other factors, which are not discussed here, also make further, somewhat smaller contributions. To emphasize, what is being referred to are spatial modulations in the amplitude of the wave function for electrons that have *not* been inelastically scattered. Together, these amplitude-modulation effects do reduce the number of electrons that remain in the image, but they generally produce only a small contribution to the image contrast.

As indicated above, phase modulations in the exit wave result from Coulombic interactions between the incident electron and atoms in the specimen. Although the Coulomb potentials of the positively charge atomic nuclei of neutral atoms are shielded by their surrounding electron clouds, once the incident electron penetrates into this cloud it accelerates as it approaches the nucleus, and it decelerates again to its original speed as it passes by. During this process, the electron wavelength transiently becomes shorter, before returning to its initial value. The effect is similar to that of light entering and leaving a refractive medium, except that the speed of light is reduced, not increased, inside a refractive medium.

The combined effect on the transmittance function of the object is represented as the product of phase-modulation and amplitude-modulation terms:

$$T(x, y) = e^{-i\frac{2\pi}{hv}\int eV(x,\, y,\, z)\mathrm{d}z}\mathrm{e}^{-\frac{\mu(x,\, y)}{\Lambda}}, \tag{1.2}$$

where i is the imaginary number $\sqrt{-1}$, h is Planck's constant, v is the speed of the electron, e is the electron charge, $V(x, y, z)$ is the shielded Coulomb potential, $\mu(x, y)$ describes the effective absorption of electrons due to inelastic scattering and other

Box 1.1. Relationship between amplitude contrast and phase contrast

As mentioned in section 1.2.2, the amount of amplitude modulation of the transmittance function in equation (1.3) is usually approximated as being a fixed fraction of the phase modulation. Recognizing, however, that inelastic scattering is the main factor contributing to amplitude modulation, a first-order improvement would be to account for the fact that inelastic scattering can occur even when the incident electron passes an atom at some distance. In other words, inelastic scattering is known to be 'delocalized' by as much as 1 nm or more (Egerton 2017). A better approximation might thus be to consider the amplitude-modulation term in equation (1.3) to be the convolution product of a spatially extended point-spread function, which accounts for this delocalization, and the shielded Coulomb potential, describing the phase modulation. Although that is surely a crude approximation as well, it would be better than to say that the point-spread function is a Dirac delta function, i.e. that the amplitude and phase modulations are related by a constant factor.

The consequences expected from the convolution model are that amplitude contrast should play a diminishing role at higher resolution, rather than continuing to be a constant proportion of phase contrast at all resolutions. As mentioned in section 1.2.2, however, the improvement afforded by accounting for a diminishing role of amplitude contrast at higher resolution may not be very significant, since the effect of amplitude contrast is small to begin with.

factors, and Λ is the mean-free-path for all events that effectively result in absorption of electrons.

When the arguments of the exponential functions in equation (1.2) are small enough to ignore higher order terms when expanding the exponential functions, the transmittance function of the object is well approximated by the following linear expansion:

$$T(x, y) = 1 - i\frac{2\pi}{hv}\int eV(x, y, z)\mathrm{d}z - \frac{\mu(x, y)}{\Lambda}. \tag{1.3}$$

When equation (1.3) is valid, the specimen is said to be a mixed, weak-phase, weak-amplitude object. As is further emphasized in section 1.2.4, even the image intensity—and not just the exit wave function—is linear in the projected structure in this favorable case. Note that the imaginary part of the transmittance function in equation (1.3) corresponds to a weak phase modulation and the real part corresponds to a weak amplitude modulation of what previously was a plane wave. The amplitude-modulation term is often approximated as being a fixed amount, between 0.07 and 0.1 times the phase-modulation term. Although that is expected to be a rather poor approximation (see box 1.1 for more on that point), a more accurate approximation is hardly needed in practice.

1.2.3 The image wave function, and thus the image intensity, suffers from imperfections in the microscope optics

The image wave function can be thought of as being a magnified (enlarged) version of the exit wave, but there is much more to it than that. The full story is too lengthy to develop here, but more can be found in many places, including Goodman (1968) and chapter 3 of Glaeser *et al* (2007).

In short, the effect of the objective aperture, image defocus, and lens aberrations can be factored into (i) a microscope-specific 'coherent point-spread' function, $h(x, y)$, and (ii) an enlarged (magnified) version of the object-specific transmission function $T(x, y)$. The coherent point-spread function can be regarded as the image wave function for the case when the exit wave is a Dirac delta function, i.e. a single point. The image wave function is not just a simple scalar product of these two factors, however; instead, it is the convolution product. Readers who need to know more about the mathematical operation of convolution are again referred to chapter 3 of Glaeser *et al* (2007), among other sources.

If the objective aperture is large its effect on the image can be ignored, as is done in the following. However, the combined effects of defocus and spherical aberration play extremely important roles in cryo-EM. Their effect can be represented mathematically by multiplying the Fourier transform of the exit wave function, also known as the scattered (diffracted) wave function, by a wave-aberration function, $e^{i\gamma(s)}$. The wave-aberration function adds a resolution-dependent phase error to the scattered wave function, represented by

$$\gamma(s) = 2\pi\left[\frac{C_s}{4}\lambda^3 s^4 - \frac{\Delta Z}{2}\lambda s^2\right], \tag{1.4}$$

where ΔZ is the amount by which the image is defocused and the effect of spherical aberration is represented by the constant C_s. The meanings of two other symbols in equation (1.4) are as follows: the two-dimensional spatial frequency variable in the Fourier transform is represented by its magnitude, s, and the electron wavelength is again represented by λ.

What is measured experimentally, of course, is the image intensity, which is the square of the image wave function. In brief, the image wave function is the inverse Fourier transform of the product of the scattered electron wave and the wave-aberration function. The Fourier transform of this product produces the convolution product of the Fourier transforms of the scattered wave function and of the wave-aberration function, respectively, the latter being the coherent point-spread function, $h(x, y)$.

Rather than describing the square of this convolution product, it is more useful to instead pass directly to the simpler result obtained by computing the Fourier transform of the resulting image intensity. In order to simplify the mathematical notation, we begin by describing the case that the image intensity has been normalized to have an average value of 1.0. We further emphasize that the object must be thin enough that its transmittance function can be described in terms of projections of the shielded Coulomb potential and the effective absorption of electrons, as in equation (1.3). The resulting 2D Fourier transform of the image intensity, $\mathcal{F}\{I(x, y)\}$, is then given by

$$\mathcal{F}\{I\}(s_x, s_y) = \delta(s) - 2\boldsymbol{F}(s_x, s_y)\sin\gamma(s) - 2\mathcal{F}\{\mu\}(s_x, s_y)\cos\gamma(s), \tag{1.5}$$

where s is the magnitude of the spatial frequency variable in the Fourier transform and s_x, s_y denote its two orthogonal components, $\delta(s)$ is the Dirac delta function, $\boldsymbol{F}(s_x, s_y)$ is the 2D Fourier transform of the projection of the shielded Coulomb potential, often referred to as the structure factor of the object, $\sin\gamma(s)$ is called the 'phase-contrast transfer function', often shortened—perhaps inappropriately—to just 'the contrast transfer function (CTF)', $\mathcal{F}\{\mu\}$ can be understood as a second kind of structure factor of the object, but this time for the amplitude modulations in the exit wave, and $\cos\gamma(s)$ represents the transfer function for amplitude contrast. (Estimation of the CTF is covered in section 4.3, along with applying a CTF correction that accounts for both amplitude and phase contrast.) The structure factor, $\boldsymbol{F}(s_x, s_y)$, is shown in bold font to draw attention to the fact that it has real and imaginary parts and thus it is often represented as a vector in the complex plane.

The same theory of image formation sketched above can be used to describe a more perfect form of phase contrast, usually referred to as Zernike phase contrast. In the ideal case, the image would be exactly in focus, i.e. $\Delta Z = 0$, and there would be no spherical aberration. Instead, the wave-aberration function written in equation (1.4) would be replaced by a constant phase shift equal to $\frac{\pi}{2}$ at all spatial frequencies except when $s = 0$. Such a phase shift reverses the roles of the real and imaginary parts in equation (1.3), thus converting phase modulations into amplitude modulations. An optical device that applies a constant phase shift to the scattered wave, but not to the unscattered wave, or—equivalently—applies a phase shift to the unscattered wave but not to the scattered wave, is commonly referred to as a 'phase plate'. More will be said about this in section 3.4.

1.2.4 Intermediate summary: the image intensity is linear in the projected Coulomb potential of the object

A large number of points have been made above so it is worthwhile to summarize what they are and to elaborate upon them a bit before continuing.

Unstained biological macromolecules are usually described as being weak-phase objects as far as high-energy electrons are concerned and the 'weak-phase object' approximation is a very useful simplification to make. In particular, and to an excellent degree of accuracy, the weak-phase approximation informs us that the phase of the exit wave is linear in the projected Coulomb potential of the object. One should still keep in mind, however, that a more complete description of the exit wave, as was first written in equation (1.1), includes an additional contribution of weak amplitude modulation and this is included in equation (1.2).

At the same time, additional complexity is introduced in the form of a convolution operation when image defocus and other optical aberrations are accounted for. Fortunately, this convolution product is converted to the more easily understood scalar product when one calculates the Fourier transform of the image intensity. Computing the Fourier transform of the image intensity is, in any case, something that is normally done during data processing.

The most important point to summarize is that the image intensity remains linear in the projected Coulomb potential of the object, even though the image intensity is the square of the image wave function. Retaining this linear dependence may initially seem to be a logical contradiction, but the algebraic approximation,

$$(1 - \epsilon)^2 \cong 1 - 2\varepsilon, \tag{1.6}$$

when ϵ is small, immediately explains the condition under which linearity is preserved upon squaring the image wave function.

Before continuing, it is also worthwhile to draw attention to the fact that the word 'phase' is used in cryo-EM with three rather different meanings. We have so far used the word to describe the exit wave as electrons leave the specimen and we have used the word to describe the wave-front aberration associated with defocus and optical aberrations. In what follows we will use the same word again to refer to the ratio of imaginary and real parts of the calculated Fourier transform of the image intensity, more specifically to the arctangent of that ratio. Since the intended meaning of the word 'phase' is usually clear from the context, using the same word to mean different things causes no problems for those who are already familiar with this situation. Since that is not expected to be the case for most of the intended readers of this ebook, however, further exposition about the three possible meanings of the word 'phase' is provided in box 1.2.

1.2.5 Structure-factor phases, as well as amplitudes, are retained in the computed Fourier transforms of image intensities

Returning now to a further discussion of equation (1.5), recall that the Fourier transform of the image intensity is linear in the Fourier transform of the Coulomb potential of the biological macromolecule being imaged. Since the Coulomb potential plays the same role for electron scattering that is played by the electron charge density for x-rays, its Fourier transform is referred to as the (complex-valued) molecular structure factor, $\boldsymbol{F}(s_x, s_y)$, the same terminology that is used in x-ray crystallography.

Unlike the case of x-ray crystallography, however, the Fourier transform of a cryo-EM image is performed in a computer. As a result, both the real and the imaginary parts of the structure factor are retained. Said in another way, the phase of the structure factor is retained when one calculates the Fourier transforms of image, unlike the case when a diffraction pattern is recorded.

The amplitudes of structure factors are also retained in the computed Fourier transforms of images, of course. However, structure-factor amplitudes that are recovered from images are, in general, inherently less accurate than those obtained in x-ray crystallography. This is true, in part, because the signal obtained by merging data from a large number of individual molecules is necessarily noisier than the signal obtained from the even-larger number of molecules in a crystal. A second reason why image amplitudes are less accurate is that the CTF further reduces the already limited amount of signal that is present. (Section 4.3 presents more information on how the CTF is estimated and how images are then corrected for the systematic error that it produces.) A third reason is that the high-resolution amplitudes in Fourier transforms of images are systematically reduced by an overall 'envelope' function due to multiple factors that are absent in a diffraction pattern. (The causes for the envelope function and how a correction is made to compensate for it are described in section 4.7.)

Box 1.2. The word 'phase' is used in three completely different contexts in cryo-EM.

The word 'phase' is used in the description of (i) the exit wave formed as an electron passes through a thin specimen, (ii) the electron wave in the back focal plane of the objective lens, and (iii) the Fourier transform of the object. As is mentioned in section 1.2.4, the intended meaning of the word 'phase' is always clear from the context, for those already familiar with the subject. Since this degree of prior familiarity is not expected for readers of this ebook, however, further exposition about the three possible meanings of the word is provided here.

There is only a single meaning of the expression 'the phase of a sinusoidal function', of course. The word 'phase' refers to the offset between the origin of the coordinate system and the origin of the sinusoid. For example, a sine function is equal to zero if its argument is zero, and its values are positive for small, positive values of its argument. The origin of the sine function need not coincide with the origin of the chosen coordinate system, however, and the amount of offset (if any) is referred to as the phase of the sine wave. The same definition (meaning) applies in the three contexts discussed above, but it may require some explanation to clarify that it is the case.

In the case of the exit wave shown in figure 1.5, the phase of the wave front varies from point to point, after the electron has passed through the specimen. We thus say that there is a spatial modulation of the phase of the exit wave. This spatial modulation of phase is in contrast to the phase of the incident electron wave, which is everywhere the same at all points on a given plane when the electron first approaches the specimen.

In the case of the electron wave-front at the back focal plane, what had been a plane wave incident on the specimen is transformed by a perfect lens into a spherical wave converging to a point in the back focal plane. In other words, the phase is now constant everywhere on the surface of any given sphere whose center is located on the back focal plane. Lens aberrations, including defocus, can be modeled as causing a systematic phase shift (or phase error) with respect to the ideal case of a converging spherical wave.

The meaning of the word 'phase' in the context of the Fourier transform of an object is a bit more difficult to grasp. This is because the expressions 'the phase of the Fourier transform' or 'the phase of the structure factor' are used as jargon, which actually refers to the phase of one or another sinusoidal (i.e. Fourier) component of the object. In the case of cryo-EM images, the image intensity represents a projection of the Coulomb potential of the object, which, mathematically, can be represented as the sum of many sinusoidal components, each of which has its own amplitude and phase.

A further point of potential confusion arises when the real and imaginary parts of a particular value of the Fourier transform of an image is represented by a vector in the complex plane. The angle that this vector makes with the real axis in such an 'Argand diagram' is, in fact, the phase of the corresponding Fourier component of the image intensity. When one records a diffraction pattern, all information is lost about the ratio of the real and the imaginary parts of the diffracted wave function. In other words, phase information is lost when one measures diffraction intensities. The same information is retained, however, when one computes the Fourier transform of image intensities, provided that the phase (and amplitude) modulations in the exit wave are quite weak (see section 1.2.4).

The phases obtained from images, on the other hand, are generally much more accurate than those obtained in crystallography. To illustrate, the expected phase error for a given Fourier component is only 45 degrees as soon as just one scattered electron contributes the image at that spatial frequency (Henderson 1992), and the expected phase error rapidly decreases as more scattered electrons contribute. In crystallography, on the other hand, phase information is recovered only in an indirect way and thus phase determination is more prone to experimental error than it is in cryo-EM.

On balance, cryo-EM benefits tremendously from the fact that the map quality is more sensitive to having accurate phases than it is to having accurate amplitudes. As a result, the interpretability of 3D density maps is usually significantly better in cryo-EM than what it is in crystallography, for a given value of the resolution of the map.

1.2.6 The projection theorem: the Fourier transform of an image corresponds to a 2D 'central' section within the 3D Fourier transform of the object

As is summarized in equation (1.5) and in section 1.2.5, Fourier transforms of 2D images provide experimental measurements of both the amplitudes and the phases of the structure factor. These values are restricted to spatial frequencies that lie on a plane, however. Since this plane passes through the origin of 3D Fourier space, as is shown schematically in figure 1.6, it is commonly referred to as a central section. The

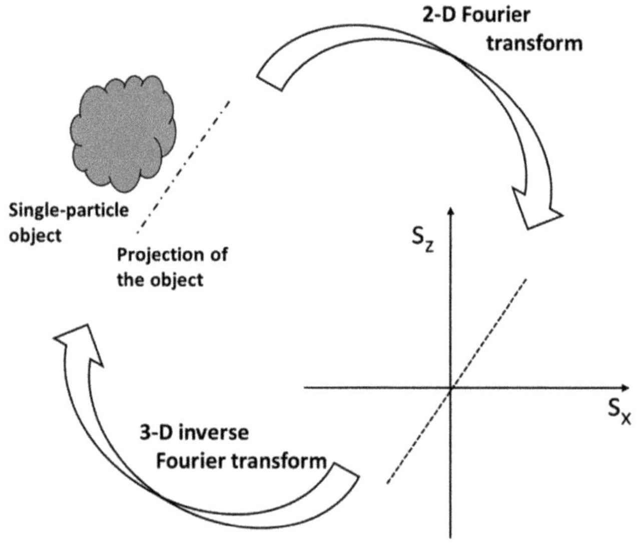

Figure 1.6. Schematic diagram showing the relationship between a 2D projection of a 3D object, which is in real space, and a 2D central section in Fourier space, which corresponds to the 2D Fourier transform of the 2D projection. While not shown here, but as can be imagined, each such central section corresponds to a different 2D projection, obtained in a different direction. Once data have been combined from enough, different projections, it is possible to then reconstruct the full 3D object in real space. As is indicated in this diagram, one way to do this is to compute a 3D inverse Fourier transform, using the 3D set of amplitudes and phases that had been assembled in Fourier space.

fact that the values of amplitude and phase are confined to a central section is stated here simply as a fact. However, a mathematical derivation—known at the projection theorem—can be found in many places, including section 4.3 of Glaeser *et al* (2007).

The usual statement that images of an object correspond to projections of its structure is only an approximation, however. In particular, this statement may no longer remain valid as the resolution of the image increases, or if the object is thick. The breakdown of the projection approximation can be understood as being due to the defocus value of the image no longer being the same for features at the top of a particle as it is at the bottom of a particle (Downing and Glaeser 2018) or, alternatively, in terms of curvature of the Ewald sphere (DeRosier 2000). Although such effects are ignored for now, correcting for the effect of curvature of the Ewald sphere, discussed in section 4.8 along with other important corrections, has recently become of practical interest.

1.2.7 The 3D object can be reconstructed from multiple projections

Although the information contained in a single image is often enormously informative, it still falls far short of what is needed to reconstruct the full, three-dimensional structure of an object. Consider, for example, that a 3D object consists of $N \times N \times N = N^3$ volume elements, i.e. 'voxels'. A single image of this object, on the other hand, contains no more than $N \times N = N^2$ picture elements, i.e. 'pixels'. As a result, the number of unknowns—the density values of each of the voxels—greatly exceeds the number of observations, and thus one image cannot possibly provide enough information to determine the 3D structure of an object.

On the other hand, images of an object that is held in many different orientations provide information that is distributed within multiple central sections. By combining all of that information it is possible to ultimately measure all N^3 values of the structure factor in three dimensions, as opposed to just the N^2 values that can be recovered from a single central section. Once a full set of data have been merged in this way, a 3D inverse Fourier transform can be used to reconstruct the original, 3D structure of the object, as is indicated in figure 1.6.

For example, if an object is tilted by a succession of angles about a single axis of rotation, as is shown in figure 1.7(A), one will generate a set of central sections that are also rotated by the same amount. If the tilt angle between successive images is small enough, the measurements will sample all points in the 3D Fourier transform, as is illustrated schematically in figure 1.7(B). The number of images that are required to completely sample all values of the structure factor is (Crowther *et al* 1970)

$$N = \pi \frac{D}{d}, \tag{1.7}$$

where d is the resolution to which one wants to collect data and D is the size of the object. Alternatively, the angular increment between successive images needs to be

$$\Delta\theta = \frac{d}{D}. \tag{1.8}$$

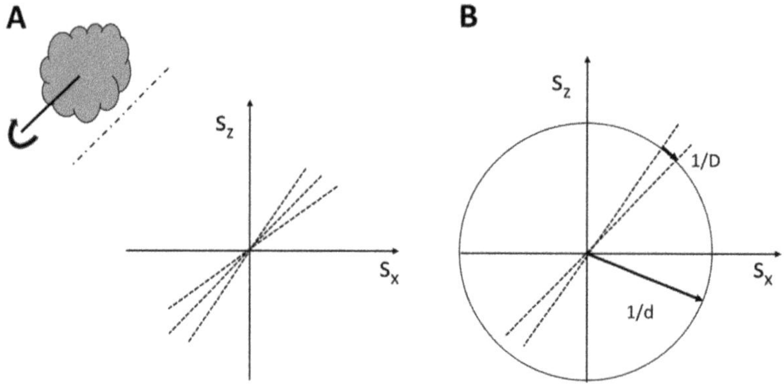

Figure 1.7. Key points involved in deriving the Crowther formula for the number of images needed to achieve a 3D reconstruction at a resolution *d*, see equation (1.7). (A) Only three central sections are shown out of a much larger set. It is assumed that the sample is rotated about just a single axis between recording successive projections and that the amount of rotation is always the same. (B) Only two central sections are shown here for simplicity, but it is imagined that there are *N* of them that are rotated by the same amount relative to one another. As is known from crystallography, the structure factor of an object only needs to be sampled at points that are uniformly spaced by a distance of $1/D$, where *D* is the size of a unit cell, in order to collect 100% of the information needed for a 3D reconstruction. This sampling will thus be complete, at a spatial frequency (i.e. resolution) $1/d$, if the angular interval between successive projections, and thus between their corresponding central sections, subtends an arc of $1/D$ at that resolution.

Single-axis electron tomography (Frank 1992) does, in fact, use repeated images of the same object that are recorded over a large range of tilt angles.

As is discussed in section 3.2, however, radiation damage requires that data, corresponding to different projection images, must be recorded from separate macromolecules if one hopes to obtain a high-resolution reconstruction. Merging data from a large number of different molecules is, in fact, an essential aspect of single-particle cryo-EM.

After data have been merged from a sufficient number of central sections, it is possible, in principle, to reconstruct a 3D density map of the object by simply computing a 3D inverse Fourier transform. In practice, however, the 3D reconstruction step is not as simple as that. In fact, a number of quite sophisticated algorithms and many different software packages are used to obtain the 3D reconstruction, as is summarized in section 4.4.

1.2.8 Similarities and differences between sub-tomogram averaging and single-particle cryo-EM

From a mathematical point of view, it is better to fractionate the exposure that is allowed within a fixed electron budget and use each aliquot to record a series of images as a tilt series (Hegerl and Hoppe 1976). Doing so recovers three-dimensional information that is otherwise lost if the allowed electron exposure is used to record a single projection. Not surprisingly, this result remained controversial after it first appeared (Saxberg and Saxton 1981, Hoppe and Hegerl 1981). Simulations (McEwen *et al* 1995) eventually confirmed that the result was indeed correct and resolved some of the ambiguities in wording that had led to earlier misunderstanding. As a result, other things being equal, sub-tomogram averaging would appear to be the preferred way to merge image data from single particles.

From a practical point of view, however, beam-induced specimen motion (doming) causes increasingly severe loss of resolution, perpendicular to the tilt axis, as the tilt angle increases. In addition, the thickness of material that the electron beam passes through often increases as the tilt angle increases, thereby reducing the image quality due to issues discussed in section 2.4. As a result, it is unlikely that tomographic data collection will generally replace the currently preferred method of using untilted specimens for single-particle data collection.

Nevertheless, there are special cases when sub-tomogram averaging is clearly the preferred way to obtain high resolution structures. This is certainly the case when one seeks to determine the structure of a macromolecular complex in its native, cellular environment (Zhang 2019, Lucic *et al* 2005). In addition, sub-tomogram averaging is perhaps the only option to recover the high-resolution structure of a complex that is present in a large number of copies, assembled in two or more layers, in which they have little or no predictable amount of translation and orientation relative to one another. Recovery of the high-resolution structure of the HIV envelope protein (Schur *et al* 2016) is one such example where sub-tomogram averaging was used to great advantage.

References

Atherton J *et al* 2019 Structural determinants of microtubule minus end preference in CAMSAP CKK domains *Nat. Commun.* **10** 5236

Boyaci H, Chen J, Jansen R, Darst S A and Campbell E A 2019 Structures of an RNA polymerase promoter melting intermediate elucidate DNA unwinding *Nature* **565** 382–5

Clare D K, Vasishtan D, Stagg S, Quispe J, Farr G W, Topf M, Horwich A L and Saibil H R 2012 ATP-triggered conformational changes delineate substrate-binding and -folding mechanics of the GroEL chaperonin *Cell* **149** 113–23

Crowther R A, Derosier D J and Klug A 1970 Reconstruction of 3 dimensional structure from projections and its application to electron microscopy *Proc. R. Soc. Lond.* A **317** 319–40

DeRosier D J 2000 Correction of high-resolution data for curvature of the Ewald sphere *Ultramicroscopy* **81** 83–98

Downing K H and Glaeser R M 2018 Estimating the effect of finite depth of field in single-particle cryo-EM *Ultramicroscopy* **184** 94–9

Dubochet J, Lepault J, Freeman R, Berriman J A and Homo J C 1982 Electron microscopy of frozen water and aqueous solutions *J. Microsc.* **128** 219–37

Egelman E H 2007 The iterative helical real space reconstruction method: surmounting the problems posed by real polymers *J. Struct. Biol.* **157** 83–94

Egerton R F 2017 Scattering delocalization and radiation damage in STEM-EELS *Ultramicroscopy* **180** 115–24

Fan C, Sukomon N, Flood E, Rheinberger J, Allen T W and Nimigean C M 2020 Ball-and-chain inactivation in a calcium-gated potassium channel *Nature* **580** 288–93

Fernandez I S, Bai X C, Hussain T, Kelley A C, Lorsch J R, Ramakrishnan V and Scheres S H W 2013 Molecular architecture of a eukaryotic translational initiation complex *Science* **342** 1240585

Fitzpatrick A W P, Falcon B, He S, Murzin A G, Murshudov G, Garringer H J, Crowther R A, Ghetti B, Goedert M and Scheres S H W 2017 Cryo-EM structures of tau filaments from Alzheimer's disease *Nature* **547** 185–90

Frank J 1992 *Electron Tomography: Three-dimensional Imaging with the Transmission Electron Microscope* (New York: Plenum)

Frank J 2002 Single-particle imaging of macromolecules by cryo-electron microscopy *Annu. Rev. Biophys. Biomol. Struct.* **31** 303–19

Fujii T, Iwane A H, Yanagida T and Namba K 2010 Direct visualization of secondary structures of F-actin by electron cryomicroscopy *Nature* **467** 724–9

Gao Y, Cui Y, Fox T, Lin S, Wang H, de Val N, Zhou Z H and Yang W 2019 Structures and operating principles of the replisome *Science* **363** eaav7003

Glaeser R M, Downing K, Derosier D, Chiu W and Frank J 2007 *Electron Crystallography of Biological Macromolecules* (Oxford: Oxford University Press)

Goodman J W 1968 *Introduction to Fourier Optics* (San Francisco, CA: McGraw-Hill)

Greber B J, Toso D B, Fang J and Nogales E 2019 The complete structure of the human TFIIH core complex *eLife* **8** e44771

Grimm R, Typke D, Barmann M and Baumeister W 1996 Determination of the inelastic mean free path in ice by examination of tilted vesicles and automated most probable loss imaging *Ultramicroscopy* **63** 169–79

Halic M, Becker T, Pool M R, Spahn C M, Grassucci R A, Frank J and Beckmann R 2004 Structure of the signal recognition particle interacting with the elongation-arrested ribosome *Nature* **427** 808–14

Han Y, Reyes A A, Malik S and He Y 2020 Cryo-EM structure of SWI/SNF complex bound to a nucleosome *Nature* **579** 452–5

He S, Wu Z, Tian Y, Yu Z, Yu J, Wang X, Li J, Liu B and Xu Y 2020 Structure of nucleosome-bound human BAF complex *Science* **367** 875–81

Hegerl R and Hoppe W 1976 Influence of electron noise on 3-dimensional image reconstruction *Z. Naturforsch.* A **31** 1717–21

Henderson R 1992 Image contrast in high-resolution electron microscopy of biological macromolecules: TMV in ice *Ultramicroscopy* **46** 1–18

Henderson R 1995 The potential and limitations of neutrons, electrons and x-rays for atomic-resolution microscopy of unstained biological molecules *Q. Rev. Biophys.* **28** 171–93

Henderson R, Baldwin J M, Ceska T A, Zemlin F, Beckman E and Downing K H 1990 Model for the structure of bacteriorhodopsin based on high-resolution electron cryo-microscopy *J. Mol. Biol.* **213** 899–929

Ho C M, Li X, Lai M, Terwilliger T C, Beck J R, Wohlschlegel J, Goldberg D E, Fitzpatrick A W P and Zhou Z H 2020 Bottom-up structural proteomics: cryoEM of protein complexes enriched from the cellular milieu *Nat. Methods* **17** 79–85

Hoppe W and Hegerl R 1981 Some remarks concerning the influence of electron noise on 3D reconstruction *Ultramicroscopy* **6** 205–6

Hua T *et al* 2020 Activation and signaling mechanism revealed by cannabinoid receptor-Gi complex structures *Cell* **180** 655–65

Jobe A, Liu Z, Gutierrez-Vargas C and Frank J 2019 New insights into ribosome structure and function *Cold Spring Harb. Perspect. Biol.* **11** a032615

Kellogg E H, Hejab N M A, Poepsel S, Downing K H, DiMaio F and Nogales E 2018 Near-atomic model of microtubule-tau interactions *Science* **360** 1242–6

Kühlbrandt W and Wang D N 1991 Three-dimensional structure of plant light-harvesting complex determined by electron crystallography *Nature* **350** 130–4

Li F *et al* 2020 Ion transport and regulation in a synaptic vesicle glutamate transporter *Science* **368** 893–7

Li X M, Mooney P, Zheng S, Booth C R, Braunfeld M B, Gubbens S, Agard D A and Cheng Y F 2013 Electron counting and beam-induced motion correction enable near-atomic-resolution single-particle cryo-EM *Nat. Methods* **10** 584–90

Liang B, Li Z, Jenni S, Rahmeh A A, Morin B M, Grant T, Grigorieff N, Harrison S C and Whelan S P J 2015 Structure of the L protein of vesicular stomatitis virus from electron cryomicroscopy *Cell* **162** 314–27

Liao M, Cao E, Julius D and Cheng Y 2013 Structure of the TRPV1 ion channel determined by electron cryo-microscopy *Nature* **504** 107–12

Liu H, Jin L, Koh S B, Atanasov I, Schein S, Wu L and Zhou Z H 2010 Atomic structure of human adenovirus by cryo-EM reveals interactions among protein networks *Science* **329** 1038–43

Liu Y T, Jih J, Dai X, Bi G Q and Zhou Z H 2019 Cryo-EM structures of herpes simplex virus type 1 portal vertex and packaged genome *Nature* **570** 257–61

Lucic V, Forster F and Baumeister W 2005 Structural studies by electron tomography: from cells to molecules *Annu. Rev. Biochem.* **74** 833–65

Ma M, Stoyanova M, Rademacher G, Dutcher S K, Brown A and Zhang R 2019 Structure of the decorated ciliary doublet microtubule *Cell* **179** 909–22 e912

McEwen B F, Downing K H and Glaeser R M 1995 The relevance of dose fractionation in tomography of radiation-sensitive specimens *Ultramicroscopy* **60** 357–73

Miyazawa A, Fujiyoshi Y, Stowell M and Unwin N 1999 Nicotinic acetylcholine receptor at 4.6 angstrom resolution: transverse tunnels in the channel wall *J. Mol. Biol.* **288** 765–86

Murata K, Mitsuoka K, Hirai T, Walz T, Agre P, Heymann J B, Engel A and Fujiyoshi Y 2000 Structural determinants of water permeation through aquaporin-1 *Nature* **407** 599–605

Nguyen T H D, Tam J, Wu R A, Greber B J, Toso D, Nogales E and Collins K 2018 Cryo-EM structure of substrate-bound human telomerase holoenzyme *Nature* **557** 190–5

Nogales E, Wolf S G and Downing K H 1998 Structure of the $\alpha\beta$ tubulin dimer by electron crystallography *Nature* **391** 199–203

Patel A B, Louder R K, Greber B J, Grunberg S, Luo J, Fang J, Liu Y, Ranish J, Hahn S and Nogales E 2018 Structure of human TFIID and mechanism of TBP loading onto promoter DNA *Science* **362** eaau8872

Peet M J, Henderson R and Russo C J 2019 The energy dependence of contrast and damage in electron cryomicroscopy of biological molecules *Ultramicroscopy* **203** 125–31

Punjani A and Fleet D J 2020 3D variability analysis: directly resolving continuous flexibility and discrete heterogeneity from single particle cryo-EM images *bioRxiv* https://doi.org/10.1101/2020.04.08.032466

Punjani A, Rubinstein J L, Fleet D J and Brubaker M A 2017 CryoSPARC: algorithms for rapid unsupervised cryo-EM structure determination *Nat. Methods* **14** 290–6

Rayment I, Holden H M, Whittaker M, Yohn C B, Lorenz M, Holmes K C and Milligan R A 1993 Structure of the actin–myosin complex and its implications for muscle contraction *Science* **261** 58–65

Rice W J, Cheng A C, Noble A J, Eng E T, Kim L Y, Carragher B and Potter C S 2018 Routine determination of ice thickness for cryo-EM grids *J. Struct. Biol.* **204** 38–44

Samatey F A, Imada K, Nagashima S, Vonderviszt F, Kumasaka T, Yamamoto M and Namba K 2001 Structure of the bacterial flagellar protofilament and implications for a switch for supercoiling *Nature* **410** 331–7

Saxberg B E H and Saxton W O 1981 Quantum noise in 2D projections and 3D reconstructions *Ultramicroscopy* **6** 85–90

Scheres S H 2016 Processing of structurally heterogeneous cryo-EM data in RELION *Methods Enzymol.* **579** 125–57

Schilbach S, Hantsche M, Tegunov D, Dienemann C, Wigge C, Urlaub H and Cramer P 2017 Structures of transcription pre-initiation complex with TFIIH and mediator *Nature* **551** 204–9

Schur F K M, Obr M, Hagen W J H, Wan W, Jakobi A J, Kirkpatrick J M, Sachse C, Krausslich H G and Briggs J A G 2016 An atomic model of HIV-1 capsid-SP1 reveals structures regulating assembly and maturation *Science* **353** 506–8

Sirohi D, Chen Z, Sun L, Klose T, Pierson T C, Rossmann M G and Kuhn R J 2016 The 3.8 Å resolution cryo-EM structure of Zika virus *Science* **352** 467–70

Staus D P, Hu H, Robertson M J, Kleinhenz A L W, Wingler L M, Capel W D, Latorraca N R, Lefkowitz R J and Skiniotis G 2020 Structure of the M2 muscarinic receptor-β-arrestin complex in a lipid nanodisc *Nature* **579** 297–302

Sugita Y, Matsunami H, Kawaoka Y, Noda T and Wolf M 2018 Cryo-EM structure of the Ebola virus nucleoprotein–RNA complex at 3.6 Å resolution *Nature* **563** 137–40

Taylor K A and Glaeser R M 1974 Electron diffraction of frozen, hydrated protein crystals *Science* **186** 1036–7

von der Ecken J, Heissler S M, Pathan-Chhatbar S, Manstein D J and Raunser S 2016 Cryo-EM structure of a human cytoplasmic actomyosin complex at near-atomic resolution *Nature* **534** 724–8

Yin J, Chen K M, Clark M J, Hijazi M, Kumari P, Bai X C, Sunahara R K, Barth P and Rosenbaum D M 2020 Structure of a D2 dopamine receptor-G-protein complex in a lipid membrane *Nature* **584** 125–9

Yonekura K, Braunfeld M B, Maki-Yonekura S and Agard D A 2006 Electron energy filtering significantly improves amplitude contrast of frozen-hydrated protein at 300 kV *J. Struct. Biol.* **156** 524–36

Yuan Z *et al* 2017 Structural basis of Mcm2–7 replicative helicase loading by ORC-Cdc6 and Cdt1 *Nat. Struct. Mol. Biol.* **24** 316–24

Zernike F 1955 How I discovered phase contrast *Science* **121** 345–9

Zhang P 2019 Advances in cryo-electron tomography and subtomogram averaging and classification *Curr. Opin. Struct. Biol.* **58** 249–58

Zhang W *et al* 2020 Novel tau filament fold in corticobasal degeneration *Nature* **580** 283–7

Zhang X, Yan C, Zhan X, Li L, Lei J and Shi Y 2018 Structure of the human activated spliceosome in three conformational states *Cell Res.* **28** 307–22

Chapter 2

Sample preparation

2.1 Overview

Following the broad overview that was presented in chapter 1, this chapter goes back to the very beginning and addresses what is involved in preparing grids for cryo-EM.

At the beginning of a single-particle EM project samples are often examined first in negative stain, a process described in section 2.2. Although negative staining is not always a reliable way to get a first look at the specimen (vesicles collapse as they are dried, for example), it is a quick, mature, and relatively safe way to prepare macromolecules for electron microscopy. Although it is a rather low-resolution method, using negatively stained samples at the beginning has four advantages: the samples can be examined at room temperature with minimal equipment required, they are relatively immune to radiation damage, the high contrast means that a small number of micrographs will be sufficient to provide initial class averages and even a preliminary 3D reconstruction, and macromolecular complexes are—perhaps surprisingly—less likely to be damaged when samples are prepared in negative stain.

Once it is established that the biochemical preparation looks promising enough, grids containing unstained samples are then prepared by the methods described in section 2.3. This is perhaps the first point at which grid preparation may become more challenging, even though many mature protocols do exist for preparing samples for cryo-EM, and vitrification often proves to be a reliable process. Nevertheless, the fact is that biological macromolecules are at risk when they diffuse and come into contact with the air–water interface (AWI), a hazard that they normally do not face in bulk solution. This is an issue because all samples that are used for high-resolution electron microscopy must be very thin. The reasons why that is so are explained in section 2.4. Fortunately, as section 2.5 describes, there are many ways to overcome the hazard associated with the AWI, although none of them is a universal solution. Furthermore, as section 2.5 explains, improving the current methods of grid preparation is now an active area of research, and grid preparation is bound to become less challenging as time goes on.

2.2 Initial screening of samples in negative stain

Patricia Grob
Howard Hughes Medical Institute, University of California, Berkeley, CA 94720, USA
Department of Molecular and Cell Biology, Howard Hughes Medical Institute, University of California, Berkeley, CA 94720, USA

2.2.1 Introduction

As explained in section 1.1, an exponentially increasing number of high-resolution cryo-EM structures is now deposited in the PDB every year (www.rscb.org). In this context does negative-stain electron microscopy (NS-EM), generally considered a low-resolution technique, still have a place? The answer is yes! In the increasingly competitive cryo-EM field it has become an integral part of high-throughput pipelines to design and prepare macromolecular complexes for high-resolution structure determination.

2.2.2 Negative staining for TEM

The negative-staining method for electron microscopy (NS-EM) was introduced in 1959 by Brenner and Horne for the study of viruses (Brenner and Horne 1959). It is a relatively simple and easy ensemble of preparation techniques, described in more detail in section 2.2.4, that makes the observation of biological materials by EM possible by (i) preserving lower resolution features such as the tertiary and quaternary structure of specimens against radiation damage, (ii) protecting these same features against dehydration artifacts, and (iii) improving contrast (Kellenberger and Kistler 1979, Harris 1997, Bremer *et al* 1992). Negative staining is achieved by what is essentially a buffer exchange, performed while the sample is on a TEM grid coated with a carbon support film, in which a heavy-atom salt solution replaces the original sample buffer. The excess liquid is subsequently blotted away and the remainder is allowed to dry, thus producing a solid film, thin enough to let electrons through. The heavy atoms form a shell encasing the biological molecules. The shell of heavy-atom salt scatters electrons much more strongly than does the biological molecule itself and it can sustain more electron irradiation, thus providing high contrast relative to the proteins and resulting in images that have a relatively high signal-to-noise ratio (SNR).

While this technique provides high contrast images, its main drawback comes from the fact that the signal originates from the heavy atoms in the stain surrounding the protein, thus producing a 'negative' image of the shape of the macromolecules. It is further limited in resolution by how well the stain can follow the protein shape, and it cannot reveal the details of the macromolecule's atomic structure. In this sense NS-EM is complementary to the cryo-EM technique, which produces images through the scattering of electrons by the lighter atoms of the biological macromolecules, which are themselves preserved in their native state by being embedded in vitreous ice. The resulting signal is weaker and noisier than it is in NS-EM, but it contains much higher-resolution information.

Staining artifacts are unavoidable due to the necessity of drying the solution before introducing the specimen into the microscope vacuum. Staining results are usually quite variable, as well, depending on how the solution spreads on the support. If the film of solid stain is too thin, the extent to which it covers the macromolecules can be uneven and 'one-sided'. In such cases, the side in contact with the carbon film may be stained better than the one protruding out. Another well known artifact of negative staining is the flattening of structures upon drying, stretching the features out in the carbon film plane and squeezing them down in its perpendicular direction. The resulting 2D images and 3D reconstructions are therefore dilated by up to 25% relative to the original structure (Kellenberger and Kistler 1979). However, these artifacts are well known and documented, and they can be somewhat reduced by optimizing the staining protocol (e.g. see 'stain additives' in section 2.2.4.1) (Bremer *et al* 1992, Haschemeyer and Myers 1972).

Knowing the limitations of the negative-staining technique, one can still use it as a low-resolution, qualitative method to assess samples destined for cryo-EM.

2.2.3 Purpose of negative staining when starting a project

Negative staining usually plays a crucial role in the identification and early analysis of potential targets for high-resolution cryo-EM studies. Rapid identification of high-quality recombinant proteins and protein complexes, or of purified complexes from endogenous sources, has become essential to the cryo-EM pipeline.

The main uses of NS-EM are for (i) optimization of sample purification prior to cryo-EM and (ii) identification of targets for proteomics (Takizawa *et al* 2017, Arnold *et al* 2018, Arthur and Ciferri 2019, Ho *et al* 2020, Brillault and Landsberg 2020). Also, worth mentioning is the fact that it is the standard way in which new students are introduced to EM. NS-EM, on its own, also plays an important role in completing the picture while trying to understand complex biochemical processes. Not all macromolecular complexes will yield a sub-nanometer resolution reconstruction, yet even a low-resolution one can give some mechanistic insights into its function, in particular for dynamic complexes. Several book chapters (Harris 1997, Haschemeyer and Myers 1972) and reviews (Bremer *et al* 1992, Ohi *et al* 2004, De Carlo and Harris 2011, Scarff *et al* 2018) cover the many techniques and applications of negative staining.

NS-EM is a relatively high-throughput, easy technique, which makes use of materials and tools that are readily available and cheap. It is a well-established technique, and thus the expertise, equipment, and materials needed are usually accessible on most biology research campuses. The specimen preparation only requires affordable consumables and reagents (see the negative-stain protocol, section 2.2.4.2) and no specialized equipment, and a low-end electron microscope is sufficient for imaging the negatively stained samples. It is an easy technique to learn for beginners, which can yield information without spending too much time gaining expertise in either specimen preparation or imaging. Characterizing a sample by NS-EM is often a prerequisite to gain access to cryo-EM facilities.

Negative staining requires very little sample and the prepared EM grids can be kept and re-used practically indefinitely as opposed to frozen hydrated samples, which are generally single-use. A few μl of material at a 10–50 μM concentration are sufficient to prepare a single NS-EM grid, which will be more than enough material to collect an entire dataset. With the high SNR imparted by the stain, very little data are needed to provide immediate feedback about how to prepare and optimize macromolecular complexes for structural characterization (see sections 2.2.4.3, 2.2.5 and figure 2.3). A few hours between specimen preparation and data collection on a lower voltage microscope are sufficient to provide relevant information on the specimen. With additional data processing, 24 h are sufficient to provide feedback for biochemistry, which can save several days of work when time-sensitive samples or cell cultures are involved.

Its relatively high throughput makes NS-EM an ideal tool to screen large numbers of experimental conditions when reconstituting a complex *in vitro*, such as buffer composition, protein concentration, incubation conditions, comparing fractions from a purification, etc. A series of conditions can be quickly prepared in parallel, screened within a few minutes with a low-end electron microscope, and the results can be compared to provide immediate feedback for specimen optimization, with or without the aid of 2D data processing (see section 2.2.5.3 and figure 2.3).

Although NS-EM images are limited in resolution, 2D classification of the data can provide significant amounts of structural information at an early stage of a project. It complements analytical techniques that are commonly used in biochemistry, such as SDS-PAGE analysis, mass spectrometry, size-exclusion chromatography, or density gradient centrifugation. NS-EM is particularly useful in providing information about the homogeneity of the sample, its oligomerization state, flexibility, and structural preservation, in addition to the particle's size and concentration (see section 2.2.5.3). Negatively stained material can also produce an initial 3D model when none is available from previous work. Although such initial models are at low resolution, they often facilitate the initial steps in cryo-EM data processing. In general, without being an end in itself, structural NS-EM provides an intermediate step that is taken after the initial reconstitution or isolation of a target complex, using traditional, biochemical techniques. Information from NS-EM is thus invaluable in guiding the final optimization for high-resolution cryo-EM, which will require a larger amount of a very well-behaved and homogeneous specimen (figure 2.3(B)).

2.2.4 Techniques for the preparation of negatively stained samples

2.2.4.1 Designing your negative-stain experiment

The main goal of NS-EM is to efficiently provide information on a specimen and its optimization. For that purpose, it should be approached as a part of the biochemistry pipeline (figure 2.3(B)) with specific questions about the process in mind. Common examples would be identifying fractions from a purification step, the level of homogeneity of a mixture or the effect of a modification on a complex. Some excellent reviews illustrate the many uses of negative staining for this purpose (Takizawa *et al* 2017, Ohi *et al* 2004).

Know your sample

First, one needs to make sure the complexity of the problem does not exceed the amount of information yielded by the technique. One advantage of NS-EM is that very little material is needed. 0.1 pmoles of sample is enough for one negative-stain EM grid, *provided it is pure and concentrated enough* (see data interpretation, section 2.2.5). Prior knowledge of the protein properties from the biochemistry is useful to guide the staining procedure and choose conditions compatible with the most stable and homogeneous macromolecular complexes (e.g. concentration, pH, ions). The main properties of the staining reagents are summarized in the next section and table 2.1.

Second, the specimen needs to be compatible with the NS-EM technique. The optimal specimen size is in the 150 kDa–MDa size range (it can be even more for polymers). The optimal concentration is typically about 10–50 nM, and the specimen should be in a buffer compatible with the stain, ideally 20 mM HEPES or PIPES. High salt concentrations and phosphate buffers should be avoided. The specific properties of the different staining reagents are summarized in the next section.

Know your stain

A variety of reagents are available as stains, with different properties. A short list of commonly used stains and their properties can be found in table 2.1. More exhaustive lists and their specific uses can be found elsewhere (Harris 1997, Bremer *et al* 1992, Haschemeyer and Myers 1972, Scarff *et al* 2018). In order to obtain meaningful information from the experiment, knowing and testing the different stains and protocols is also essential. The negative-staining results are always sample-specific. Therefore, in practice it is a good idea to test a few conditions by trial-and-error and

Table 2.1. List of the most commonly used staining solutions and their properties.

Stain	Concentration used	pH	Charge	Properties
Uranyl acetate	1%–3%	4.5	Positive	Very high contrast, fine grain, positive staining
Uranyl formate	0.75%–1.5%	4–5.5 (adjust with NaOH/KOH)	Positive	High contrast, high resolution for small molecules, positive staining
Sodium/potassium phosphotungstate	2%	5–8.0 (adjust with NaOH/KOH)	Negative	Low contrast, disrupts phospholipid membranes
Methylamine tungstate	2%	6–7	Negative	Low contrast, no positive staining
Ammonium molybdate	2%	5.3–8.0 (adjust with NaOH/ NH_4OH)	Negative	High contrast

compare them by EM when preparing a sample for the first time. The qualities to look for in a stain are resolution, contrast, embedding qualities, and compatibility with the specimen. One also should avoid negative-staining reagents that react with parts of the specimen and produce artifacts such as disrupting structures or creating a 'positive staining' effect (see table 2.1).

The optimal staining solution has properties compatible with the specimen's, results in a staining depth matching the specimen size (see stain additives and figure 2.1), and resolves the level of detail needed for the experiment.

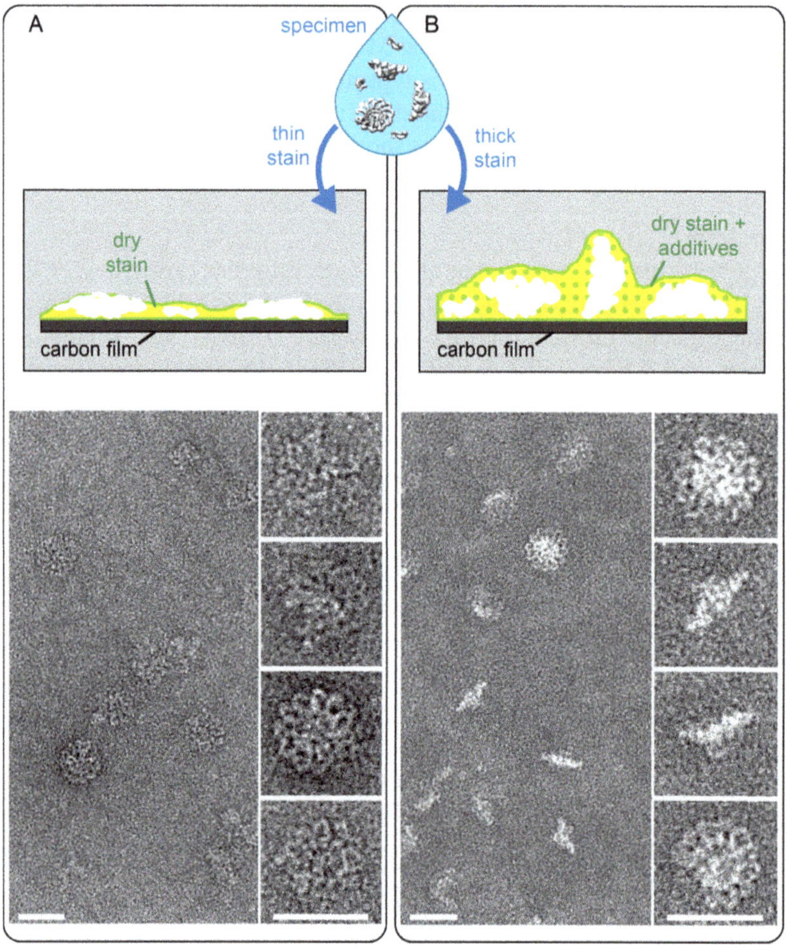

Figure 2.1. Effect of the stain layer thickness on the specimen and on the contrast. The same specimen is embedded in (A) a thin and (B) a thick layer of stain. Thin stain is well suited for small molecules, while thicker stain obtained, for example, by using embedding additives helps preserve large macromolecular structures in different orientations on the carbon film. The micrographs and particle galleries illustrate the resulting contrast for the 1.4 MDa NAIP5-NLRC4 inflammasome complex (Tenthorey *et al* 2017). Note that both top (top and bottom-most particles) and side views (middle rows) are visible in the thick stain layer (right panels), but only top views are present in the thin stain (left). Scale bars are 20 nm. The NAIP5-NLRC4 complexes were produced by J Tenthorey in R Vance's lab at UC Berkeley.

In most cases uranyl stains with their high resolution and high density give the best results, with some exceptions when the specimen is sensitive to low pH or interacts directly with uranium ions. In particular uranyl ions react with carboxyl groups in proteins and phosphate groups in lipids and nucleic acids producing a 'positive staining' effect and potentially obscuring parts of a structures containing these groups (e.g. DNA in a protein/DNA complex) (Harris 1997, Haschemeyer and Myers 1972). Uranyl acetate and formate also have a slight fixative effect that helps stabilize protein complexes (Zhao and Craig 2003).

However, when specimens are pH sensitive, positive staining masks parts of the structure, or the use of radioactive materials is not possible, alternative stains should be considered (table 2.1).

Stain additives

These can be added directly to the sample or in a wash buffer applied to EM grid, after incubation with the specimen and prior to applying the staining solution, to improve sample preservation and stain coverage (section 2.2.4.2, figures 2.1 and 2.2).

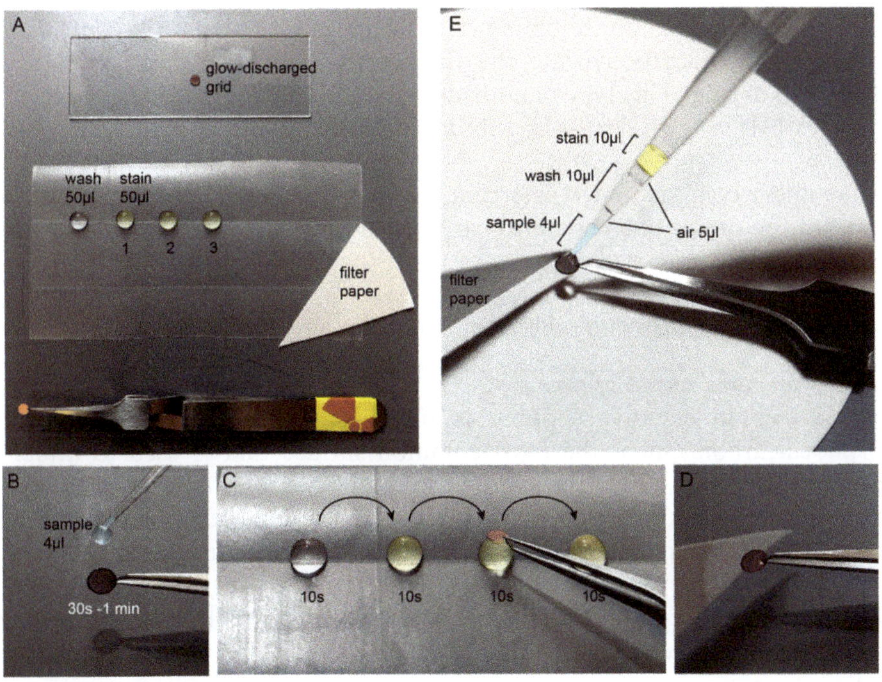

Figure 2.2. Layout and procedure for two commonly used, negative-staining protocols. (A)–(D) Classic droplet technique. (A) Suggested layout for staining one specimen with wash buffer and stain solution drops deposited onto Parafilm. (B)–(D) Procedure. (B) Apply 4 μl of specimen on a glow discharge treated EM grid and let it incubate for a desired length of time. (C) Wash, then stain in the successive drops (specimen side of the grid facing down). (D) Blot the last stain drop from the side, then let dry. The whole procedure takes less than 10 min, including drying time. (E) Rapid flushing variation of the negative-staining technique: for quick application of concentrated samples. Preload a pipet tip with all the solutions in successive order (reverse order from application). Eject the whole contents of the pipet tip onto the grid while blotting from the opposite side, then dry.

Some detergents can be added when poor spreading of the stain on the EM carbon film substrate is observed, if there is poor penetration of stain in the complex, or if some are required for the specimen stability (Harris 1997). Nonidet P-40 (NP-40) and n-octyl-β-D-glucopyranoside (OG) are commonly used at 0.01%–0.02% w/v concentration. Mild detergents employed for the isolation and preservation of integral membrane protein complexes such as digitonin or n-dodecyl-D-maltoside (α- or β-DM) can also aid staining. Always keep the detergent concentration below the critical micelle concentration (CMC) to avoid forming micelles, which would add to the NS-EM data heterogeneity.

Several carbohydrates can serve as embedding agents to preserve 3D structures from flattening and allow coverage of larger complexes by the stain. They can also prevent some drying effects at the AWI that could result in denaturation of parts of the protein or complex exposed (visible as aggregation or 'blurred-out' densities). Trehalose (1%–5%) added to stain (uranyl or ammonium molybdate; De Carlo and Harris 2011, Chiu *et al* 2011) or aurothioglucose and tannic acid are most traditionally used. Sucrose, glucose, and to some extent glycerol and are not recommended due to their instability under the electron beam resulting in 'fuzzy', grainy images or bubbling (Haschemeyer and Myers 1972). The higher the percentage of carbohydrate the thicker the stain layer will remain after blotting and drying. This type of embedding is suitable for large macromolecular complexes (MDa, virus capsids, tubular assemblies) but may obscure smaller complexes.

Sometimes a combination of detergent and carbohydrate is needed to preserve the specimen from drying and denaturing artifacts during staining. These could also be used for cryo-EM grids if needed.

2.2.4.2 General negative staining protocol

Materials prepared ahead of time

The first step in negative staining is preparing the EM substrates. A thin, amorphous-carbon film is the most common surface that is used to attach the macromolecules and spread the staining solution. Carbon film coated TEM grids (300 or 400 mesh Cu or Au depending on the applications) can either be purchased commercially, or prepared in-house ahead of time if the equipment and materials are available. The preparation requires a carbon evaporator, carbon fibers or rods, and either a collodion solution or mica sheets to serve as a support, depending on the method used (Haschemeyer and Myers 1972, Scarff *et al* 2018, Booth *et al* 2011).

After choosing the specific stain(s) for your experiment, staining solutions are simply prepared by diluting the proper weight of reagent powder in ultrapure water (see table 2.1 for recommended concentrations), and adjusting the solution pH as prescribed. All solutions should be filtered before use for EM. Detailed protocols can be found elsewhere (Harris 1997, Haschemeyer and Myers 1972, Arthur and Ciferri 2019, Brillault and Landsberg 2020, Ohi *et al* 2004, Scarff *et al* 2018).

Negative staining

Many variations of the negative-staining protocols have been described (Harris 1997, Haschemeyer and Myers 1972, Arthur and Ciferri 2019, Brillault and Landsberg 2020, Ohi *et al* 2004, Scarff *et al* 2018). As mentioned before, the appropriate method and conditions are sample-dependent and require some testing for each new experiment (section 2.2.4.1). They all consist in the same basic steps.

Right before depositing the sample, the carbon film surface used as support should be charged to render it hydrophilic. This step is done by placing the coated grids, carbon side up, in the plasma chamber of a glow discharge unit and treating them following the protocol specific to the equipment used.

A drop of sample solution is applied to the pre-treated grid surface (see figures 2.2(A)–(D)) and the specimen is allowed to attach to the substrate through its charges for a few seconds to minutes depending on the concentration of the specimen. Specialized techniques can be used when the sample concentration is very low (e.g. GraFix, Stark 2010; streptavidin grids, Han *et al* 2012) or needs to be kept very high (e.g. rapid flushing method figure 2.2(E), Imai *et al* 2015).

When the specimen buffer is not compatible with the stain, a buffer exchange or 'wash' step can be added at this point. Additives can also be introduced at this step (section 2.2.4.1 and figures 2.1 and 2.2). The excess buffer is subsequently blotted off or directly exchanged with the staining solution. There are many variations of the staining method; the two methods illustrated in figure 2.2 are among the most commonly used. After the excess solution has been removed and the grid thoroughly dried, the stain-embedded specimen can be stored indefinitely or transferred to an electron microscope for evaluation.

2.2.4.3 Imaging

The grid preparation only takes a few minutes. When imaging a new sample, a few conditions (1–10) can be prepared in parallel (e.g. different protein concentrations and/or staining solutions) to minimize the waiting time and screen them by EM more efficiently (if not limited by the amount of specimen).

NS-EM imaging is relatively straightforward. As indicated previously, a low-end, low voltage 80–120 kV side-entry electron microscope with room temperature holder and data collection camera (CCD, CMOS, or direct electron detector) is recommended. The lower voltage of the microscope and using a high electron dose both contribute to the high SNR, often referred informally as 'contrast', in the images. After evaluating the specimen and staining qualities at low and intermediate magnifications, the specimen can be exchanged rapidly or data collected at higher magnification, with a relatively high electron dose (Arthur and Ciferri 2019). Since a few thousand particles are sufficient for processing the NS-EM data, 10–200 images are usually enough for a dataset depending on the particle distribution on the carbon film, which usually takes only 1–3 h to collect. Automated data collection software (Arthur and Ciferri 2019, Mastronarde 2005, Stagg *et al* 2006, Suloway *et al* 2005) are helpful to collect, archive, and connect the data with the processing pipeline, but not necessary.

2.2.5 Use of data processing to provide feedback to optimize samples for cryo-EM

2.2.5.1 Need for data processing

Samples that are initially produced, either to isolate or to reconstitute macromolecular complexes, are rarely completely homogeneous in structure. Some information about these initial samples is evident from the micrographs, such as the protein concentration, the presence of large oligomers, or the propensity of the molecules to either aggregate or fall apart. However negative-stain EM images are still quite noisy and, in general, raw 2D data require some processing and analysis to provide useful information about the 3D molecular complex in order to further optimize its preparation. In order to do so, particle data must be extracted from the images and processed in order to interpret it in terms of sources of variability.

The steps and techniques used are similar to the ones used to process cryo-EM data, however, less data is necessary thanks to the higher SNR in NS-EM images. Multiple examples of applications are found in the literature and some are compiled in Bremer *et al* (1992), Takizawa *et al* (2017), and Ohi *et al* (2004). The details and the mathematical description of the methodologies are software specific and can be found elsewhere (Punjani *et al* 2017, Samso *et al* 2002, Frank 1996, Grigorieff 2007, Scheres 2012, Tang *et al* 2007, Hohn *et al* 2007, Van Heel 1987, van Heel *et al* 1996).

2.2.5.2 Processing steps

After sorting the images acquired to reject those with too much drift, astigmatism, or bad staining, particles are picked. The extracted particle images are pre-processed to produce the dataset that will be analyzed in 2D or 3D (figure 2.3(A), more details in sections 4.2 and 4.3).

Different data processing software uses different algorithms, but the overall method is the same (Punjani *et al* 2017, Samso *et al* 2002, Frank 1996, Grigorieff 2007). Briefly, the 2D data particles are iteratively aligned to each other, their pixel intensities compared, and grouped according to their degree of similarity. The resulting groups or classes of data provide both statistics of their grouping and formation of class-average images. The SNR of each average depends on the number of images assigned to the corresponding class. The more evenly distributed the views, and the more heterogeneous the dataset, the more data are required to obtain an accurate and reliable representation. The high SNR from NS-EM data helps these types of methods reach statistically meaningful averages (section 3.2).

In some cases when 2D sorting proves challenging (e.g. heterogeneous data), 3D classification and/or 3D projection matching can be used to sort out the data, in particular when a 3D structure of a similar complex is available. It can also be necessary to determine which 3D structure is the target when the 2D class averages are difficult to interpret (see the following section) (Punjani *et al* 2017, Scheres 2012).

2.2.5.3 How to interpret the data

The main use of negative-stain EM, together with 2D/3D data processing, is judging the heterogeneity of a sample. Difficulties arise from the complexity of the samples

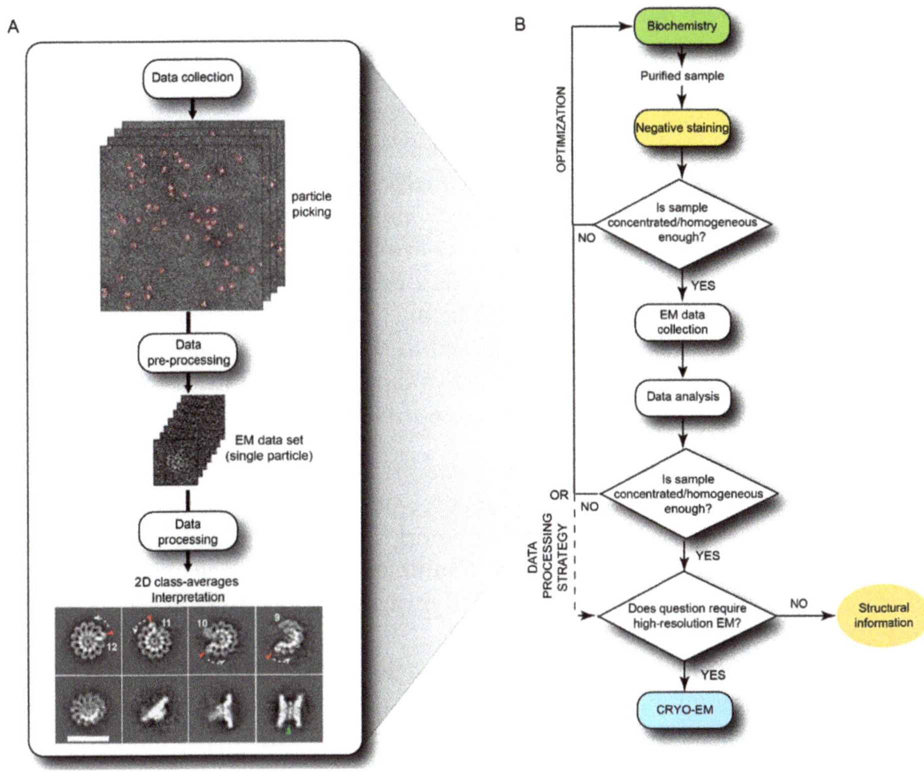

Figure 2.3. Negative staining and data analysis workflow integrated in the specimen optimization feedback loop for cryo-EM. (A) The basic steps of sample evaluation by negative-stain EM and data analysis, applied to the NAIP5-NLRC4 inflammasome complex (Tenthorey *et al* 2017). Introducing several mutations at the N-terminus and the interaction surface of the NLRC4 CARD domains helped to reduce the number of oligomerization states found when expressing and assembling the 1.4 MDa complex *in vitro*. NS-EM was used to fine-tune the mutations and reconstitute a functional complex bound to its pathogen target, the bacterial protein flagellin, ultimately causing it to behave well enough to elucidate its ligand-activated mechanism by high-resolution cryo-EM. At the bottom are representative class averages showing the characteristic top (top row), tilted, and side views (bottom row). The top class averages reveal the inflammasome particles formed with different oligomeric states, mostly as 12- or 11-mers, with higher-flexibility regions appearing blurred-out. This analysis lead to using a data processing strategy like that described in section 4.5.3, in order to separate the oligomeric states and mask the more variable regions during processing of data collected by cryo-EM. (B) Integration of NS-EM/data analysis in the specimen optimization process flowchart for high-resolution cryo-EM. The scale bar is 20 nm.

and from the TEM data being noisy two-dimensional ‘projections’ of three-dimensional structures. 2D and 3D alignment and classification methods are used to approximate representative views or 3D models matching the data. The resulting averages are then sorted in terms of sources of heterogeneity: (i) different views of the complex, (ii) different compositions or oligomeric states, or (iii) different conformations (Ohi *et al* 2004).

Apparent heterogeneity due to particle orientation

Before any information regarding other sources of heterogeneity can be extracted, the data need to be sorted between the different orientations that a complex can adopt on the carbon support film. The resulting projections can look drastically different and are sometimes hard to interpret as different 'views' versus different conformations or oligomeric states of a complex. *A priori* knowledge of the 3D structure of the complex is useful when interpreting 2D class averages, modeled from a previous study or a known homologous structure. When none is available it is possible to obtain a 3D reconstruction from 2D negative-stain data, as long as the views are evenly distributed, or by collecting 0°–60° tilted-pair data (Punjani *et al* 2017, Radermacher *et al* 1987). Each 2D class average can then be attributed to different views of the 3D structure, visually or by using multi-reference matching with projections of the current 3D model. The sorted views can then be further analyzed if necessary.

Heterogeneity due to different oligomeric states

For each view of the complex, differences in class-average features can be interpreted as different conformations or compositions, depending on the position or presence/absence of certain densities. Examples of applications are described in two excellent reviews by Ohi *et al* and Takizawa (Takizawa *et al* 2017, Ohi *et al* 2004) and can be summarized as follows:

- I*dentifying symmetries*: NS-EM and single-particle analysis can be used to identify different symmetries of complexes in a mixed population (also demonstrated by the class averages in figure 2.3(A)).
- *Measuring binding constants*: If 3D model information is available, projection matching of the 2D classes to the different views of the complex with different compositions can also be used to quantify the classes assigned to each view. The number of particles assigned to each class corresponding to a certain orientation, e.g. with or without a certain subunit, can be used to estimate binding constants corresponding to the experimental conditions. The values can be used to change the ratio of different components, for example.
- *Extracting a complex from a mixture*: This technique can also be useful when forming a complex by mixing proteins *in vitro*. This can be applied to an initial test for a reconstitution, or for a complex that is too fragile to undergo any further purification.
- *Labeling components of a complex*: This can be achieved using covalent protein tags (GFP, MBP, etc), immuno-labeling or gold. Comparison of the tagged versus untagged complex (visually or using difference maps) will point to the tagged subunit location (Takizawa *et al* 2017, Ciferri *et al* 2012).

Conformational heterogeneity

In this case, the different classes (sorted by view) show a given density in different positions relative to the rest of the complex. These differences could represent discreet conformational states, which can then be quantified as in the previous section, or a continuum of states, which would appear blurred out in the class averages. In any case different buffer and incubation conditions can be tested by

NS-EM to reduce the conformational landscape (e.g. non-hydrolysable nucleotide addition, cross-linking step) (Takizawa *et al* 2017).

2.2.5.4 Feedback loop

If the goal is high-resolution cryo-EM, the bottleneck may be the biochemistry. Once the different sources of heterogeneity have been attributed, the information can be used to optimize either the sample or the data processing approach.

Some degree of *in silico* purification is possible (Punjani *et al* 2017, Scheres 2012), but NS-EM and single-particle analysis will help estimate whether the level of heterogeneity is tolerable for high-resolution cryo-EM. Sometimes the initial NS analysis provides enough information to devise strategies for the computational sorting of the heterogeneity. Guided local masking and 3D classification can be used to obtain high-resolution structures from heterogeneous datasets. If parts of the structure are disordered, *in silico* masking (sections 4.5.3 and 4.5.5) or *in vitro* labeling strategies can help identify/localize domains while others can be resolved.

If the level of heterogeneity is still too high for cryo-EM, the biochemistry can be optimized. Possible solutions include using mutations, pull down strategies (Han *et al* 2012, He *et al* 2013), and cross-linking (Stark 2010). This process can be repeated multiple times until the desired specimen homogeneity is reached and NS-EM can be integrated into the cryo-EM optimization feedback loop as illustrated in figure 2.3. Eventually NS-EM and data processing will provide essential information to help devise new strategies to isolate or reconstitute increasingly challenging complexes (section 2.5).

Additionally, NS-EM can help complete our understanding of the mechanisms of action of some macromolecular complexes by employing lower resolution data and labeling (see figure 2.3(B)). Ultimately combining different resolution EM techniques offers more possibilities to understand these complex systems.

2.3 Standard method of making grids for cryo-EM

Melanie D Ohi

Department of Cell and Developmental Biology and Life Sciences Institute, University of Michigan, 210 Washtenaw Avenue, Ann Arbor, MI 48109, USA

The central idea of cryo-EM is to cryogenically freeze a biological sample in a thin layer of non-crystalline amorphous ice so that it is preserved in a near-native state (Taylor and Glaeser 1973, 1974, Adrian *et al* 1984, Dubochet *et al* 1988). The goal is to obtain a vitrified ice layer that is as thin as possible to minimize the amount of noise generated in the images, but thick enough to fully embed the sample in the vitrified ice layer, thus keeping the protein hydrated and avoiding the destructive AWI. The vitrified grid is then imaged in the electron microscope with the ultimate goal of determining an atomic-resolution 3D structure.

Almost all aspects of cryo-EM, including the development of more stable and powerful microscopes, automated image collection, sensitive direct electron detectors (DEDs), computation capabilities, and image processing algorithms, have improved over the last decade (Binshtein and Ohi 2015, Nogales and Scheres

2015, Takizawa *et al* 2016, Earl *et al* 2017, Cheng 2018). However, the steps used for vitrifying biological samples for cryo-EM have, up until very recently, not changed much from the approach introduced by Dubochet's group in the 1980s (Dubochet *et al* 1988). While the method is conceptually easy to grasp, successful execution remains a major experimental bottleneck and arguably one of the most frustrating steps of the entire 'sample-to-structure' cryo-EM pipeline. In many ways, finding the best, reproducible vitrification conditions for high-resolution cryo-EM imaging is analogous to the trial-and-error involved in growing highly ordered crystals for x-ray crystallography analysis. This means that each lab often has their own elaborately developed ethos for what steps and methods are required to obtain the best vitrified grids, whether these steps and/or tweaks are scientifically grounded or not.

There are now a number of helpful resources that review the general steps of vitrifying a biological sample for cryo-EM analysis using plunge-freezing devices (for examples, see Iancu *et al* 2006, Grassucci *et al* 2007, Dobro *et al* 2010, Sgro and Costa 2018), and practical hands-on training classes are offered at the National Cryo-EM Facilities. Although the protocols are all generally the same for traditional plunge freezing, there are a number of choices that can be made by the practitioner that can profoundly affect the success of the method. The next subsections outline the general steps and decisions that need to be made for successfully vitrifying samples for imaging by cryo-EM. Although the focus of this section will be on the steps for the standard plunge-freezing approach used for single-particle cryo-EM analysis, new developments in methods and technologies used for preparing grids will also be discussed when appropriate.

2.3.1 Grids and support films

The first step of vitrifying a sample requires choosing a grid and support film. There are many choices about what grids to use and it is not always clear which choice will be best for each sample. The grid (3 mm), which is composed of a metal mesh of varying sizes, can be made from a variety of materials, with the most common choices for single-particle cryo-EM being copper or gold (figure 2.4(A)). While gold grids are more fragile to handle than copper grids, they are more stable for imaging (Russo and Passmore 2016). This makes them a great choice for collecting images that can be used for high-resolution structure determination. However, copper grids are still commonly used and also do not limit the ability to determine high-resolution structures, in particular when using DED cameras and the most up-to-date motion correction software.

The metal grids are coated with a thin film that can be either continuous or 'holey'. Holey grids make it possible to suspend the sample in a thin layer of vitrified ice across the holes, thereby reducing background noise (figure 2.4(B)). Holey films can have holes that are either regular in size and evenly spaced made with micro-fabrication techniques, or 'lacey'—with varying sizes and spacings between holes (figure 2.4). Continuous films, on the other hand, can help when trying to vitrify more

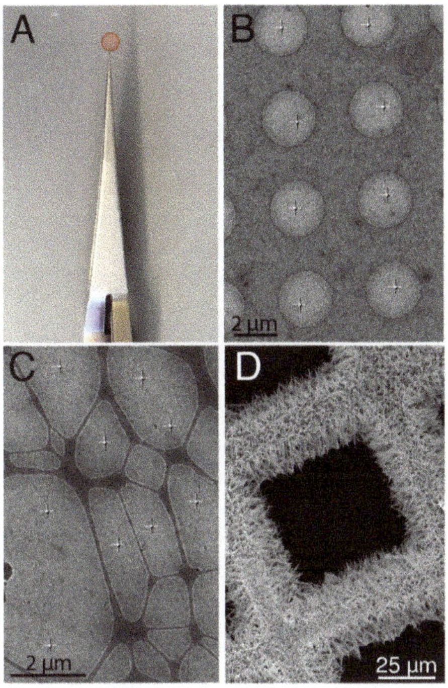

Figure 2.4. Types of grids used to vitrify samples in cryo-EM. (A) A 3 mm copper grid. (B) A Quantifoil grid. (C) A Lacy carbon grid. Images courtesy of Rachel Torrez, Dr Clarissa Durie, and Dr Jeongmin Chung at the University of Michigan. (D) Scanning electron microscopy image of a self-wicking grid. Scale bar = 25 μm. Image provided courtesy of Dr Hui Wei at the National Resource for Automated Molecular Microscopy at the Simons Electron Microscopy Center.

dilute samples or samples that are reticent to being suspended across holes, as is explained more fully in section 2.5. These films are often overlayed on the holey grids.

There are many choices for the types of film substrates to coat the grids and some of the more popular options include carbon, gold, streptavidin monolayer crystals, graphene, and graphene oxide (Russo and Passmore 2016, Wang *et al* 2008, Han *et al* 2016, Palovcak *et al* 2018, Doerr 2019, Zheng *et al* 2020, Han *et al* 2020). Once again, more on this is covered in section 2.5. Briefly, however, graphene, graphene oxide, and streptavidin monolayer crystals all have the added advantage of being amenable to various bioactive functionalizations that can make it easier to trap dilute particles on the grid for vitrification and, in some cases, also reduce the issues that arise from particles interacting with the AWI (Han *et al* 2012, Benjamin *et al* 2016, Liu *et al* 2019, Wang *et al* 2020). While grids with various support films can now be purchased commercially, it is not uncommon for cryo-EM labs to take the time and effort to buy the grids and then add the various types of support films 'in-house'.

2.3.2 Plasma cleaning or 'glow discharging' grids

Grids, in particular when covered with a carbon film, become hydrophobic over time, which causes the sample to not spread evenly across the holes or the continuous

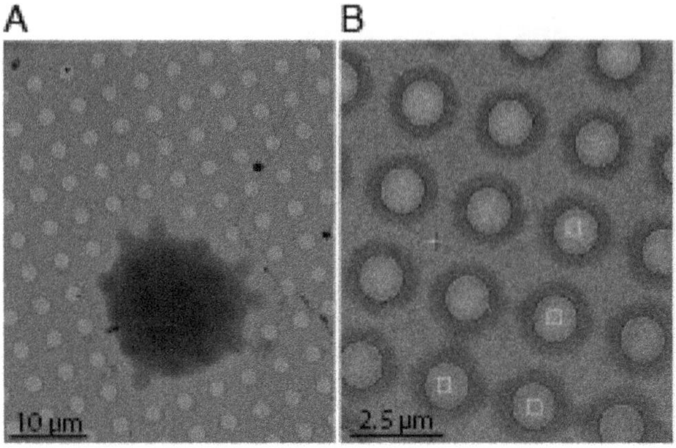

Figure 2.5. Glow discharging the grid before vitrification helps the sample to spread evenly. (A) Example of a grid that was not properly glow discharged before adding the sample and plunging. Scale bar = 10 μm. (B) Example of a grid that was properly glow discharged. Images courtesy of Drs Kelly Manthei and Clarissa Durie at the University of Michigan.

support film. This leads to uneven vitrified ice that can include bulges in the center of the grid squares (figure 2.5(A)).

To overcome this issue, grids are first plasma cleaned shortly before use, a process often referred to as 'glow discharging'. Plasma cleaning is done using a specialized piece of equipment that creates a modest vacuum, which is the condition needed for the remaining gas to be ionized. The gas is usually simply air, but more precise mixtures of specific gases can also be used. The ionization of the gas creates radicals that interact with the surface of the support film, rendering it hydrophilic. This, in turn, allows for the sample to spread evenly across the grid (figure 2.5(B)). The evenly spread liquid is then blotted off before plunging into cryogen to create a thin layer (~100–200 nm thick) of vitrified ice. Glow discharging a grid for too long can, however, lead to the support film breaking, thus making it unusable for vitrifying a sample.

There has been a recent focus on developing more efficient and reproducible approaches for vitrifying samples, with many of these methods using smaller amounts of sample than needed for traditional plunging methods. However, obtaining the small volumes of liquid to evenly spread across even a plasma-cleaned grid turned out to be a major challenge. This led to the development of self-wicking grids which are composed of a copper grid that has nanowires grown on the grid bars (Wei *et al* 2018) (figure 2.4(D)). These special grids produce a thin layer of liquid suspended across holey support films, in times that can be as short as tens of milliseconds from application. Self-wicking grids represent an important auxiliary component of a newly developed cryo-plunger called Spotiton (Jain *et al* 2012), now sold commercially as the Chameleon (TPP Labtech) (figure 2.6(A)), or used with an open source instrument that can be built by laboratories called Back-it-up (Tan and Rubinstein 2020) (figure 2.6(B)).

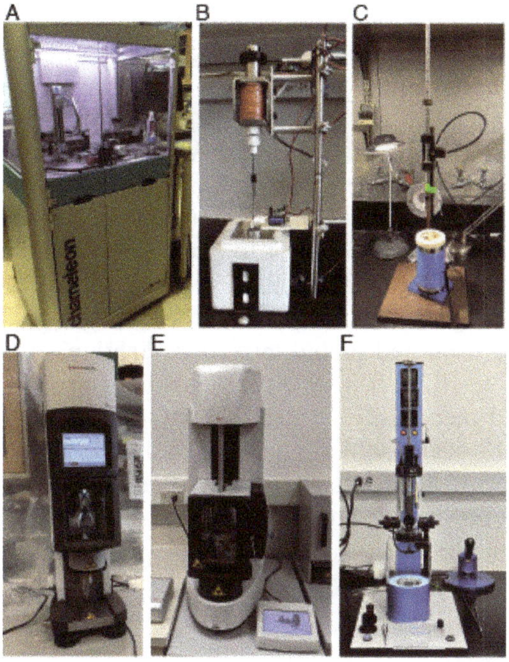

Figure 2.6. Images of types of vitrifying machines used for preparing vitrified samples. (A) Chameleon (TPP Labtech). Image courtesy of Rachel Torrez at the University of Michigan. (B) Back-it-up. Image courtesy of Dr John Rubenstein at the University of Toronto. (C) Manual Plunger. Image courtesy of Dr Gabe Lander at the Scripps Institute. (D) Thermo Fisher Vitrobot™. Image courtesy of Dr Clarissa Durie at the University of Michigan. (E) Leica's EM GP2™ Plunge Freezer. Image courtesy of Dr Matt Jaremko at Cold Spring Harbor Laboratory. (F) Gatan's Cryoplunge™ system. Image courtesy of Dr Daniela Nicastro at University of Texas Southwestern.

2.3.3 Types of apparatus used for plunge freezing

Plunge freezing, regardless of the type of apparatus used, involves placing a plasma-cleaned grid in fine-tip forceps, adding sample (usually ~3 μl) to the grid, blotting away excess sample using filter paper, and then rapidly plunging the grid into liquid ethane to vitrify the sample. There are two common plunge-freezing apparatus that can be found in most cryo-EM facilities. These include manual plunge freezers which rely on gravity to plunge the grid into cryogen, and automated plunge freezers that use robotics to help control humidity, as well as blotting force and blotting times.

The design of manual plungers has changed very little since their initial use by cryo-electron microscopists in the 1980s, and they are often still built 'in-house' (Dubochet *et al* 1988) (figure 2.6(C)). These instruments require more user experience to obtain reproducible results, but they are a very cost-effective approach for vitrifying samples for cryo-EM. There are also a variety of commercially available automated plunge freezers that each have slightly different approaches for controlling humidity, blotting, blot-force, and blot-times. Currently the most commonly used instruments are Thermo Fisher's Vitrobot™, Leica's EM GP2™ Plunge Freezer, and Gatan's Cryoplunge™ system (figures 2.6(D)–(F)).

While plunge freezing is still the most commonly used approach for vitrifying samples for cryo-EM imaging, there is a new generation of vitrification technology under development that aims to tackle many of plunge freezing's limitations that include wasting over 99% of the sample during the blotting process and being too slow to minimize particle interaction with the AWI. While these instruments are still not commonly available at every institution, they are poised for wide-spread adoption in the field. They could represent the next 'revolution' in cryo-EM if they prove to make it easier and quicker to determine atomic-resolution structures.

The new instruments take a variety of approaches to reproducibly deliver small volumes of samples to grids including piezoelectric dispensers (Jain *et al* 2012), an ultrasonic humidifier (Tan and Rubinstein 2020, Ashtiani *et al* 2018), pin printing technology (Arnold *et al* 2017), and instruments with proper mixing chambers to initiate reactions and then vitrify samples quickly enough for time-resolved microscopy (Feng *et al* 2017, Kaledhonkar *et al* 2018, Kontziampasis *et al* 2019). Many of these approaches and instruments are so new that they are currently not widely available for testing; however, a recent review provides a helpful summary comparing the various methods that are currently being developed for depositing small amounts of sample on a grid for vitrification (Klebl *et al* 2020a).

2.3.4 Blotting and plunging the grid using plunge freezers

For plunge freezing, whether using a manual or automated instrument, once the sample has been applied to the grid excess liquid must be blotted away in order to leave only a thin liquid film on the grid. Thinner ice is more amenable to high-resolution imaging, but if the vitrified ice is too thin it can lead to preferred orientation of the particles and/or denaturing of the complex due to interactions with the AWI (Glaeser 2018, Glaeser and Han 2017). The temperature, humidity, blotting time, and blotting pressure with which samples are plunged at are all factors that help control the thickness of the vitrified ice layer and they are parameters that must be optimized for each sample.

If using a manual plunger this information must be retained by the individual doing the plunging; however, if using an automated instrument these parameters can be saved and easily reproduced by any user. Blotting is essential to produce the thin films required for vitrification, however, we still do not fully understand how the filter paper interacts with the sample and the filter paper can leach contaminates into the sample (Armstrong *et al* 2020, Walker *et al* 1994).

In addition, blotting removes ~99.9% of the sample applied to the grids and takes at least 1 s (often more) from blotting the grid to the sample reaching the cryogen. This amount of time exacerbates exposure of the particles to the AWI. As is discussed in greater detail in sections 2.5 and 4.6, this can lead to denaturation and/or preferred orientation of the sample, as well as impeding the ability to trap reaction intermediates that happen on the millisecond timescale (Arnold *et al* 2017, Noble *et al* 2018a, D'Imprima *et al* 2019). Thus, blotting arguably remains one of the most important but least understood steps of the entire grid preparation methodology.

2.3.5 Common issues faced in making grids for cryo-EM imaging

It has now become very common for single-particle cryo-EM practitioners to expect sub-4.0 Å resolution 3D structures, an achievement that before the introduction of DEDs would have been considered a very challenging accomplishment (Binshtein and Ohi 2015, Cheng 2018). Of course, biochemistry still remains the most important component of whether it is possible to determine a high-resolution 3D structure, with the maxim 'garbage-in equals garbage-out' remaining in full force regardless of technical breakthroughs in microscopes, cameras, computational approaches, and/or vitrification protocols. However, preparing grids for imaging remains another major bottleneck that can significantly thwart scientific progress. There are many steps in vitrifying cryo-EM grids that can lead to sub-optimal results including choice of grid, film support, blotting instrument, filter paper, and blotting time.

User error in not maintaining the proper temperature of the cryogen or not using proper technique also adds variability of the vitrification process. In addition, even biochemically pure samples can refuse to go into the grid holes, aggregate, denature, and/or adopt highly preferred orientations during the vitrification process. There remains no clear consensus, even from the experts, about how to overcome these issues, except that trouble-shooting is almost always required (Carragher *et al* 2019). Thus, making cryo-EM grids that are amenable to high-resolution imaging remains a variable and uncertain process that should not be assumed to always be reproducible. Considering all the steps that can go wrong while vitrifying grids, it is amazing the process works as often as it does.

The hope is that as new approaches and instruments become more wide-spread and commercially available that this current bottleneck in the 'sample-to-structure' pipeline will be at least somewhat relieved. However, most frustratingly, even when the proper vitrification conditions have been experimentally determined, there often remains variability between grids vitrified during the same session, using what appears to be the same parameters. For this reason, it is very common that experienced practitioners will vitrify multiple grids of the same sample under the 'same' conditions and then screen these grids to find which ones are amenable for high-resolution imaging. A low-magnification 'atlas' is often enough to make a quick determination of whether a grid is good enough for further imaging or can simply be discarded. Freezing multiple grids per sample can also increase your chances of success by hedging against biochemical variability between purifications, as well as against other mishaps that can happen during the processes of clipping and loading vitrified grids into the microscope for imaging.

2.4 Requirement to make very thin specimens for cryo-EM

Robert M Glaeser
Lawrence Berkeley National Lab, University of California, Berkeley, CA 94720, USA

Specimens prepared for single-particle cryo-EM are best if they are made to be very thin, less than 100 nm, for example, or even less than 50 nm in thickness. Very thin specimens are required because a number of factors act to decrease the quality of images as the specimen thickness increases. At the same time, it can also be very hazardous for proteins to be confined within thin aqueous films. Although we do not address this latter aspect here, the topic is taken up in section 2.5.

In this section, we address only the reasons why images are best if the samples are very thin. In addition, we describe some of the options that one has for estimating what the sample thickness is, so that one can target data collection to areas that appear to be the most promising ones.

2.4.1 Inelastic electron scattering causes the image quality to deteriorate with increasing sample thickness values

As electrons pass through a specimen, a fraction of them experience inelastic scattering events, and that fraction naturally increases as the specimen thickness increases. Unfortunately, the loss of some electrons to inelastic scattering reduces the fraction of the incident electrons that ultimately contribute to the formation of useful image information, as was mentioned in section 1.2. If the sample is very thin, perhaps 0.1 or even 0.2 of the mean-free-path for inelastic scattering, the reduction in useful image information is relatively small. On the other hand, if the sample is as thick as 0.5 of the mean-free-path for inelastic scattering, then the remaining fraction of useful electrons is reduced to about 0.6 of the number that were incident at the front of the specimen. Thus, assuming that the mean-free-path for inelastic scattering is about 350 nm for 300 keV electrons, the loss of electrons due to inelastic scattering is not such a serious issue if the sample thickness is less than 50 nm, but it becomes increasingly problematic if the sample thickness is greater than about 100 nm.

The inelastically scattered electrons will continue to make an unwelcome contribution to the random shot-noise (statistical fluctuations) in the measured image intensity, unless they are removed with an energy-filter lens. As a result, using an energy-filter lens can increase the SNR, even though doing so does not compensate for the loss of some signal due to inelastic scattering. In practice, the benefit of decreasing the noise by removing inelastically scattered electrons is likely to be marginal, provided that the sample thickness is a small fraction of the mean-free-path for inelastic scattering.

It nevertheless is widely acknowledged that particles are more easily seen if the ice is not much thicker than the particles themselves; see, for example, such statements in the section ‘Specimen preparation for single-particle cryo-EM’ in Cheng *et al* (2015). This effect is likely to be particularly important if images are recorded at lower values of defocus. In addition, there is some evidence that the quality of the contrast transfer function (CTF) fit (Rice *et al* 2018), as well as the resolution achieved (Kim *et al* 2018), is somewhat better with very thin samples, even for relatively large particles.

The adverse consequences of increasing the sample thickness therefore seem—to this author, anyway—to be more detrimental than can be accounted for by the reduction in the SNR due to inelastic scattering. Perhaps a continual rearrangement of water molecules during irradiation, which causes the Thon rings from vitreous ice to be so much weaker than those from amorphous carbon (McMullan *et al* 2015) may cause more severe effects when the ice is thick than when it is thin. This possibility does not seem to have been investigated, however. In fact, the quantitative extent to which the image quality deteriorates, due to factors in addition to the reduction of SNR caused by inelastic scattering, remains unknown at this point.

2.4.2 The projection approximation may fail if the sample is too thick

Leaving aside the effects of inelastic scattering and deterioration of image quality, increasing sample thickness also causes unwanted complications for data processing. The first issue is that the amount of defocus will vary from one particle to another, depending upon the height of each particle within a thick ice film. Applying the same CTF correction to all particles causes an envelope function that attenuates the signal at higher resolution, just as would happen if the objective lens current were unstable, or if the energy spread of the electrons were much larger than what it really is (Jensen and Kornberg 2000). To avoid such an unwanted loss of signal, therefore, it is necessary to correct the CTF for each particle, something that was first introduced in the context of tomography (Galaz-Montoya *et al* 2016) and subsequently implemented in software used for single-particle cryo-EM (Zivanov *et al* 2018, Su 2019). More is said in section 4.3.1 about the causes for variation in defocus across the field of view and the benefit of carrying out per-particle CTF correction.

If a particle itself is quite thick, however, that will introduce an additional effect, which can be described either as being due to curvature of the Ewald sphere (DeRosier 2000) or, equivalently, to the depth of field being similar to, and even less than the size of the particle (Downing and Glaeser 2018). Since the depth of field decreases as the square of the resolution, it may start to be beneficial to correct for curvature of the Ewald sphere, as is done in Zivanov *et al* (2018), when the resolution is in the range of 0.2–0.3 nm. Much more is said about this in sections 4.8.4 and 4.8.5.

Both of the ‘thick sample’ effects mentioned here can be seen as being due to failure of the projection approximation that is used in the usual description of image formation, summarized, for example, in section 1.2. This failure of the projection approximation is due primarily to the fact that the CTF changes by a significant amount throughout the thick specimen. Multiple elastic scattering (often referred to as dynamical scattering, see chapter 15 of Glaeser *et al* 2007) is another reason why the projection approximation can fail, but that is not likely to be a problem in the case of single-particle cryo-EM.

2.4.3 Areas of a grid where the sample is obviously too thick can, and should be, avoided during data collection

It is best to avoid recording images in areas where the ice thickness is too great. There are multiple ways in which one can estimate how thick the sample is. Thus, if one experiences difficulty in seeing/boxing particles, which can easily happen at relatively low values of defocus, or if the map resolution is less than desired, the situation may be improved by first estimating the sample thickness, and then limiting data collection to just the areas that are very thin.

Perhaps the quickest and most accurate way to estimate the ice thickness is to measure the average values of the intensities for images recorded with and without an energy filter, respectively, both of which are taken at a relatively low value of defocus (Rice *et al* 2018). The logarithm of the ratio of these two image intensities is then proportional to the sample thickness divided by the mean-free-path for inelastic scattering. Although estimates of the inelastic mean-free-path vary considerable in the literature, this author recommends to use a value of 350 nm for 300 keV electrons, based on the measurements of Grimm *et al* (1996) at 120 keV. This value is close to, but slightly less than what is reported in Rice *et al* (2018).

An alternative, and equally fast, method to estimate the ice thickness is to measure the average intensities in different areas of images recorded in 'SEARCH' mode. These images should be recorded with a very large defocus value, of course. In this case the idea is to compare the intensity measured in an empty hole to that in an ice-filled hole. Figure 2.7, reprinted from Agard *et al* (2014), illustrates how this is done. If there are no empty holes one can expose an ice-filled hole to the electron beam for a long time beforehand in order to remove all water by beam-induced desorption (effectively by sublimation). Once again, one can use a value of 350 nm for the mean-free-path of 300 keV electrons, although in this case there is poor

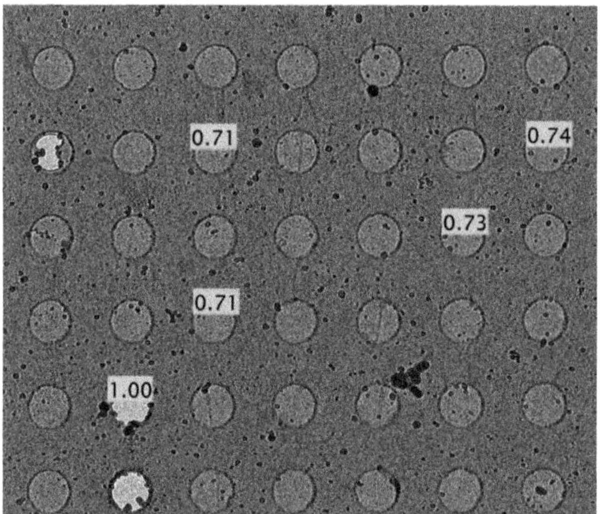

Figure 2.7. Example of a low-magnification image of a cryo-EM specimen recorded with the low-dose software of the electron microscope set to the SEARCH condition. Most of the holes in the Quantifoil carbon film shown here are filled with a thin film of vitrified sample, but three of them—recognizable by their greater transparency to electrons—are partially or completely empty. The electron intensity, as registered by the camera, can be used to compute the apparent transmission of electrons through a hole 'filled' with sample, relative to that of an open hole. The numbers in the white boxes in this figure, for example, are the ratio of the local intensity, relative to that in an open hole. We thank Dr Roseann Csencsits for writing the DigitalMicrograph script used to compute the apparent electron transmission, and Dr Alison Killilea for providing this example of estimating the ice thickness. This figure is reproduced with permission from Agard *et al* (2014). Copyright (2014), with permission from Elsevier.

theoretical justification for using the same value as the one that is appropriate for energy-filtered images.

Either way, having a reliable and reproducible way to know the relative ice thickness is more important than to have an accurate, absolute value of thickness. As an initial rule of thumb, images are likely to be best if the fractional transmission—measured in either way described above—is between 0.8 and 0.9. Images will almost certainly become markedly worse if the transmission ratio is 0.7 or less and they are also likely to become poor when the ratio is higher than about 0.9. As a caution, it is not certain whether the optimal ice thickness values are the same for all types of particles. Some authors, such as Cheng *et al* (2015) for example, report that images of membrane proteins are better if particles are in thicker ice than when the ice is too thin. As a result, one must ultimately correlate the empirically determined electron transparency, whatever the thickness value may correspond to, with the image quality and then use that correlation as a guide for further data collection.

2.4.4 Areas where the sample is much too thin, perhaps even air-dried, can sometimes be avoided just on the basis of their subjective appearance

Depending upon how much previous experience one has had, it is sometimes possible to recognize that the sample may have become air-dried, even though the ambient humidity may have remained close to 100% prior to vitrification. If the contrast of the particles is 'too good to be true', for example, the explanation may be that the particles are no longer surrounded by (i.e. embedded in) ice. (See section 6.5 of Glaeser *et al* (2007) for an explanation of contrast matching, which occurs as long as the proteins are still embedded in ice.) Furthermore, the background may sometimes become markedly more granular than usual if a sample is air-dried, and a residue of precipitated salts and other small-molecule solutes is often left behind. Finally, one can always use very high electron exposures to see whether bubbling occurs (Cheng *et al* 2014), which is usually a good indication that organic material is still embedded in ice. Two possible examples of air-dried areas of a sample are shown in figure 2.8. In other cases, for which the sample is not yet completely dry, a meniscus of water may appear to surround individual particles and clusters of particles, which causes the image to be slightly darker close to the particle than it is in the larger gaps between the particles.

All of these descriptions raise the question of why there is anything at all remaining in a hole if the sample has been air-dried. This point is not an issue if one had used some type of support film, of course, such as a thin carbon film. In other cases, however, one possible explanation may be that there is a continuous, denatured-protein skin across the entire area of each hole—see the cartoon in figure 2.9 of Glaeser (2018), which then serves as a support film for bound macromolecules plus any residue of other solutes that remains after air drying. It is also possible that such a thin membrane of denatured protein is what prevents films of water from rupturing when they are so thin that particles are seen to be gathered near the rim of a hole but are absent in the middle—see figure 2.13 in section 2.5.

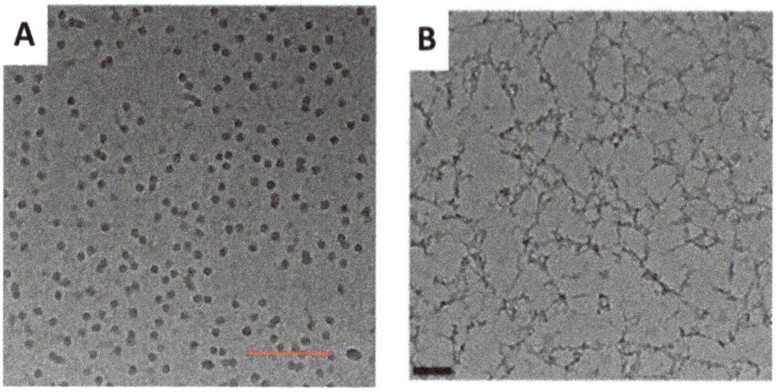

Figure 2.8. Examples of specimens that may have been air-dried before the grid was plunged into liquid ethane. (A) Unidentified particles that the authors suspected to have been freeze dried; note that the contrast appears to be too high for the particles to be embedded in vitreous ice. Reproduced from figure 2D of Grassucci *et al* (2007) with permission from Springer Nature, copyright 2007. Scale bar = 250 nm. (B) Sample of a ryanodine receptor, taken from a gallery of examples, showing common cryo-EM artifacts encountered in preparing cryo-EM grids, which is presented in Cabra and Samso (2015). Reproduced with permission. Not only does this material not have the globular shape expected for a detergent-solubilized membrane protein, but the contrast is higher than expected if the material were embedded in vitreous ice. Scale bar = 10 nm.

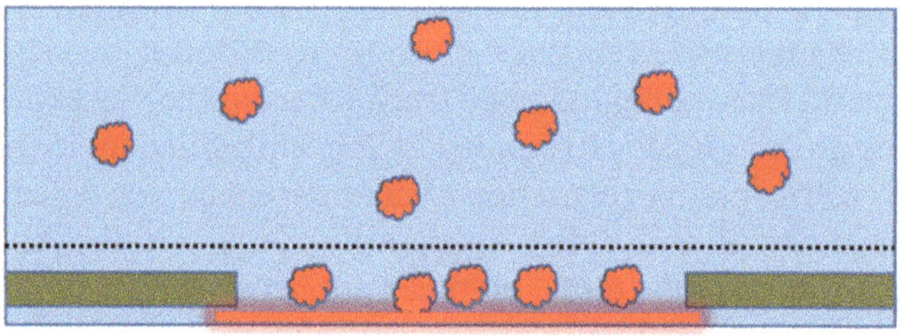

Figure 2.9. Cartoon illustrating the concept that a sacrificial denatured-protein skin may form over the hole on a support film when sample is applied to an EM grid, and additional protein molecules may then adsorb to the denatured monolayer. Such a denatured-protein skin might also serve as a continuous support film onto which solutes and debris would be deposited if an area of the specimen becomes air-dried prior to vitrification. This figure is reproduced from figure 3 in Glaeser (2018), copyright (2018), with permission from Elsevier.

It may be more difficult to recognize areas of a grid where a uniform distribution of particles remains, but the ice thickness is still less than the particle diameter. Particles in such areas are nevertheless at risk of being flattened or otherwise deformed, or they may even experience partial unfolding upon contract with the AWI. Also difficult to recognize are cases in which fluid shear (Armstrong *et al* 2020) may possibly damage the particles, in particular as the sample thickness becomes thinner and thinner. This latter point is mentioned simply to caution that there may be factors, other than those that are intuitively obvious, which damage particles when the ice thickness is not great enough. Thus, once again, if particles do not seem to be well preserved in areas where the sample is thin, it may be helpful to use one of the ways described above to measure the relative ice thickness and limit data collection to areas that were previously found to be optimal.

2.5 Current strategies for optimizing preparation of cryo-grids

Rebecca F Thompson
School of Molecular and Cellular Biology, Faculty of Biological Sciences, University of Leeds, Leeds, UK

The standard methods for preparing single-particle cryo-EM grids described in section 2.3 are undoubtedly successful, with the vast majority of structures solved by single-particle cryo-EM to date using this blotting paper-based 'plunge freezing' approach to prepare the grids. However, some problems can be encountered. These primarily relate to the behavior of single particles in the thin film environment, which is required for imaging (section 2.4), and the resulting irreproducibility of this process.

2.5.1 Behavior of particles in the thin film environment

Throughout this section we will refer to the specimen carriers on which the thin-specimen film is formed and vitrified as cryo-EM grids. These grids typically comprise a 3 mm metal disk (typically copper or gold) with square windows, over which a continuous support (typically carbon or gold) with perforations (holes) over which the thin film containing the specimen is formed (figure 2.10(A)). The grid may have an additional thin, continuous support over the top, discussed in section 2.5 (figure 2.10(B)). This creates at least one AWI and, optionally, a second support-water interface (SWI), with which particles may interact. In this section, we will discuss the practical implications of interactions with the AWI, and the measures that can be taken to ameliorate any negative effects.

In an ideal situation our macromolecular complex would have no affinity for the AWI, SWI, or itself. The particles would be suspended in a thin film just thicker than the particles' largest dimension, well distributed in the hole, and adopt random orientations (figure 2.10(A)). In order to generate an accurate, high-resolution reconstruction by single-particle cryo-EM, there is a requirement for particles to be distributed in a mono-disperse fashion with a range of angular orientations in the thin film being imaged (discussed further in section 1.2.7). The appearance of particles on a cryo-EM grid in reality is often far from this theoretical ideal situation (Drulyte *et al* 2018, Carragher *et al* 2019). Particles aggregating (figure 2.10(C)), adopting a preferred orientation (figure 2.10(D)), partitioning to the thick continuous support of the grid (figure 2.10(E)), as well as particles appearing morphologically different from the state that would be predicted from biochemical analysis (for example, missing a subunit, or becoming denatured as discussed below) (figure 2.10(F)) can all be barriers to structure determination.

By the use of cryo-electron tomography (cryo-ET), we can visualize particle partitioning by reconstructing a 3D model of the thin film environment. This is achieved by acquiring a series of projection images of the sample from various angular orientations. The subsequent image tilt-series is then aligned and computationally merged to form a 3D reconstruction known as a tomogram (Beck and Baumeister 2016, Wan and Briggs 2016). Analysis of the reconstructed tomogram can provide details on the interfaces of the thin film, the location of the particles within the film, and

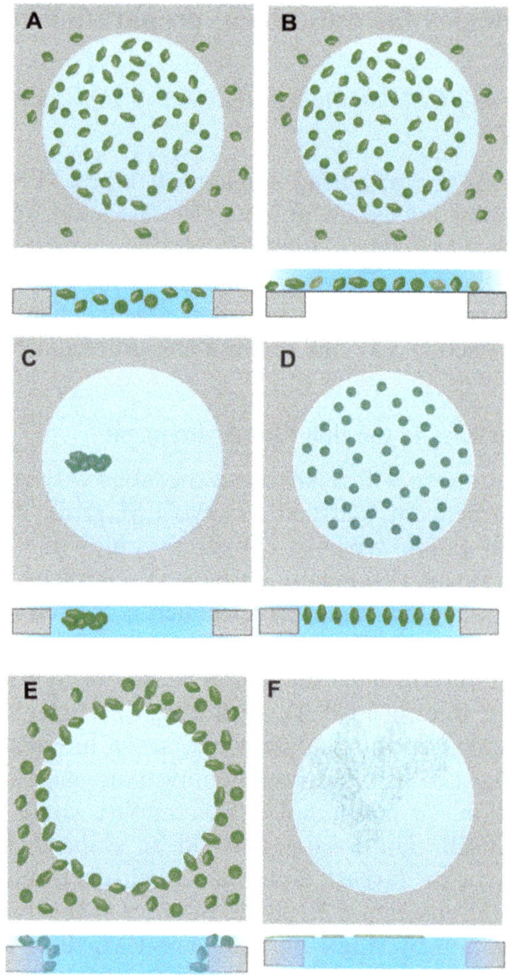

Figure 2.10. Schematic of particle distribution on cryo-EM grids. For each example a hole in the grid support (gray) is shown from the top and rotated 90° so as to show the particle distribution within the ice layer. Typically, this hole would be 1–2 μm in diameter. Hypothetical macromolecular complexes are in green and the vitreous ice in blue. (A) An ideal situation where particles are randomly distributed within the ice layer. (B) An ideal situation where a thin continuous support film (black line) is present over the top of the grid support. (C) Particle aggregation. (D) Particle preferred orientation. (E) Particle partitioning to the thick grid support where we cannot image. (F) Particle denaturation to a thin film at the air–water interface. Note that this denatured layer cannot be seen in raw cryo-EM images, the depiction here is a schematic.

any contamination on the surface of the film. It has been observed by cryo-ET that, for most of specimens examined, the majority interact with and partition to the AWI (Bharat and Scheres 2016, Noble *et al* 2018b) (figure 2.11). Where a thin continuous film is used, particles sometimes appear to preferentially partition to the SWI as opposed to the AWI.

How do particles behave at the AWI and SWI, and how does this affect our cryo-EM experiment? In a generalized model of particle behavior in the thin film

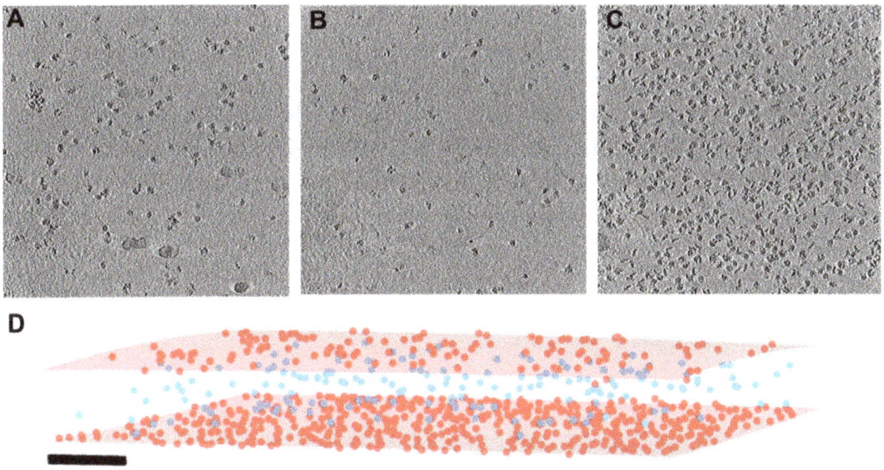

Figure 2.11. Cryo-electron tomography can be used to show where particles are partitioned in the ice layer. In this example the ribosomes have been vitrified on cryo-EM grids with a 2 μm hole using the Vitrobot. (A)–(C) show sections through the reconstructed tomogram at the top AWI (A), middle (B), and bottom AWI (C) of the tomogram. In these images we can see that both AWI contain more particles compared with the middle of the tomogram. A schematic of the particle locations in the ice later (D) shows the location of the AWI (red) and the particles partitioned to the interface (red circles) and particles not associated with the AWI (blue). Scale bar = 50 nm. Data kindly provided by David Klebl in the Stephen Muench group, University of Leeds.

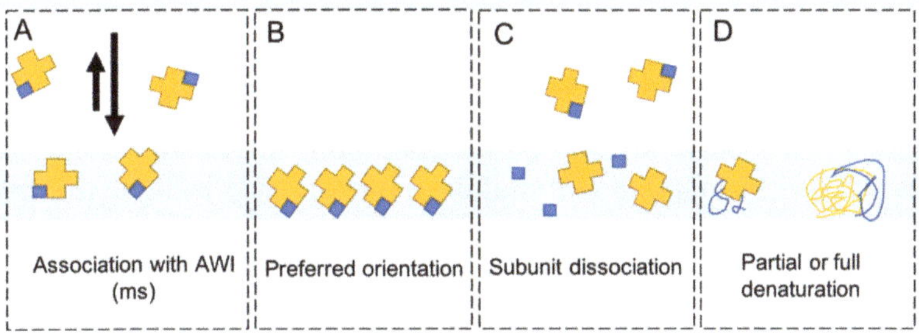

Figure 2.12. Potential effects of macromolecular-complex interactions at the AWI. (A) Proteins associate with the AWI within a millisecond of thin film formation. They may stay associated or desorb back away from the interface. (B) Once associated with the AWI, they may adopt a preferred orientation. (C) Forces at the AWI can cause structural changes to the complex, such as subunit dissociation. (D) The macromolecular complex may become subject to partial or full denaturation.

(figure 2.12), the initial stage involves diffusion-limited interactions with the AWI/SWI. This is very rapid; calculations show that a protein may interact with the AWI tens of times within a single millisecond (Taylor and Glaeser 2008, Naydenova *et al* 2019). After interacting with the interface, secondary effects can occur. The protein may desorb back away from the interface, and/or adopt a preferred orientation, or forces present at the AWI may cause dissociation of complexes and even partial or full denaturation of macromolecules (Glaeser 2018). Partial denaturation and exposure of hydrophobic regions may be a mechanism by which particles self-associate and aggregate in the thin film environment, complicating or even preventing structure determination. Upon denaturation, the protein may form a network at the AWI, which may influence how other particles interact subsequently with the AWI. The way in which macromolecules interact with the AWI, and their behavior across time scales from milliseconds to seconds, may be highly variable for each specimen, which may go some way to explain why such specimen-dependent cryo-EM grid optimization is often required (Klebl *et al* 2020a).

This all raises the question, in the event that the distribution or denaturation of particles in the thin film prevents single-particle structure determination: what strategies are available to researchers to alter their behavior during the sample preparation process?

2.5.2 Approaches to alter particle behavior in the thin film

Before trying to optimize cryo-EM grid preparation, it is worth considering if the source of the problems can or should be tackled by biochemical means first. If a protein preparation is of poor quality no amount of optimization on the cryo-EM grid side is likely to yield a successful result. Biochemical tools including functional assays, gel analysis, size-exclusion chromatography (SEC), SEC coupled with multi-angle laser light scattering (SEC-MALLS), circular dichroism, and thermal stability assays can all be helpful in determining if a protein preparation is sufficiently optimized to proceed with cryo-EM. A well-suited preparation of macromolecular complex should be pure with as few contaminants as possible and at an appropriate concentration to yield a sufficient number of particles in each micrograph. Efforts should be made to ensure complexes are compositionally homogeneous and stable on time scales relevant to cryo-EM sample preparation. Tools such as ProteoPlex can be used to identify buffer conditions under which macromolecular complexes have maximal thermal stability, which may include the presence of particular buffer constituents (e.g. salt, pH) or additives (e.g. small molecules, lipids, or metal ions) (Chari *et al* 2015). Some buffer components can interfere with the process of vitrification or subsequent structure determination by reducing the SNR. These include sucrose, glycerol, and high (~>350 mM) concentrations of salts, and therefore should be avoided where possible. It should, however, be noted that high-quality reconstructions can still be achieved in 'sub-optimal' buffers if the ice thickness is very well optimized, such as in the case of the HIV intasome structure (Passos *et al* 2020). Negative-stain analysis (as discussed in section 2.2) can also be an invaluable tool to quickly and cheaply check the quality of preparations before proceeding to cryo-EM, although in a small number of cases there is no straightforward relationship between the appearance in the negative stain and cryo-EM. For labile macromolecular complexes, cross-linking approaches, such as GraFix (Kastner *et al* 2008), may be used to increase complex stability and homogeneity, however, these approaches are not always successful (Carragher *et al* 2019).

Despite best efforts, even the most highly optimized biochemical sample preparations may yield unpredictable results during cryo-EM analysis. There is a large difference between the environment of a macromolecular complex in an Eppendorf tube and the thin film environment formed during cryo-EM grid preparation. The surface area to volume ratio is many orders of magnitude larger in a cryo-EM thin film than in an Eppendorf tube, and the protein is exposed to either one or two AWIs in the thin film, depending on whether a continuous support is used. As discussed above, these AWI interactions can manifest in a range of unhelpful ways. Where we do observe a difference between our biochemical/negative-stain analysis and how the complexes appear in cryo-EM, we can use a range of different tools to fine-tune particle behavior and interface interactions to overcome sub-optimal particle distributions in the thin film environment (Drulyte *et al* 2018).

One basic parameter that requires optimization is ice thickness, both in terms of generally achieving ice that is in the thickness range suitable for high-resolution imaging, but also because ice thickness and morphology can affect particle distribution. As discussed in section 2.3, for traditional paper-based blotting sample preparation methods varying the ice thickness is primarily achieved by adjusting the time over which the grid is exposed to the blotting paper ('blot time'). The morphology of the ice can affect particle crowding; this is particularly apparent with larger particles (e.g. viruses and ribosomes) but can be observed for a wide range of protein samples (figure 2.13). Ice thickness and morphology can also be tuned by varying grid type. Parameters that are varied can include using grids with smaller or larger diameter holes, or by using 'lacey' carbon films with an irregular array of holes. Lacey carbon grids can be a convenient way to screen different hole sizes on the same grid. Additives such as surfactants (molecules that can change the surface tension properties of the thin film and form layers at the AWI), whose use is discussed in more detail below, can also alter the morphology of the thin film formed. This in turn means that grid type or blotting parameters may need to be altered to achieve optimal results if surfactants are used. Adjusting the charge on the grid is usually achieved by plasma treatment or 'glow

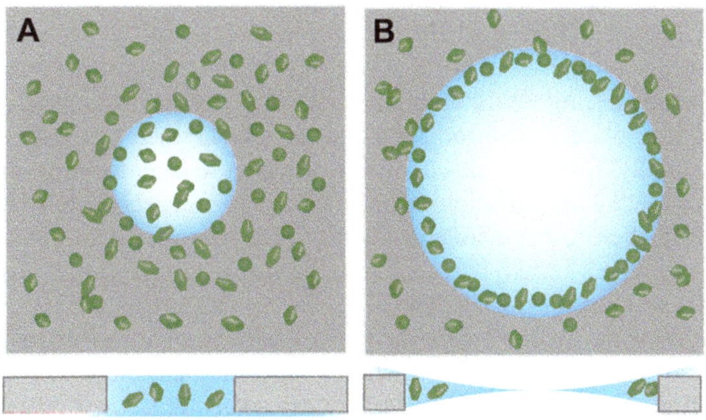

Figure 2.13. Schematic to show ice morphology affecting particle crowding. (A) Small hole with even ice across the hole. (B) Larger hole where the ice thins toward the center, physically excluding the particles from the center of the hole.

discharge', but can also be altered chemically, such as with poly-L-lysine (Jin *et al* 2019) or PEG-treated gold grids (Meyerson *et al* 2014). These approaches can affect both the formation of the thin film and partitioning of particles onto the support film or into the holes.

From a structural biologist's perspective the biggest danger of macromolecular-complex interaction at the AWI is denaturation. Full denaturation is catastrophic to the experiment at hand. While it is hard to unambiguously identify a thin monolayer of denatured proteins, the 'symptoms' of such an occurrence are often quite obvious, with few discernible intact particles. The researcher is usually able to determine that further optimization of sample preparation is required in this case. Partial denaturation may result in the researcher collecting large amounts of data and spending time and computational resources attempting to reconstruct before being able to identify that partial denaturation is complicating or precluding structure determination, or that the binding partner of interest has become disassociated from the complex. In one reported case, ~90% of yeast fatty acid particles collected within a dataset were partially denatured at the AWI (D'Imprima *et al* 2019). This study was able to show that the partial denaturation was not observed in negatively stained samples, indicating that the damage was being caused during the cryo-EM grid preparation process.

One generic approach to alter the properties of the AWI and potentially ameliorate harmful interactions is the use of surfactants. Surfactants can change the surface tension properties of the thin film and form layers at the AWI, this can reduce particle access to the hydrophobic side of the interface and thus protect them from potential denaturation and aggregation. Surfactants are necessary when preparing membrane proteins to shield the hydrophobic regions of the protein and stabilize individual particles, but adding surfactants to soluble complexes is also a common approach to improving the quality of cryo-EM grids. Many different kinds of surfactants have been reported to assist with cryo-EM sample preparation when added below their CMC (where applicable) including a range of detergents (e.g. LMNG, CHAPS, DDM, NP40) (Snijder *et al* 2018, Chen *et al* 2019) and fluorinated lipids (Johnson and Chen 2017). Surfactants which interact with the specimen of interest (Chen *et al* 2019) may also work via the same mechanism of preventing access to the hydrophobic AWI, or perhaps change the surface chemistry of the complex, altering its interactions with the AWI. Interestingly, the use of surfactants does not seem to produce a generic effect, since many macromolecules show dramatically different results when varying concentrations and types of surfactant are used. This variation means a cryo-EM researcher may have to screen many conditions before being able to identify an optimal buffer condition. A word of caution: researchers should be aware that the surfactant may itself have a negative impact on the stability of the complex.

Interestingly, denatured proteins themselves may be able to form a layer akin to that formed by detergent or lipid surfactants. Upon association with the AWI, it has been shown that protein films comprised of completely denatured proteins can form at the interface. These films have unique characteristics (including dimensions, charge, viscosity, and elasticity) depending on the proteins from which they have formed (Martin *et al* 2002). As further proteins then interact with this layer, they are prevented from accessing the hydrophobic environment but interact instead with the denatured-protein network. While it has not been possible to date to unambiguously identify the formation of this denatured-protein layer in the context of a cryo-EM grid, given the evidence from general observations on protein–interface interactions it seems highly likely this is occurring at least for some macromolecules in the context of cryo-EM grid preparation (Yoshimura *et al* 1994).

Another common strategy to protect macromolecular complexes from interactions at the AWI is to use grids with a continuous support such as to sequester particles at the SWI. Given the current evidence, it is thought that interactions with the SWI may be less hazardous and detrimental to the integrity of protein samples than interactions at the AWI. Despite this reasoning, it should be noted at this stage that there may be additional forces detrimental to protein structure present at the SWI, however, these are not understood as well as those at the AWI.

Due to the requirements for a thin film environment (section 2.4), an ideal continuous support should be as thin as possible, whilst sequestering the particles away from the AWI. For larger particles such as ribosomes and viruses, in particular where they have a natural affinity for the continuous carbon support, a thin (2–3 nm) continuous film of carbon can be used. Typically, such an approach can also be used to increase the number of particles compared with using holey supports. However, thin amorphous-carbon supports are not ideal for many specimens, as they introduce a significant additional layer through which the electrons must be transmitted. This extra obstacle can lead to a deterioration of the SNR achieved in the image, making identifying and aligning particles more challenging.

An alternative approach to achieving a continuous carbon film is through the use of graphene, a single-atom thick layer of carbon with attractive properties in terms of mechanical stability and conductance. Pristine graphene is highly hydrophobic and so unsuitable for cryo-EM grids, however, graphene can be treated by the process of hydrogenation, rendering it hydrophilic (Russo and Passmore 2014). Grids produced using this method can increase particle numbers in the field of view, just as continuous carbon supports can, but with very little effect on the SNR compared to grids with no continuous support. Graphene can also be made hydrophilic via non-covalent aromatic functionalization (Pantelic *et al* 2014, D'Imprima *et al* 2019). In addition, grids with a continuous graphene oxide support have been used to solve a large number of structures to a high resolution (Boland *et al* 2017), however, there can be local variations in the graphene film on this type of grid (e.g. graphene coverage and state of oxidation) owing to their method of production, which can make selecting areas to collect data from more challenging.

Although the continuous supports described above are often useful for a range of different proteins and macromolecular complexes, they are only tunable to a limited extent. Continuous supports can be engineered with specific characteristics based on the requirements of the sample. One development concerning continuous graphene films is the ability to pattern a range of different chemical functionalization's across a single grid. This recent advance opens up the possibility of testing multiple environments in order to identify an optimal condition for a particular macromolecular complex (Naydenova *et al* 2019). Another example that seeks to stabilize particles by reducing interactions at the AWI is the development of engineered grids known as 'affinity' grids. This approach can be as straightforward as immobilizing antibodies to a carbon support and using their affinity for the macromolecule of interest to sequester particles away from the AWI (Yu *et al* 2014). Other approaches pair chemically functionalized graphene oxide with covalent affinity tags (Wang *et al* 2019). A further novel approach takes advantage of the high affinity between streptavidin and biotin. This method utilizes a monolayer of streptavidin crystals to bind a recombinant biotin-tagged macromolecule. This approach has the advantage that the streptavidin crystal can be mathematically subtracted from the images before processing (Wang *et al* 2008); it also may help in assessing the quality of the image (Han *et al* 2017). Despite these advantages, the process of producing streptavidin-functionalized grids can be challenging and a range of different strategies for biotin-labeling may need to be trialed before success, depending on the macromolecule of interest.

While the focus of this section has been on the negative aspects of the AWI, it should be noted that sometimes there may also be beneficial effects. By using electron tomography to calculate the concentration of particles in the thin film and comparing this with solution concentration, for example, it has been shown for a range of specimens that the traditional filter paper-based blotting approach to forming a thin film creates an environment where the AWI concentrates particles by up to 30×, although the range of the concentration effect is highly specimen-dependent. For some complexes which are hard to purify or concentrate this by-product of the grid preparation process is actually an essential step in creating a cryo-EM grid with a viable number of particles in each micrograph (Klebl *et al* 2020a).

2.5.3 New technologies for sample preparation

The majority of cryo-EM structures solved to date have made use of commercially available devices such as the Vitrobot (Thermo Fisher Scientific), EP GP (Leica Microsystems), or CP3 (Gatan) to carry out a filter paper-based removal of excess specimen followed by plunge freezing (section 2.3). Remarkably, this method has not developed significantly in concept since its invention by Dubochet and colleagues in the 1980s (Dubochet *et al* 1988) and, as discussed in section 2.3, it can result in highly variable results. In order to address both problems of reproducibility and efficiency in terms of amount of sample required, new approaches are currently being developed, which are summarized in table 2.2.

Some of the technologies developed focus on improving reproducibility of thin film production, such as Spotiton, VitroJet, and cryoWriter (table 2.2). These systems are united in their approach of dispensing specific small volumes of sample to the grid, reducing the sample volume required. Interestingly, each of these systems uses a different approach to produce the thin films, with the use of self-wicking grids, pin printing very small volumes, and controlled evaporation being used for the Spotiton, VitroJet, and cryoWriter, respectively.

Approaches that apply droplets to the grid, such as Shake-it-off, the surface acoustic waves device, and time-resolved EM device (table 2.2) are all able to apply

Table 2.2. Cryo-EM sample preparation devices currently under development.

Device	Approach	Sample volume required	Fastest reported time from sample application to vitrification	Typical results	References
CryoWriter (under commercialization by NUONEX)	Microcapillary dispensing system to apply specimen to grid in a spiral pattern before plunge freezing. Series of flexible modules before sample dispensing can assist with a range of different sample preparation types, including working with single-cell lysate followed by affinity isolation of untagged proteins using fragment antigen binders from small quantities of cell lysate. Evaporation assists sample thinning. Can also be used to prepare negative-stain grids.	3–20 nl per grid	~350 ms from sample application to vitrification, ~1 s to 'write' a whole grid	A grid with a range of different ice thicknesses, depending on area of grid 'written'	(Arnold *et al* 2017, Schmidli *et al* 2019)

(Continued)

Table 2.2. (*Continued*)

Device	Approach	Sample volume required	Fastest reported time from sample application to vitrification	Typical results	References
Shake-it-off	An inexpensive piezoelectric sprayer applies sample in droplet form (droplet size not experimentally determined, but likely 10–100 nm) to self-wicking grids before plunge freezing into the cryogen.	~0.5–1 μl (Actual volume dispensed is in nl range so multiple grids can be prepared with 1 μl)	~100 ms	Grid with random dispersion of droplets, some of which have appropriate ice thickness at their periphery	(Rubinstein *et al* 2019)
Back-it-up	A variant on the Shake-it-off device, which uses standard EM grids rather than self-wicking grids. Once sample is applied using the piezoelectric sprayer, the liquid is wicked through the grid using glass fiber filter before the grid is plunged into the cryogen. Replacing the self-wicking grids used with Shake-it-off with through-grid wicking increases the amount of ice suitable for high-resolution data collection and makes the process less sensitive to variations in the volume or type of sample applied.	~0.5–1 μl (as for Shake-it-off)	135 ms	Significant (~1/3) areas of the grid with ice suitable for high-resolution data collection.	(Tan and Rubinstein 2020)

Spotiton	A piezo droplet dispenser applies small volumes of sample to self-wicking grids, which reduce sample thickness to a thin film, as monitored by an optical system, before vitrification. Two independent streams of sample can be used to achieve time-resolved applications.	5 μl, 30–40 nl deposited per grid	~133 ms	Grid with strip of squares with optimal ice thickness	(Jain *et al* 2012, Dandey *et al* 2018, 2020, Wei *et al* 2018, Darrow *et al* 2019, Klebl *et al* 2020b, Liu *et al* 2020)
Chameleon (SPT Labtech)	Commercialization of the Spotiton device with the addition of automation around grid handling, glow discharge and reporting.	2 μl, then ~10 nl per grid	54 ms		
Surface acoustic waves	Surface acoustic waves to deliver the sample from a microfluidic chip in the form of an aerosol (6–8 μm droplets) to a standard EM grid before vitrification.	Not reported	52.6 ms	Grid with random dispersion of sample 'mist', some of which have appropriate ice thickness	(Ashtiani *et al* 2018)
Vitrojet (cryoSOL)	Pin printing of sample in overlapping spiral pattern produces thin film of specimen in a humidity and temperature-controlled environment. Integrated glow discharge module and jet vitrification reduce number of manual handling steps.	0.5 μl for 'stock', sub nl for each grid	Sample deposition 1–2 s. Transfer of the grid between the process chamber and the vitrification chamber <80 ms. Vitrification of the sample <1 ms.	A grid with potentially many grid squares of optimized ice thickness, depending on area of grid sample is applied.	(Ravelli *et al* 2019)

(*Continued*)

Table 2.2. (*Continued*)

Device	Approach	Sample volume required	Fastest reported time from sample application to vitrification	Typical results	References
Time-resolved EM device	Primarily designed to perform time-resolved experiments by rapidly mixing (for example) protein and substrate before vitrification on a millisecond timescale, this device either uses gas pressure or voltage assisted mechanism to generate droplets (~ 5–75 μm) which land on the grid.	33 μl stock, ~ 1 μl per grid	10 ms (or slower, depending on machine set up)	A grid with a random dispersion of droplets, at the edge of which are areas suitable ice thickness for imaging.	(Kontziampasis *et al* 2019, Klebl *et al* 2020a)

sample and vitrify faster than is achieved with devices that dispense set volumes. These high-speed devices allow the possibility of time-resolved electron microscopy experiments, which may provide unique insights into macromolecule function. Despite these unique advantages, these methods come at the cost of ice consistency and reproducibility, with no active monitoring of ice thickness on these devices and more variable results across and between each grid, although technological advances may help to ameliorate these short comings (Klebl *et al* 2020a). Back-it-up is an example of a hybrid device that combines spraying of droplets with subsequent blotting to achieve more consistent ice thickness. By using a glass fiber filter opposed to traditional filter paper, some of the negative effects of blotting paper (Armstrong *et al* 2020) may be overcome.

Since these devices are still under development, but some are at various stages of commercialization, it is likely that these methodologies may become more reliable, reproducible, and accessible in the coming years. These emerging technologies may enable cryo-EM to tackle new research questions, such as introducing elements of time resolution, as well as improving the reproducibility of the production of thin films for vitrification. However, many of the problems discussed in section 2.5.1 are introduced by interactions of particles at the AWI, which as discussed above can occur within a millisecond. This means that, even with the devices which can vitrify samples on the fastest times (6 ms fastest reported at time of writing), we often are unable to outrun the potential aberrant effects taking place at the AWI by speed alone. Combining these devices with other tools such as affinity grids and engineered continuous supports (discussed in section 2.5.3), may provide additional routes to improve the quality and reproducibility of cryo-EM samples in the future. The area of sample preparation is now developing quickly and exciting future developments may once more transform the types of research question that can be tackled using cryo-EM.

References

Adrian M *et al* 1984 Cryo-electron microscopy of viruses *Nature* **308** 32–6

Agard D, Cheng Y F, Glaeser R M and Subramaniam S 2014 Single-particle cryo-electron microscopy (cryo-EM): progress, challenges, and perspectives for further improvement *Advances in Imaging and Electron Physics* vol 185 ed P W Hawkes (Amsterdam: Academic)

Armstrong M, Han B G, Gomez S, Turner J, Fletcher D A and Glaeser R M 2020 Microscale fluid behavior during cryo-EM sample blotting *Biophys. J.* **118** 708–19

Arnold S A, Albiez S, Bieri A, Syntychaki A, Adaixo R, McLeod R A, Goldie K N, Stahlberg H and Braun T 2017 Blotting-free and lossless cryo-electron microscopy grid preparation from nanoliter-sized protein samples and single-cell extracts *J. Struct. Biol.* **197** 220–6

Arnold S A, Müller S A, Schmidli C, Syntychaki A, Rima L, Chami M, Stahlberg H, Goldie K N and Braun T 2018 Miniaturizing EM sample preparation: opportunities, challenges, and 'visual proteomics' *Proteomics* **18** e1700176

Arthur C P and Ciferri C 2019 High-throughput protein analysis using negative stain electron microscopy and 2D classification *Methods Mol. Biol.* **2025** 477–85

Ashtiani D, Venugopal H, Belousoff M, Spicer B, Mak J, Neild A and de Marco A 2018 Delivery of femtolitre droplets using surface acoustic wave based atomisation for cryo-EM grid preparation *J. Struct. Biol.* **203** 94–101

Beck M and Baumeister W 2016 Cryo-electron tomography: can it reveal the molecular sociology of cells in atomic detail? *Trends Cell Biol.* **26** 825–37

Benjamin C J *et al* 2016 Nonfouling NTA-PEG-based TEM grid coatings for selective capture of histidine-tagged protein targets from cell lysates *Langmuir* **32** 551–9

Bharat T A M and Scheres S H W 2016 Resolving macromolecular structures from electron cryo-tomography data using subtomogram averaging in RELION *Nat. Protoc.* **11** 2054–65

Binshtein E and Ohi M D 2015 Cryo-electron microscopy and the amazing race to atomic resolution *Biochemistry* **54** 3133–41

Boland A *et al* 2017 Cryo-EM structure of a metazoan separase–securin complex at near-atomic resolution *Nat. Struct. Mol. Biol.* **24** 414–8

Booth D S, Avila-Sakar A and Cheng Y 2011 Visualizing proteins and macromolecular complexes by negative stain EM: from grid preparation to image acquisition *J. Vis. Exp.* **58** e3227

Bremer A, Henn C, Engel A, Baumeister W and Aebi U 1992 Has negative staining still a place in biomacromolecular electron microscopy? *Ultramicroscopy* **46** 85–111

Brenner S and Horne R W 1959 A negative staining method for high resolution electron microscopy of viruses *Biochim. Biophys. Acta* **34** 103–10

Brillault L and Landsberg M J 2020 Preparation of proteins and macromolecular assemblies for cryo-electron microscopy *Methods Mol. Biol.* **2073** 221–46

Cabra V and Samso M 2015 Do's and don'ts of cryo-electron microscopy: a primer on sample preparation and high quality data collection for macromolecular 3D reconstruction *J. Vis. Exp.* **95** 52311

Carragher B, Cheng Y, Frost A, Glaeser R M, Lander G C, Nogales E and Wang H-Y 2019 Current outcomes when optimizing 'standard' sample preparation for single-particle cryo-EM *J. Microsc.* **276** 39–45

Chari A *et al* 2015 ProteoPlex: stability optimization of macromolecular complexes by sparse-matrix screening of chemical space *Nat. Methods* **12** 859–65

Chen J, Noble A J, Kang J Y and Darst S A 2019 Eliminating effects of particle adsorption to the air/water interface in single-particle cryo-electron microscopy: bacterial RNA polymerase and CHAPSO *J. Struct. Biol.: X* **1** 100005

Cheng N Q, Wu W M, Watts N R and Steven A C 2014 Exploiting radiation damage to map proteins in nucleoprotein complexes: the internal structure of bacteriophage T7 *J. Struct. Biol.* **185** 250–6

Cheng Y, Grigorieff N, Penczek P A and Walz T 2015 A primer to single-particle cryo-electron microscopy *Cell* **161** 438–49

Cheng Y 2018 Single-particle cryo-EM-How did it get here and where will it go *Science* **361** 876–80

Chiu P L, Kelly D F and Walz T 2011 The use of trehalose in the preparation of specimens for molecular electron microscopy *Micron* **42** 762–72

Ciferri C, Lander G C, Maiolica A, Herzog F, Aebersold R and Nogales E 2012 Molecular architecture of human polycomb repressive complex 2 *eLife* **1** e00005

Dandey V P *et al* 2018 Spotiton: new features and applications *J. Struct. Biol.* **202** 161–9

Dandey V P *et al* 2020 Time-resolved cryoEM using Spotiton *Nat. Methods* **17** 897–900

Darrow M C, Moore J P, Walker R J, Doering K and King R S 2019 Chameleon: next generation sample preparation for cryoEM based on Spotiton *Microsc. Microanal.* **25** 994–5

De Carlo S and Harris J R 2011 Negative staining and cryo-negative staining of macromolecules and viruses for TEM *Micron* **42** 117–31

DeRosier D J 2000 Correction of high-resolution data for curvature of the Ewald sphere *Ultramicroscopy* **81** 83–98

D'Imprima E, Floris D, Joppe M, Sanchez R, Grininger M and Kuehlbrandt W 2019 Protein denaturation at the air-water interface and how to prevent it *eLife* **8** e42747

Dobro M J *et al* 2010 Plunge freezing for electron cryomicroscopy *Methods Enzymol.* **481** 63–82

Doerr A 2019 Graphene-on-gold grids for cryo-EM *Nat. Methods* **16** 578

Downing K H and Glaeser R M 2018 Estimating the effect of finite depth of field in single-particle cryo-EM *Ultramicroscopy* **184** 94–9

Drulyte I, Johnson R M, Hesketh E L, Hurdiss D L, Scarff C A, Porav S A, Ranson N A, Muench S P and Thompson R F 2018 Approaches to altering particle distributions in cryo-electron microscopy sample preparation *Acta Crystallogr.* D **74** 560–71

Dubochet J, Adrian M, Chang J, Homo J, Lepault J, McDowall A W and Schultz P 1988 Cryo-electron microscopy of vitrified specimens *Q. Rev. Biophys.* 21 129–228

Earl L A *et al* 2017 Cryo-EM: beyond the microscope *Curr. Opin. Struct. Biol.* **46** 71–8

Feng X *et al* 2017 A fast and effective microfluidic spraying-plunging method for high-resolution single-particle cryo-EM *Structure* **25** 663–70

Frank J 1996 *Three-dimensional Electron Microscopy of Macromolecular Assemblies* 1st edn (San Diego, CA: Academic)

Galaz-Montoya J G, Hecksel C W, Baldwin P R, Wang E, Weaver S C, Schmid M F, Ludtke S J and Chiu W 2016 Alignment algorithms and per-particle CTF correction for single particle cryo-electron tomography *J. Struct. Biol.* **194** 383–94

Glaeser R M 2018 Proteins, interfaces, and cryo-EM grids *Curr. Opin. Colloid Interface Sci.* **34** 1–11

Glaeser R M, Downing K, Derosier D, Chiu W and Frank J 2007 *Electron Crystallography of Biological Macromolecules* (Oxford: Oxford University Press)

Glaeser R M and Han B G 2017 Opinion: hazards faced by macromolecules when confined to thin aqueous films *Biophys. Rep.* **3** 1–7

Grassucci R A, Taylor D J and Frank J 2007 Preparation of macromolecular complexes for cryo-electron microscopy *Nat. Protoc.* **2** 3239–46

Grimm R, Typke D, Barmann M and Baumeister W 1996 Determination of the inelastic mean free path in ice by examination of tilted vesicles and automated most probable loss imaging *Ultramicroscopy* **63** 169–79

Grigorieff N 2007 FREALIGN: high-resolution refinement of single particle structures *J. Struct. Biol.* **157** 117–25

Han B G *et al* 2012 Electron microscopy of biotinylated protein complexes bound to streptavidin monolayer crystals *J. Struct. Biol.* **180** 249–53

Han B G *et al* 2016 Long shelf-life streptavidin support-films suitable for electron microscopy of biological macromolecules *J. Struct. Biol.* **195** 238–44

Han B G, Watson Z, Cate J H D and Glaeser R M 2017 Monolayer-crystal streptavidin support films provide an internal standard of cryo-EM image quality *J. Struct. Biol.* **200** 307–13

Han Y *et al* 2020 High-yield monolayer graphene grids for near-atomic resolution cryoelectron microscopy *Proc. Natl Acad. Sci. USA* **117** 1009–14

Harris J R 1997 *Negative Staining and Cryoelectron Microscopy: The Thin Film Techniques* (Oxford: BIOS Scientific)

Haschemeyer R H and Myers R J 1972 Negative staining *Principles and Techniques of Electron Microscopy* ed M A Hayat (New York: Van Nostrand)

He Y, Fang J, Taatjes D J and Nogales E 2013 Structural visualization of key steps in human transcription initiation *Nature* **495** 481–6

Ho C M *et al* 2020 Bottom-up structural proteomics: cryoEM of protein complexes enriched from the cellular milieu *Nat. Methods* **17** 79–85

Hohn M *et al* 2007 SPARX, a new environment for Cryo-EM image processing *J. Struct. Biol.* **157** 47–55

Iancu C V *et al* 2006 Electron cryotomography sample preparation using the Vitrobot *Nat. Protoc.* **1** 2813–9

Imai H *et al* 2015 Direct observation shows superposition and large scale flexibility within cytoplasmic dynein motors moving along microtubules *Nat. Commun.* **6** 8179

Jain T, Sheehan P, Crum J, Carragher B and Potter C S 2012 Spotiton: a prototype for an integrated inkjet dispense and vitrification system for cryo-TEM *J. Struct. Biol.* **179** 68–75

Jensen G J and Kornberg R D 2000 Defocus-gradient corrected back-projection *Ultramicroscopy* **84** 57–64

Jin M, Han W, Liu C, Zang Y, Li J, Wang F, Wang Y and Cong Y 2019 An ensemble of cryo-EM structures of TRiC reveal its conformational landscape and subunit specificity *Proc. Natl Acad. Sci. USA* **116** 19513–22

Johnson Z L and Chen J 2017 Structural basis of substrate recognition by the multidrug resistance protein MRP1 *Cell* **9** 1075–85

Kaledhonkar S *et al* 2018 Time-resolved cryo-electron microscopy using a microfluidic chip *Methods Mol. Biol.* **1764** 59–71

Kastner B *et al* 2008 GraFix: sample preparation for single-particle electron cryomicroscopy *Nat. Methods* **5** 53–5

Kellenberger E and Kistler J 1979 The physics of specimen preparation *Advances in Structure Research by Diffraction Methods* ed W Hoppe and R Mason (Wiesbaden: Vieweg)

Kim L Y, Rice W J, Eng E T, Kopylov M, Cheng A, Raczkowski A M, Jordan K D, Bobe D, Potter C S and Carragher B 2018 Benchmarking cryo-EM single particle analysis workflow *Front. Mol. Biosci.* **5** 50

Klebl D P, Monteiro D C F, Kontziampasis D, Kopf F, Sobott F, White H D, Trebbin M and Muench S P 2020a Sample deposition onto cryo-EM grids: from sprays to jets and back *Acta Crystallogr.* D **76** 340–49

Klebl D P *et al* 2020b Need for speed: examining protein behaviour during cryoEM grid preparation at different timescales *Structure* **3** 1238–48

Kontziampasis D, Klebl D P, Iadanza M G, Scarff C A, Kopf F, Sobott F, Monteiro D C F, Trebbin M, Muench S P and White H D 2019 A cryo-EM grid preparation device for time-resolved structural studies *IUCrJ* **6** 1024–31

Liu N *et al* 2019 Bioactive functionalized monolayer graphene for high-resolution cryo-electron microscopy *J. Am. Chem. Soc.* **141** 4016–25

Liu Y *et al* 2020 FACT caught in the act of manipulating the nucleosome *Nature* **577** 426–31

Martin A H, Grolle K, Bos M A, Cohen M A Stuart and van Vliet T 2002 Network forming properties of various proteins adsorbed at the air/water interface in relation to foam stability *J. Colloid Interface Sci.* **254** 175–83

Mastronarde D N 2005 Automated electron microscope tomography using robust prediction of specimen movements *J. Struct. Biol.* **152** 36–51

McMullan G, Vinothkumar K R and Henderson R 2015 Thon rings from amorphous ice and implications of beam-induced Brownian motion in single particle electron cryo-microscopy *Ultramicroscopy* **158** 26–32

Meyerson J R, Rao P, Kumar J, Chittori S, Banerjee S, Pierson J, Mayer M L and Subramaniam S 2014 Self-assembled monolayers improve protein distribution on holey carbon cryo-EM supports *Sci. Rep.* **4** 7084

Naydenova K, Peet M J and Russo C J 2019 Multifunctional graphene supports for electron cryomicroscopy *Proc. Natl Acad. Sci. USA* **116** 11718–24

Noble A J *et al* 2018a Reducing effects of particle adsorption to the air–water interface in cryo-EM *Nat. Methods* **15** 793–5

Noble A J *et al* 2018b Routine single particle cryoEM sample and grid characterization by tomography *eLife* **7** 32

Nogales E and Scheres S H 2015 Cryo-EM: a unique tool for the visualization of macromolecular complexity *Mol. Cell* **58** 677–89

Ohi M, Li Y, Cheng Y and Walz T 2004 Negative staining and image classification—powerful tools in modern electron microscopy *Biol. Proced. Online* **6** 23–34

Palovcak E *et al* 2018 A simple and robust procedure for preparing graphene-oxide cryo-EM grids *J. Struct. Biol.* **204** 80–4

Pantelic R S, Fu W, Schoenenberger C and Stahlberg H 2014 Rendering graphene supports hydrophilic with non-covalent aromatic functionalization for transmission electron microscopy *Appl. Phys. Lett.* **104** 134103

Passos D O *et al* 2020 Structural basis for strand-transfer inhibitor binding to HIV intasomes *Science* **14** 810–4

Punjani A, Rubinstein J L, Fleet D J and Brubaker M A 2017 CryoSPARC: algorithms for rapid unsupervised cryo-EM structure determination *Nat. Methods* **14** 290–6

Radermacher M, Wagenknecht T, Verschoor A and Frank J 1987 Three-dimensional reconstruction from a single-exposure, random conical tilt series applied to the 50S ribosomal subunit of *Escherichia coli J. Microsc.* **146** 113–36

Ravelli R B G, Nijpels F J T, Henderikx R J M, Weissenberger G, Thewessem S, Gijsbers A, Beulen B W A M M, López-Iglesias C and Peters P J 2019 Automated cryo-EM sample preparation by pin-printing and jet vitrification *bioRxiv*: **561** 651208

Rice W J, Cheng A C, Noble A J, Eng E T, Kim L Y, Carragher B and Potter C S 2018 Routine determination of ice thickness for cryo-EM grids *J. Struct. Biol.* **204** 38–44

Rubinstein J L, Guo H, Ripstein Z A, Haydaroglu A, Au A, Yip C M, Di Trani D M, Benlekbir S and Kwok T 2019 Shake-it-off: a simple ultrasonic cryo-EM specimen-preparation device *Acta Crystallogr.* D **75** 1063–70

Russo C J and Passmore L A 2014 Controlling protein adsorption on graphene for cryo-EM using low-energy hydrogen plasmas *Nat. Methods* **11** 649–52

Russo C J and Passmore L A 2016 Ultrastable gold substrates: properties of a support for high-resolution electron cryomicroscopy of biological specimens *J. Struct. Biol.* **193** 33–44

Samso M, Palumbo M J, Radermacher M, Liu J S and Lawrence C E 2002 A Bayesian method for classification of images from electron micrographs *J. Struct. Biol.* **138** 157–70

Scarff C A, Fuller M J G, Thompson R F and Iadanza M G 2018 Variations on negative stain electron microscopy methods: tools for tackling challenging systems *J. Vis. Exp.* **132** e57199

Scheres S H 2012 RELION: implementation of a Bayesian approach to cryo-EM structure determination *J. Struct. Biol.* **180** 519–30

Schmidli C, Albiez S, Rima L, Righetto R, Mohammed I, Oliva P, Kovacik L, Stahlberg H and Braun T 2019 Microfluidic protein isolation and sample preparation for high-resolution cryo-EM *Proc. Natl Acad. Sci. USA* **116** 15007–12

Sgro G G and Costa T R D 2018 Cryo-EM grid preparation of membrane protein samples for single particle analysis *Front. Mol. Biosci.* **5** 74

Snijder J, Ortego M S, Weidle C, Stuart A B, Gray M D, McElrath M J, Pancera M, Veesler D and McGuire A T 2018 An antibody targeting the fusion machinery neutralizes dual-tropic infection and defines a site of vulnerability on Epstein-Barr virus *Immunity* **17** 799–811

Stagg S M *et al* 2006 Automated cryoEM data acquisition and analysis of 284742 particles of GroEL *J. Struct. Biol.* **155** 470–81

Stark H 2010 GraFix: stabilization of fragile macromolecular complexes for single particle cryo-EM *Methods Enzymol.* **481** 109–26

Su M 2019 goCTF: geometrically optimized CTF determination for single-particle cryo-EM *J. Struct. Biol.* **205** 22–9

Suloway C *et al* 2005 Automated molecular microscopy: the new Leginon system *J. Struct. Biol.* **151** 41–60

Takizawa Y, Binshtein E, Erwin A L, Pyburn T M, Mittendorf K F and Ohi M D 2017 While the revolution will not be crystallized, biochemistry reigns supreme *Protein Sci.* **26** 69–81

Tan Y Z and Rubinstein J L 2020 Through-grid wicking enables high-speed cryoEM specimen preparation *Acta Cryst.* D **76** 1092–103

Tang G *et al* 2007 EMAN2: an extensible image processing suite for electron microscopy *J. Struct. Biol.* **157** 38–46

Taylor K A and Glaeser R M 1973 Hydrophilic support films of controlled thickness and composition *Rev. Sci. Instrum.* **44** 1546–7

Taylor K A and Glaeser R M 1974 Electron diffraction of frozen, hydrated protein crystals *Science* **186** 1036–7

Taylor K A and Glaeser R M 2008 Retrospective on the early development of cryoelectron microscopy of macromolecules and a prospective on opportunities for the future *J. Struct. Biol.* **163** 214–23

Tenthorey J L *et al* 2017 The structural basis of flagellin detection by NAIP5: a strategy to limit pathogen immune evasion *Science* **358** 888–93

Van Heel M 1987 Angular reconstitution: *a posteriori* assignment of projection directions for 3D reconstruction *Ultramicroscopy* **21** 111–23

van Heel M, Harauz G, Orlova E V, Schmidt R and Schatz M 1996 A new generation of the IMAGIC image processing system *J. Struct. Biol.* **116** 17–24

Walker M *et al* 1994 Electron cryomicroscopy of acto-myosin-S1 during steady-state ATP hydrolysis *Biophys. J.* **66** 1563–72

Wan W and Briggs J A G 2016 Cryo-electron tomography and subtomogram averaging *Methods Enzymol.* **579** 329–67

Wang F, Liu Y, Yu Z, Li S, Feng S, Cheng Y and Agard D A 2020 General and robust covalently linked graphene oxide affinity grids for high-resolution cryo-EM *Proc. Natl Acad. Sci. USA* **117** 24269–73

Wang F *et al* 2020 Amino and PEG-amino graphene oxide grids enrich and protect samples for high-resolution single particle cryo-electron microscopy *J. Struct. Biol.* **209** 107437

Wang L, Ounjai P and Sigworth F J 2008 Streptavidin crystals as nanostructured supports and image-calibration references for cryo-EM data collection *J. Struct. Biol.* **164** 190–8

Wei H, Dandey V P, Zhang Z, Raczkowski A, Rice W J, Carragher B and Potter C S 2018 Optimizing 'self-wicking' nanowire grids *J. Struct. Biol.* **202** 170–4

Yoshimura H, Scheybani T, Baumeister W and Nagayama K 1994 Two-dimensional protein array growth in thin layers of protein solution on aqueous subphases *Langmuir* **9** 3290–5

Yu G *et al* 2014 Single-step antibody-based affinity cryo-electron microscopy for imaging and structural analysis of macromolecular assemblies *J. Struct. Biol.* **187** 1–9

Zhao F Q and Craig R 2003 Capturing time-resolved changes in molecular structure by negative staining *J. Struct. Biol.* **141** 43–52

Zheng L *et al* 2020 Robust ultraclean atomically thin membranes for atomic-resolution electron microscopy *Nat. Commun.* **11** 541

Zivanov J, Nakane T, Forsberg B O, Kimanius D, Hagen W J H, Lindahl E and Scheres S H W 2018 New tools for automated high-resolution cryo-EM structure determination in RELION-3 *eLife* **7** e42166

Chapter 3

Data collection

3.1 Overview

The strategies adopted for recording high-resolution images of unstained biological macromolecules differ in three key aspects from those used in many areas of materials science. Perhaps foremost among these differences is the fact that cryo-EM images of biological macromolecules must be recorded with extremely low electron exposures, which may be thousands of times, or even far more than that, lower than what is used for many other materials.

High sensitivity to radiation damage is the reason why cryo-EM exposures must be so low, as is explained in section 3.2. These exposures are so low, in fact, that it can be challenging to even see biological macromolecules in the raw images and high-resolution features are certainly not visible in images of individual particles. As a result, it is necessary to merge the noisy data obtained from a large number of images. These numbers currently stand in the range of hundreds of thousands to a few million particles in order to bring the signal-to-noise ratio (SNR) up to the level at which atomic resolution is achieved. Collecting so many images would be a daunting task if it were not for the development of various automated data collection capabilities, which are described in section 3.3.

In addition, cryo-EM images have almost no contrast when they are taken precisely in focus. Instead, they must be intentionally defocused by amounts that are considered to be unacceptable in materials science applications. As is explained in section 3.4, the reason for doing so is to generate enough phase contrast in the image intensity to be able to identify particles, and thus to extract their images, in preparation for merging the data. Ideally one might prefer to use a phase plate to generate phase contrast, as can be done in light microscopy. While significant progress has been made in that direction, again described in section 3.4, this remains an area that is actively under development.

Finally, cryo-EM specimens suffer from a unique problem that is referred to as 'beam-induced motion', something that does not occur for specimens that are far less

doi:10.1088/978-0-7503-3039-8ch3

sensitive to radiation damage. Section 3.5 describes how this problem can be substantially mitigated by recording images as a series of 'frames' and then aligning them to one another in order to compensate for beam-induced motion. The development of camera technology that made this data collection strategy possible was the key breakthrough needed to launch the cryo-EM 'revolution'.

3.2 Radiation damage in cryo-EM

John L Rubinstein
The Hospital for Sick Children, Toronto, ON, Canada
The University of Toronto, Toronto, ON, Canada

3.2.1 Introduction

Irradiation of biological specimens with a high-energy electron beam causes significant damage to the specimens. It is therefore not obvious that electron microscopy is capable of providing high-resolution insight into biological structures. Indeed, the fundamental limitation for electron microscopy of biological molecules is radiation damage. The atoms in organic molecules are held together by covalent bonds with dissociation energies of a few electron volts (e.g. ~4 eV between carbon atoms). In contrast, electrons in the electron beam of most modern instruments used for transmission electron microscopy (TEM) have energies between 100 and 300 keV. Consequently, each electron has the kinetic energy needed to break nearly 10^5 covalent bonds in the specimen. The beauty of the single-particle cryo-EM technique lies in the numerous ways that investigators have found to get around this limitation, wringing as much information as possible out of a destructive imaging technique and harnessing the power of computation to build high-resolution maps of complicated macromolecular assemblies.

3.2.2 Interaction cross sections, elastic, and inelastic interactions

Multiple kinds of interaction can occur between a biological specimen and a high-energy electron from the electron beam. The probability of a specific type of interaction occurring depends on the cross section of that interaction. Expressed with units of area, large cross sections indicate a high probability of an event occurring while small cross sections indicate the inverse.

In order to form a high-resolution image with scattered electrons, each electron would ideally be scattered elastically, with the scattered electron leaving the sample with the same energy with which it entered, and the atom in the sample that scattered it remaining in place. An undesirable form of elastic scattering is also possible, in which the scattered electron knocks an atom out of the specimen while conserving kinetic energy. However, the resulting knock-on damage occurs with an extremely low cross section for biological specimens (Glaeser *et al* 2007) and it can be ignored in the discussion that follows. Inelastic events are a more insidious type of interaction. In an inelastic interaction the incident electron loses some of its energy into the specimen during scattering. A 300 keV electron could deposit up to 300 keV into the specimen, but on average each inelastic interaction deposits ~20 eV into the specimen (Henderson 1995), which is still sufficient to break numerous covalent bonds.

The ratio of cross sections for inelastic and elastic events between biological molecules and electrons with energies typical for TEM are such that inelastic events outnumber elastic events by ~3:1. Consequently, each informative elastic event is accompanied by the deposition of ~60 eV of energy into the specimen (Henderson 1995). This unfavorable situation is still more than 1000× better than the equivalent situation with x-ray scattering, where radiation damage is thought to occur through similar mechanisms and each informative elastic scattering event is accompanied by deposition of ~80 keV of energy into the specimen (Henderson 1995). Both inelastic and elastic interactions have higher cross sections at lower energies, so that more elastic scattering, more inelastic scattering, and more radiation damage occur for the same electron exposure with lower-energy electron microscopes. There are several factors that dictate the choice of optimal electron energy, as discussed in section 3.2.4.

3.2.3 Cryoprotection and primary, secondary, and tertiary radiation damage

Primary radiation damage occurs through the breaking of covalent bonds by the incident electron. This radiolytic event creates reactive species that can migrate through the specimen, causing subsequent changes to the specimen in a process known as secondary radiation damage. Finally, the release of hydrogen, resulting from radiolysis of both biological molecules and water, leads to the generation of hydrogen gas. The hydrogen can form bubbles that distort the specimen, which is known as tertiary damage. Further details of these processes are reviewed in Baker and Rubinstein (2010).

Cryo-EM is typically performed close to liquid-nitrogen temperature (−195.8 °C or 77.4 K) to maintain specimens in the vitreous ice in which they are trapped during sample preparation. However, cooling specimens also has a dramatic effect on the apparent radiation sensitivity of specimens. Cooling specimens during cryo-EM does not prevent or reduce the probability of the initial insult to the specimen, the breaking of covalent bonds. However, as shown in the 1970s (Taylor and Glaeser 1976, Hayward and Glaeser 1979), cooling specimens reduced fading in the intensity of diffraction spots from radiation sensitive crystals during irradiation. Further improvement in cryoprotection is seen at even lower temperatures, and some early high-resolution studies of 2D crystals were performed at liquid-helium temperature (Fujiyoshi 1998).

The phenomenon of cryoprotection occurs because the reactive radiolytic products from primary damage can be trapped within the specimen at reduced temperature, decreasing their migration and reducing their ability to cause secondary damage (Henderson 1990). Cryoprotection is useful for both cryo-EM and x-ray crystallography. However, the mobility of hydrogen gas in the specimen is also affected by specimen temperature (Leapman and Sun 1995). Most hydrogen gas that is produced by radiation damage appears able to escape the specimen at liquid-nitrogen temperature. At extreme electron exposures, bubbles can be seen within specimens, probably corresponding to hydrogen that was not able to escape (Leapman and Sun 1995). In fact, the formation of bubbles has even been used to differentiate the inner body from the capsid in cryo-EM of phage particles, with the former more prone to bubbling than the latter (Wu *et al* 2012). Hydrogen appears less able to diffuse through specimens at liquid-helium temperature than at liquid-nitrogen temperature and can become trapped more easily. This trapping of hydrogen can lead to dramatic bubbling during liquid-helium temperature cryo-EM that distorts the specimen and affects contrast (Iancu *et al* 2006). For this reason, single-particle cryo-EM at liquid-nitrogen temperature is now generally accepted to be superior to cryo-EM at liquid-helium temperature (Bammes *et al* 2010, Pfeil-Gardiner *et al* 2019). An optimum may exist between those two temperatures but the advantage of working below liquid-nitrogen temperature compared to the added complication has not been fully explored (Meents *et al* 2010).

3.2.4 Radiation damage dependence on electron energy

The discussion up to this point has assumed that the amount of energy deposited into the specimen for each informative elastic interaction does not depend on the kinetic energy of the incident electron. More recently, this assumption has been challenged with measurements that suggest that 25% more elastic scatter occurs for the same amount of radiation damage with 100 keV electrons than with 300 keV electrons (Peet *et al* 2019). This analysis suggests an advantage to working with a 100 kV microscope rather than higher voltage instruments for suitably thin specimens.

However, radiation damage is not the only practical limitation in cryo-EM, which is affected by electron energy in multiple ways. Curvature of the Ewald sphere (Cohen *et al* 1984) causes complications in interpreting data at lower resolution for low energy electrons than it does for high-energy electrons. It is now known that the Ewald sphere curvature can usually be neglected at ~2 Å resolution with 300 keV electrons, but its effect will be somewhat greater for 100 keV electrons. In addition, multiple scattering events, where the same electron scatters more than once within the specimen, do not contribute meaningful information to images. These events occur with higher probability at low energy than at high energy because interaction cross sections are higher at lower energies. Further, current direct-detector device (DDD) cameras have a higher detective quantum efficiency (DQE) with 300 keV electrons than 100 keV electrons. Provided efficient detectors for 100 keV electrons become available and strategies for Ewald sphere correction prove effective (Wolf *et al* 2006, Russo and Henderson 2018a), it has been suggested that 100 kV electron microscopes will prove preferable to 300 kV electron microscopes for single-particle cryo-EM of thin specimens (Peet *et al* 2019).

At present, there is expected to be an overall advantage for cryo-EM at 300 kV rather than lower voltages. But despite this advantage, it is frequently commented, even amongst cryo-EM experts, that lower voltage cryo-EM provides more 'contrast' than higher voltage cryo-EM. The origin for this apparent paradox is that both elastic and inelastic scattering cross sections are higher at low voltage than at high voltage. Consequently, more electron scattering events occur for the same electron exposure at low voltage than at high voltage. In order to make up for the difference in scattering cross section, the experimenter must simply use an increased electron exposure to produce the same amount of specimen damage at 300 kV as that which is produced at lower voltage. At the same time, this increased exposure will provide more signal. When electron exposure is adjusted the SNR, and thus the visibility of particles (i.e. 'contrast'), should remain almost constant with voltage, while the other factors discussed above improve at higher voltage. Based on the linear energy transfer at different voltages (see page 94 of Glaeser *et al* 2007), one should use ~1.5× the exposure at 200 kV compared to 100 kV, ~1.8× the exposure at 300 kV compared to 100 kV, and ~1.2× the exposure at 300 kV compared to 200 kV. The behavior of the microscope contrast transfer function (CTF) (see both sections 1.2.3 and 4.7) also leads to a different amount of low-resolution phase contrast for the same amount of defocus at different voltages. These differences in the CTF, for the same value of defocus, appear as improved contrast at low voltage, but with all of the complications of increased defocus. To maintain CTFs that are similar at low resolution one must keep the product of defocus and electron wavelength constant. Consequently, one must use ~1.47× more defocus at 200 kV compared to 100 kV, ~1.88× more defocus at 300 kV compared to 100 kV, and ~1.27× more defocus at 300 kV compared to 200 kV.

3.2.5 Practical implications of radiation damage: image averaging in cryo-EM

Experiments by Glaeser and colleagues (Glaeser 1971) led to the recognition that electron microscopy of radiation sensitive specimens requires different imaging techniques than are used for radiation stable specimens. These experiments showed that electron diffraction patterns from radiation sensitive *l*-valine, adenosine, and uranyl acetate stained catalase crystals faded during the course of the exposure. Consequently, two conditions must be met for successful microscopy of radiation stable specimens. First, all unnecessary exposure of the specimen to the damaging effects of radiation must be avoided. Second, the electron microscope must be optimized so that as much information as possible is extracted from the specimen as it is irradiated.

The first condition required the development of so-called 'low-dose' techniques. In radiation physics, the word 'dose' refers to the energy absorbed (per unit mass). Low-dose techniques limit electron exposure in order to avoid destroying the specimen that one intends to image. Low-dose electron microscopy is accomplished by use of an electron beam blanker that can rapidly switch the electron beam on and off. Early beam blankers were mechanical shutters, while modern instruments accomplish beam blanking with electromagnetic coils that can deflect the electron beam. In addition, the electron microscope is not focused on the region of the specimen that is going to be imaged, but on a nearby region that is at approximately the same height in the microscope column. Next, the image is acquired from the region of interest by unblanking the beam and simultaneously capturing the image on the detector with a total electron exposure that avoids major distortion of the specimen, typically from 10 to 100 electrons/Å^2. Further details about low-dose imaging are described in section 3.3.

The second condition requires specific hardware to capture as much signal as possible. With the use of defocused phase contrast, a microscope with a field emission gun is needed to preserve contrast when the instrument is out of focus. A specimen cryostage is needed, not only to maintain the amorphous ice that preserves the specimen structure, but also to allow the maximum electron exposure that can be used before the effects of radiation damage become visible. Finally, a camera that efficiently detects each electron, equalizes the signal contributed by each electron, and accurately records where each electron struck the sensor allows extraction of as much signal as possible from the specimen. Due to the deformation and movement of specimens that occurs upon irradiation, a camera that records movies allows extraction of even more information (Ripstein and Rubinstein 2016); see section 3.5 for more on this point.

Even when all of the low-dose cryo-EM conditions are met, individual electron microscope images do not contain sufficient information to provide an atomic description of a biological specimen. The electron exposure needed to produce a high-resolution image of a radiation stable specimen is far beyond the 10–12 electrons/Å^2 that allow imaging of high-resolution features in radiation sensitive specimens, and even the total exposure of 50–100 electrons/Å^2 that avoids large scale distortion of these specimens. Thus electron microscope images of biological specimens are inherently noisy, even at low resolution, because radiation damage dictates that an electron exposure be used that is insufficient to describe the specimen. This phenomenon places out of reach the simple use of electron microscopy for atomic-resolution interrogation of biological structure.

To work around this predicament, cryo-EM of biological molecules makes use of image averaging to obtain high-resolution insight into biological structures. Averaging is possible when different molecules independently give the same insight into features in the specimen. In the simplest case, the specimen contains multiple copies of identical rigid structures. Thus, each image from a frozen thin-film suspension of macromolecules shows identical structures presenting different poses or views. If identical noisy images can be grouped and averaged (figure 3.1), high-resolution signal in the images adds coherently, while noise is gradually averaged out. Through this route, high-resolution structures have been obtained for 2D arrays of molecules (Henderson *et al* 1990), helical filaments (Miyazawa *et al* 2003), icosahedral viruses (Zhang *et al* 2010), ribosomes (Fernandez *et al* 2013), and smaller, low-symmetry, or asymmetric proteins (Liao *et al* 2013).

3.2.6 Resolution dependence and exposure weighting

Multiple studies of fading of diffraction spots during electron exposure showed that the highest resolution information in specimens is the first to be destroyed by radiation damage, with lower resolution information showing more resistance to

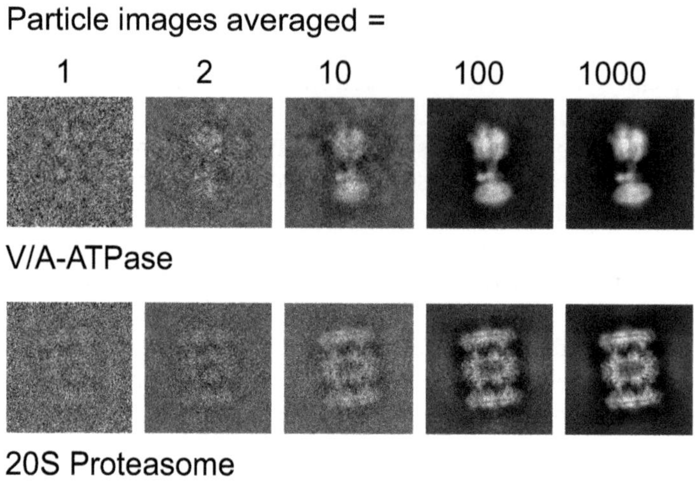

Figure 3.1. The effect of averaging on the signal-to-noise ratio of images of protein particles. Averages of 1, 2, 10, 100, and 1000 single-particle images of the *Thermus thermophilus* V/A-ATPase (top) and the *Thermoplasma acidophilum* 20S proteasome (bottom). The particles in the images are in the same orientation and the effects of the microscope contrast transfer function have been partially corrected. (Calculations and images are courtesy of Ali Punjani).

damage (figure 3.2). The intensity, I, of each diffraction spot was found to fade exponentially with

$$I_N = I_0 e^{-N/N_e},$$

where N represents the exposure in electrons/Å^2, I_0 is the initial intensity of the diffraction spot, I_N is the intensity of the diffraction spot at exposure N, and N_e is the critical exposure at which the spots decreased to $1/e$ times their initial intensity (Hayward and Glaeser 1979, Unwin and Henderson 1975). Spots are observed to fade in a resolution-dependent manner, with spots corresponding to similar resolutions displaying similar critical exposures. During imaging, information about a feature no longer increases after the feature has been destroyed, but the electron shot noise continues to increase.

Based on assumptions related to the performance of detectors, it was estimated that the optimal electron exposure to maximize the SNR in a particular resolution bin, N_{opt}, is ~2.5 N_e (Hayward and Glaeser 1979). Early experiments based on 2D crystals prescribed a total electron exposure that was aimed at obtaining the best SNR for information at around 3 Å resolution, or around 10–12 electrons/Å^2. Most single-particle cryo-EM studies followed this guideline, even when they were limited in practice to 10 or 20 Å resolution. Subsequent measurement of fading of calculated diffraction spots from images of thin catalase crystals allowed estimation of optimal exposures for resolutions between 2.9 and 86 Å (Baker *et al* 2010). These measurements enabled calculation of optimal electron exposures for intermediate-resolution cryo-EM. In addition, it was pointed out that when images were recorded as movies, each resolution range can be captured with its optimal exposure by gradually

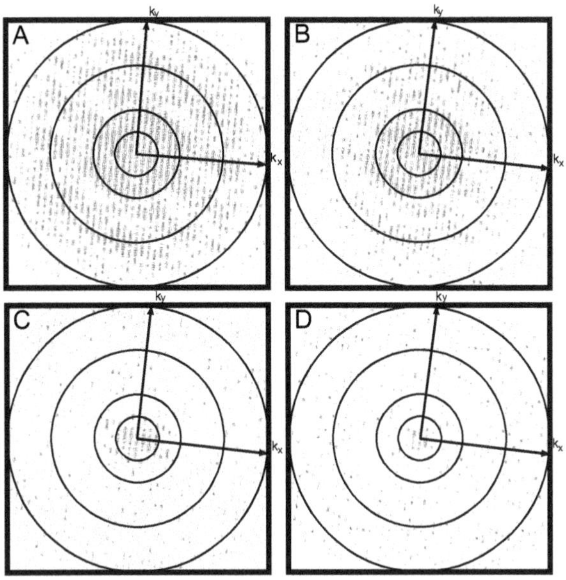

Figure 3.2. Fading of calculated diffraction spots from images of thin catalase crystals. A series of successive images were recorded on photographic film from the same area of a crystal and their diffraction patterns calculated. Images were recorded with 8 e$^-$/Å^2 with the specimen not previously exposed (A) or previously exposed to 8 e$^-$/Å^2 (B), 46 e$^-$/Å^2 (C), or 78 e$^-$/Å^2 (D). The systematic loss of calculated diffraction spots shows the fading of information from high to low resolution. Adapted from Baker *et al* (2010), copyright (2010) with permission of Elsevier.

down-weighting the higher resolution information as more frames of the movie are added to the sum (Baker *et al* 2010).

In this exposure weighting approach, high-resolution information is captured in the first few frames of the movie while lower resolution information, which can tolerate a higher electron exposure, is captured over more frames throughout the movie. This low-resolution information can thus be obtained with a high SNR, using electron exposures that are significantly higher than the 10–12 electrons/Å^2 used for high-resolution information.

These values of optimal electron exposure were subsequently measured again with a direct electron detector in movie mode, using a large icosahedrally symmetric virus particle rather than catalase crystals (Grant and Grigorieff 2015). The latter analysis showed an even more permissive total exposure can be used than was estimated from imaging catalase crystals. Figure 3.3 reproduces a plot of relative signal-to-noise as a function of electron exposure as described in Baker *et al* (2010) but using the critical exposures measured in Grant and Grigorieff (2015). Measurement of optimal exposures from either 2D crystals or icosahedral viruses suffers from the problem that disruption of specimen order during irradiation can lead to apparent loss of information even before that information is destroyed in the molecules that make up the crystal or icosahedron. Alternatively, the best weighting to use for the different frames contained in movie data can be determined at the final

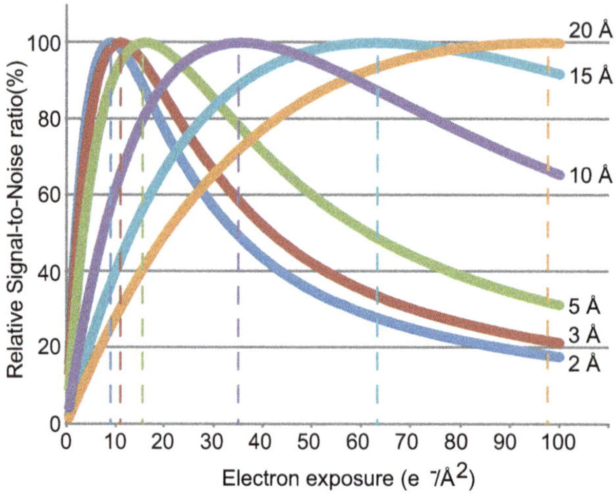

Figure 3.3. Optimal exposures for different resolutions. The relative signal-to-noise ratio is plotted as a function of electron exposure for 300 kV electrons. The plot shows that lower resolution information reaches an optimal signal-to-noise ratio at higher exposure than higher resolution information. Adapted from Baker and Rubinstein (2010), copyright (2010) with permission of Elsevier, with critical exposures measured in Grant and Grigorieff (2015).

stage of 3D reconstruction (Scheres 2014). However, experience with different approaches has shown that 3D map resolution is not particularly sensitive to the details of how this exposure weighting is carried out, as long as it is performed.

3.2.7 Radiation damage versus beam-induced motion and charging

Both progressive charging of the specimen during imaging, which can lead to movement of the image during recording, and true beam-induced mechanical motion of the specimen degrade image quality. It can be difficult to distinguish one from the other. However, carefully analysis has suggested that beam-induced motion rather than charging is responsible for most information loss in images (Russo and Henderson 2018b, 2018c).

One model of beam-induced motion (Henderson 2018, Thorne 2020) posits that movement is due to radiation-induced creep in the specimen that has trapped strain due to differences in cooling rates of the specimen and its supports, which leads to ice being compressed after it has become solid during the grid freezing process. This beam-induced motion remains a major limitation in cryo-EM. While algorithms have been developed to track the movement of specimens that occurs from one frame to the next (Li *et al* 2013, Rubinstein and Brubaker 2015, Zivanov *et al* 2019, Zheng *et al* 2017), the largest magnitude and most erratic movement usually occur with the first few electron/Å^2 exposure. This phenomenon leads to the unfortunate situation that beam-induced motion is worst early in an exposure when radiation damage is minimal and specimens should provide the most high-resolution information, while radiation damage is worst late in the exposure when beam-induced motion is easiest to correct. It has been estimated that elimination of beam-induced motion would reduce by a factor of 30-fold the number of images needed to obtain a map at 3 Å resolution, or alternatively would improve a 3 Å resolution map to 1.7 Å resolution from the same number of particle images (Henderson 2018). Therefore, while radiation damage appears to be an unavoidable consequence of imaging proteins with a highly energetic electron beam, it may be possible to provide major gains in the high-resolution signal by avoiding the fast and erratic beam-induced motion that occurs before radiation damage has set it.

3.3 Low-dose protocols for recording images

Anchi Cheng
Simons Electron Microscopy Center, National Resource for Automated Molecular Microscopy, New York Structural Biology Center, 89 Convent Street, New York, NY 10027, USA

Low-dose data collection for cryo-EM refers to (i) the strategy used to survey the grid, i.e. identify and navigate to the optimal imaging area, and (ii) the imaging conditions used to achieve minimal radiation damage. While the details of different protocols may vary, the goal remains the same: the information gathered in the final, high-magnification images should have suffered from minimal radiation damage.

3.3.1 Automated low-dose imaging

Automated data collection software (reviewed in Cheng *et al* 2016) is commonly used to efficiently carry out low-dose imaging. Such programs control the mechanical stage in order to move to the area of interest, acquire images at user-defined magnifications, and adjust the appropriate components to determine and set the defocus values for the images. Some programs also automatically select suitable areas for imaging.

The general workflow for single-particle cryo-EM is summarized in the following four steps. The electron dosage used at each step is estimated to be the minimum required by the purpose.

Step one: acquire a 'grid atlas'

A grid atlas is a low-magnification overview of part or all of the EM grid. Since there is usually no appropriate imaging condition that can cover the full 2 mm diameter open area of a standard EM grid, the software acquires multiple, low-magnification images at adjacent positions, and stitches them together to present to the user (figure 3.4).

The user will next interact with the display of this grid atlas to select suitable open areas of the grid for intermediate-magnification imaging. We refer to this selection process as 'targeting'. Very thick ice and squares with broken support films are easily excluded. However, this usually still leaves many candidate squares to work with. With experience from similar grid preparations, users are often guided by characteristics of the squares such as the average gray-level within the grid square and the extent of the ice gradient extending from the grid bar, to fine-tune the square-targeting process.

The electron dose accumulated in this step is negligible because the magnifications used are low and thus the exposures per unit area are also very low.

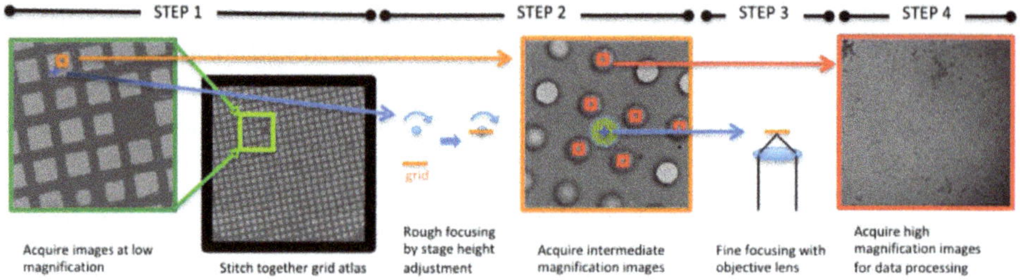

Figure 3.4. Multi-scaled imaging approach used for low-dose data collection. See the text for details of each step. The squares within the image represent the image taken at higher magnification, each framed by the matching color. The blue cross indicates a target used for focusing. The light green solid disc shown in the intermediate-magnification image in step 2 represents the footprint of the electron beam used in step 3 and step 4.

Step two: acquire intermediate-magnification images

Intermediate-magnification images provide references on which high-magnification final imaging area is selected. While it is possible to include a full grid square in a single image and thus select the final targets directly from that image, this is not regularly done. The reason is that there is a limited amount of detail provided at such a low magnification and there are also optical limitations imposed by the microscope.

Automation-software designs use, instead, appropriate workflows to get around this limitation. In some designs, such as that typically used in SerialEM (Mastronarde 2005), the software maps the whole grid square with a comprehensive series of images, each at a relatively high-magnification scale. While these images thus have enough resolution to make a good targeting selection, the area displayed in any one of them does not cover the entire grid square. A square atlas is therefore created from the individual images to be used for final target selection. In other software, such as EPU or Leginon (Suloway *et al* 2005), a double-pass targeting approach is performed by adding a step to obtain a single image at the low-magnification scale of the entire grid square. This is done so that the user, or the automated software itself, can be more selective about the area of the grid to be used at the desired intermediate-magnification image scale.

The electron dose accumulated, at a given position during step 2, is no more than 0.05 e/Å^2.

Step three: focusing

The ideal optical settings for the final image are adjusted during the step of microscope alignment. Fortunately, at least with a stable environment, almost all of the basic alignments performed at installation or maintenance time seldom need to be changed.

Normal alignment is performed in reference to the specimen height coinciding with the mechanical stage-tilt axis, i.e. at the so-called eucentric height. The objective-lens current is then adjusted to put the image in focus, after which further alignment steps can be performed, as desired. If subsequent focusing adjustments must be made during an imaging session, doing so by adjusting the specimen height rather than by adjusting the objective-lens excitation will guarantee that the previous alignments remain valid. In practice, however, alignments remain sufficiently valid even after the objective-lens focus is changed by a few micrometers. As a result, it is common practice to adjust the objective-lens focus within this limit, rather than to always focus the image by changing the specimen height.

One automated software method measures how close the specimen is to the eucentric height by observing the amount of movement between two images while tilting the specimen stage. An alternative method observes the amount of movement while tilting the electron beam as it passes through the specimen (Koster and de Ruijter 1992). The specimen-tilt method is usually performed under the intermediate-magnification condition mentioned in step 2, and it adds double the electron dose estimated previously for that step. Even more dose than that may be required if the goniometer is not ideal or if imperfect grid-flatness prevents accurate correction the first time. On the other hand, when fine focusing is performed using the beam-tilt method, the imaging is performed at similar magnification as the final exposure and thus a more significant dose is accumulated during the procedure. As a result, the location and the timing of such a focusing step is the main consideration for low-dose data acquisition.

A typical workflow first performs a rough-focusing step for each grid square, either before or during the intermediate-magnification imaging step. In order to perform the rough-focusing step the mechanical stage is moved parallel to the optical axis so as to bring the specimen to the eucentric height. This is the preferred practice, since the microscope alignment is initially optimized with the objective lens in focus at that position. The fine focusing step is usually done later, just before the final imaging begins, and that requires only a small adjustment of the objective-lens current. The software or the user then chooses a position on the intermediate-magnification image so that the footprint of the beam on the specimen during the fine focusing will not overlap with the final exposure area. Another strategy of fine focusing, which is used at some facilities, assumes that the grid is locally flat. In this case the fine focus is performed adjacent to an exposed area used in step 4 and it assumes that the desired value of focus at the next target position will not be too different.

Fine focusing typically accumulates an electron dose of 2 e/Å^2 or more, which is enough to cause substantial damage to the specimen even before an image has been recorded, as has been explained in section 3.2. As a result, focusing cannot be done at exactly the same places in the specimen that will be used to record high-resolution images.

Step four: acquire high-magnification images for data processing

The combination of steps 1, 2, and 3 is intended to provide a good selection of targets and optimal optical settings for high-magnification imaging. The content of the image is determined by the position of the target as it is mapped back to the specimen. The structural information transferred to the image is constrained by the image pixel size, the defocus, and other lens aberrations. The pixel size that is appropriate to use in this step depends on the camera characteristics, the particle size, and the purpose of the imaging session.

The concept of DQE is used to describe the efficiency with which a camera converts the input signal to its output as a function of the spatial frequency. All pixelated cameras exhibit a drop in DQE as the spatial frequency approaches the Nyquist frequency of the camera. This happens because the pixelated camera output consists of just the average value of the input across the area of each pixel. Therefore, in general, high-resolution structural information will be recorded with a higher value of the DQE if a smaller pixel size (referred to the specimen) is used (Stagg *et al* 2008). On the other hand, excessively reducing the pixel size means that much of the output band-width extends well beyond the band limit (i.e. resolution limit) of the specimen itself. Since most biological question asked in the SPA method can be answered with a structure in 2.5–3 Å resolution range, 1 Å is the most common pixel size used.

3.3.2 Improving throughput

Since the position of the camera is fixed relative to the microscope column, the images formed by the targets on the specimen need to be brought to the camera. There are two methods to do this.

The first method is to keep the beam stationary and move the specimen with the mechanical stage. Mechanical movement of the specimen guarantees that the image will be taken at the same optical settings chosen during the initial alignment. However, the accuracy of this motion is limited by the performance of the stage and microscope manufacturers can currently only commit to 2 μm accuracy. Mechanical movement also requires time for residual drift to settle after the movement. It is not unusual that 20–40 s is needed after a large movement, even with the best stage, to return to the pre-movement level of stability. One practical note, however, is that movie-recording digital cameras can align the drifting movie frames. Therefore, to return to sub-angstrom per second stability is not a requirement for high-quality data collection.

An alternative method is to deflect the electron beam to an off-axis target on the specimen and then deflect the image formed below the objective lens back by a corresponding amount. This optical operation is called beam-image shift. Beam-image shift is accurate (0.1 μm) and fast (within 1 s). It does have some drawbacks, however. To begin with, deflectors have limited range. The TFS Titan column, for example, limits the deflection to about 12 μm at the magnification typically used for taking the final image at step 4.

A further issue to be careful about is the fact that the tilt angle of the optical axis of a magnetic-field lens changes as one moves across the bore of the lens pole piece (figure 3.5). The lens aberration known as coma thus becomes a function of the magnitude and direction of the beam-image shift applied (more background about coma is presented in section 3.4 and in section 4.8). In particular, azimuthal coma increases as the beam passes further away from the optical axis of the position used during the initial, coma-free alignment step (Glaeser *et al* 2011). On TFS Titan columns, the equivalent of 0.17–0.25 mrad beam tilt is produced per μm of beam-image shift (Cheng *et al* 2018) and on TFS Talos columns the beam tilt produced is

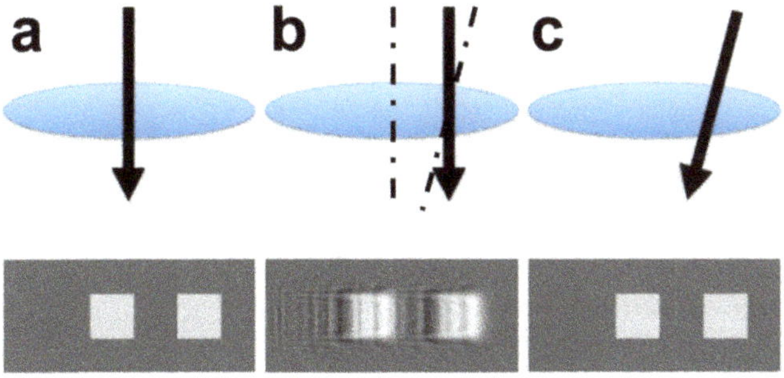

Figure 3.5. Correction of the beam-image shift coma effect by using different, pre-set values of beam tilt. The upper panel schematically shows the path of the unscattered-electron beam (arrow) as it first enters the lens. The dot-dash lines in panel (b) indicate that the optical axis of the lens changes its orientation as one moves across the face of the lens. The lower panel shows a simulation of what happens when imaging two separated squares with each of the three optical conditions that are shown schematically by the solid arrows. Column (a) shows the result for on-axis imaging. This produces images that are a perfect representation of the object. In (b) the beam is deflected to image an off-axis object, but the beam-tilt angle is not changed as the beam is shifted. The resulting coma effect produces a directional distortion of the image, as shown in the lower panel. In (c) the incident beam is tilted in order to be parallel to the optical axis at the shifted position and the images again become a perfect representation of the object.

about 30% higher. This, plus the fact that the phase error produced by coma is greater at lower high-tension values, makes this aberration more problematic in the latter instruments.

Fortunately, a compensating beam tilt can be pre-set at each location when using the beam-image shift method (Eades 2006). The correction of beam-tilt and other lens aberrations, such as stigmatism, is implemented for beam-image shift in both Leginon and SerialEM, and a detailed characterization of the performance of such corrections is available (Wu *et al* 2019). Microscope manufactures also plan in the future to implement this at a lower level. Using aberration-corrected beam-image shift has more than doubled the data collection rate in the past few years.

Further improvement of the data collection rate is also becoming available by modifying the usual settings for the condenser lenses. Thus far, all high-end microscopes are operated in a way that results in the main condenser aperture being imaged onto the specimen in an out-of-focus condition when the illumination is parallel at the specimen plane. As a result, the perimeter of the illuminated area exhibits strong Fresnel fringes. While this effect demonstrates the superb coherence of the electron source, it nevertheless presents a problem for low-dose imaging. If the beam is kept small enough to cover only the camera footprint, in order to reduce radiation damage outside that area, these fringes become a strong feature in the image. Therefore, users are forced to spread the electron beam to twice or more the diagonal distance of the camera footprint to minimize the fringe that is present in the image. However, it has been shown that it is possible to eliminate these fringes with extensive changes in the lens excitations and with corresponding re-alignment of the illumination. The result is to project a precisely in-focus image of the condenser aperture onto the specimen and still maintain parallel illumination

(Weis and Hagen 2020). Using such alignment conditions promises even better efficiency of data collection by allowing more images taken within a given area.

3.3.3 Electron exposure levels used during high-resolution data collection

The value of the electron dose that is used for the final exposure depends somewhat upon individual preferences. Most commonly, exposures of 30–45 e/Å^2 are used, even though exposures of less than 25 e/Å^2 have been shown to provide the highest resolution information (Tan *et al* 2018, Bartesaghi *et al* 2018).

Since movie frames are recorded with direct-detection cameras, the tendency is to collect datasets at a higher dose, because a greater exposure can help in the initial stages of data processing. In this case, a progressively more narrow, low-pass filter should be applied to successive movie frames before they are summed. The purpose of applying dose-dependent weighting to successive frames is to exclude the high-frequency noise components that would otherwise continue to be added, even after the most sensitive high-resolution features of the specimen are no longer present, due to radiation damage. The topic of dose weighting is covered more fully in section 3.5.

3.4 Practical considerations: defocus, stigmation, coma-free illumination, and phase plates

Radostin Danev
Graduate School of Medicine, The University of Tokyo, 113-0033 Tokyo, Japan

3.4.1 Why do we need to defocus the microscope?

Biological frozen-hydrated samples are composed mainly of light elements, such as carbon, hydrogen, oxygen, and nitrogen, and therefore behave like weak-phase objects in the electron microscope (see section 1.2). This means that they modulate predominantly the phase and much less the amplitude of the electron wave. If the microscope was an ideal aberration-free optical system and was perfectly in focus, then images of biological samples will not show any contrast. The electron detector measures the intensity of the electron wave, as represented by the number of electrons that arrive at each pixel and therefore it cannot directly measure the phase of the wave. To enable visualization of the phase modulation the wave must be optically manipulated in a way that converts the phase information into intensity modulation, which can then be observed by the camera.

The electron microscope is not a perfect optical system. It has inherent optical flaws, such as spherical and chromatic aberrations, as well as intentionally introduced aberrations, such as defocus (see section 1.2). The presence of aberrations causes partial conversion of phase modulation into intensity variations in the image, thereby it can be used as a phase-contrast mechanism. In fact, intentional defocusing is the oldest technique for phase-contrast generation in electron microscopy, and to this day it remains the most popular and easiest to use phase-contrast method. Unfortunately, it has some disadvantages. The defocus method provides only partial conversion and approximately half of the phase information is not represented in the image.

There are other, more advanced, methods for generation of phase contrast, such as phase plates. Theoretically, phase plates are better at preserving phase information but are quite challenging to realize in practice and are more difficult to use (see section 3.4.5).

3.4.2 Effects of defocus on the image and its information content

The weak-phase CTF includes the defocus as a coefficient that multiplies the square of the spatial frequency (see equation (1.4) in section 1.2). At the typical defocus values used currently in cryo-EM (>200 nm), the defocus term is the dominant aberration. The amount of defocus determines the number and periodicity of the CTF rings observed in the 2D power spectrum of images. Larger defocus values shrink the first CTF ring toward low frequencies and consequently improve the low-frequency information transfer and increase the overall image contrast (figure 3.6). However, they also increase the number and density of CTF rings in the high spatial frequency region, which makes it difficult to effectively demodulate and recover high-resolution information during data processing. Lower defocus values push the first CTF ring outwards, toward higher spatial frequencies, which causes weaker transfer of the low spatial frequencies and thereby reduces the overall contrast in the

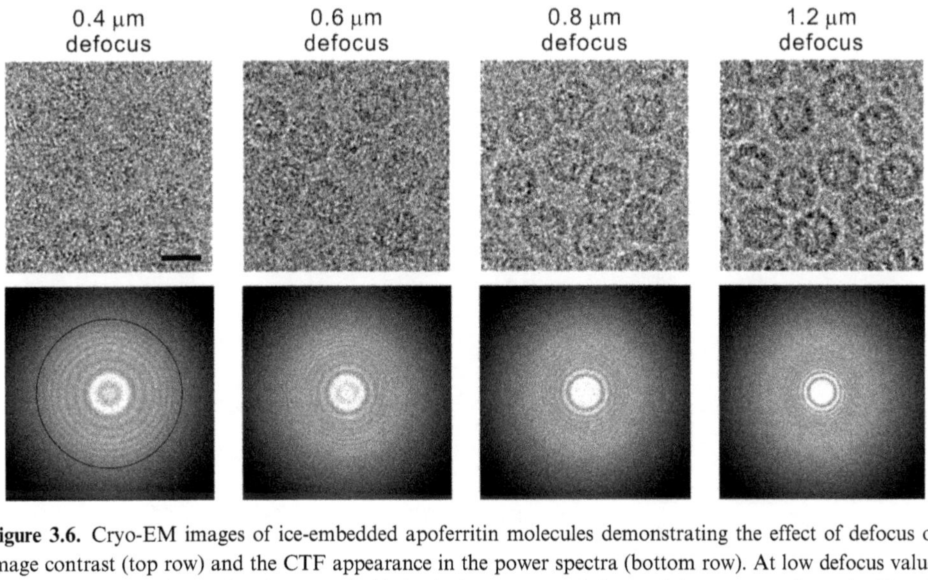

Figure 3.6. Cryo-EM images of ice-embedded apoferritin molecules demonstrating the effect of defocus on image contrast (top row) and the CTF appearance in the power spectra (bottom row). At low defocus values (top left) the molecules are barely recognizable in the image, nevertheless such images contribute significantly to the high-resolution components of the result. High defocus values (top right) improve the contrast and make easy the recognition of individual molecules but may impair the resolution. The CTF rings are well separated and easy to recognize up to high resolution at low defocus values (bottom left). Higher defocus increases the number and reduces the separation between CTF rings (bottom right). Scale bar = 10 nm. Black ring in the power spectrum = 3 Å resolution.

image (figure 3.6). Because of the lower number and wider spacing of the CTF rings, lower defocus values are more effective in preserving high-resolution information.

For optimal results it is desirable to have both high contrast and high resolution, but with the defocus method and current data processing approaches there is a practical compromise between the two. To overcome these limitations, we use a range of defocus values starting at the lowest defocus at which particles can be visually discerned in images. Depending on the size of the observed molecule, the typical value for the low end of the defocus range is between 0.5 and 1.0 μm.

3.4.3 Defocus variation is necessary to obtain uniform information coverage in reciprocal space

The zeros of the CTF create circular zones between the CTF rings where structural information is lost. If all images in a dataset are collected with the same defocus value, the reconstructed structure will lack information at the spatial frequencies corresponding to the CTF zeros and will therefore be incomplete and possibly unusable for building a molecular model. To mitigate the loss of information at the CTF zeros, images within a dataset are acquired with a range of defocus values. Varying the defocus moves the positions of the CTF zeros and helps to achieve more uniform information coverage of the frequency spectrum when the data are merged.

Figure 3.7 contains simulated images that illustrate the effect of combining different defocus values. The simulation was performed assuming that the CTF sign-reversal has been corrected by phase-flipping (see section 4.3) and demonstrates only the effect of information loss around the CTF zeros. The low defocus image (figure 3.7, left) has weaker overall contrast and emphasizes high-resolution features of the object while the high defocus image (figure 3.7, middle) has stronger contrast and carries information about larger elements. By combining data collected at different defocus values, a more uniform and balanced representation of the object can be constructed (figure 3.7, right).

Figure 3.7. Simulated defocus modulated images of a legacy electron microscope (left and middle) and their sum (right). Low defocus values (left) preserve high-resolution features in the image but reduce the overall contrast. High defocus values (middle) enhance large-periodicity features and improve the overall contrast but suppress fine details. The combination of low and high defocus (right) creates a more balanced representation of the original image. Scale bar = 10 nm.

The upper defocus limit for data acquisition depends on the type of sample and the ice thickness of the specimen. With optimal thin-ice specimens the upper defocus value should not exceed ~2 μm. In general, for most specimens a defocus range of 0.7–1.5 μm works quite well. If the sample necessitates data collection in thicker ice areas, the defocus range must be shifted toward higher values, e.g. 1.0–2.0 μm, to compensate for the loss of contrast. However, data collection in thicker ice should be avoided as much as possible because it precludes the achievement of high resolutions (<3 Å). It should be considered as a last-resort measure in cases where reasonable sample-optimization efforts have not succeeded in producing better grids (see section 2.5).

To cover the desired defocus range experimentally, the target defocus is cyclically varied in steps from image to image, e.g. 1.5, 1.4, 1.3, … 0.7, 1.5, 1.4, … etc. Some defocus variation also occurs naturally because focusing is performed on a support film area that is a few micrometers away from the acquisition area and the sample is never perfectly flat. The exact defocus value for each image is measured during the CTF fitting and CTF refinement steps of the data processing workflow and therefore precise focusing is not necessary during data acquisition. Currently, data acquisition speed is one of the top priorities during the experiment, hence focusing interruptions must be reduced to a minimum. This can be achieved by skipping the focusing step for imaging areas that are close to each other. For example, if the last focusing was performed less than 5 μm away from the current imaging position, then the defocus is presumably still accurate enough to remain within the intended defocus range. Preferably, focusing actions should be reduced by organizing the acquisition in image shift groups, as discussed below (see also section 3.3).

Recent advances in image processing enabled *in silico* correction of higher order optical aberrations and thus circumvented the requirement to collect optically perfect images in order to achieve optimal performance (see section 4.8). In practice, this means that the beam does not have to remain perfectly centered on the optical axis of the microscope. Beam-shift and image-shift coils can be used to optically move the imaging position over a region around the optical axis and collect multiple micrographs without having to move the specimen stage mechanically. This greatly reduces the number of mechanical stage-shifts and the associated stage-drift settling pauses, that are usually the most time-wasting step in the data acquisition workflow (see section 3.3). The beam-image-shift (BIS) acquisition approach has become the standard data acquisition method at most cryo-EM facilities.

The BIS acquisition method (see section 3.3) easily accommodates defocus variation. First, focusing is performed only once close to the center of the acquisition pattern. Considering that residual mechanical drift resulting from moving the stage can affect the first couple of images of the BIS series, it is advisable to start the focus variation at the high end of the defocus range. The stronger contrast would improve the accuracy of movie frame alignment and would also match the potentially lower resolution of the first couple of images. Following the initial focusing routine, after each BIS image the objective-lens defocus is reduced blindly (without explicit focus measurement) in small steps, so that by the end of the series the presumed defocus has reached the low limit of the defocus range. For example, let us assume that we are using a nine-position BIS acquisition pattern. After the stage moves to the next

BIS area, a focusing routine is carried out on the support film next to the central hole with a target defocus of 1.5 μm. Then, after each image is acquired, the objective-lens defocus is reduced by 0.1 μm. By the last, ninth image in the series, the target defocus will decrease to 0.7 μm and the desired defocus range of 0.7–1.5 μm will be covered with uniformly spaced values. As mentioned above, due to sample non-flatness the defocus of the acquired micrographs rarely matches the target defocus exactly. With the BIS acquisition scheme the deviations from the target defocus value become even larger because of the longer distances between the focusing area and some of the acquisition positions. In practice such deviations of 0.1–0.2 μm are not problematic and aid in obtaining a smooth continuous coverage of the defocus range.

3.4.4 Optical correction of astigmatism and coma aberrations

Being a real-world optical system, the electron microscope exhibits optical aberrations of varying order and magnitude. In this regard, the most important practical question is 'Are aberrations limiting the achievable resolution?'. Until recently, only defocus, astigmatism, and spherical aberration were treated as part of the numerical CTF demodulation during image processing. With the drastic resolution improvements brought about by the cryo-EM 'resolution revolution', higher order aberrations, such as coma, have become the limiting factors. Recent advances in data processing (see section 4.8) enabled numerical correction of virtually any aberration, except for chromatic aberration. Nevertheless, it is still beneficial to have the optical system aligned as well as possible to obtain a better starting point for the processing and therefore improve the chances of obtaining optimal results.

Astigmatism is a symmetrical aberration in relation to the optical axis and manifests as an almost elliptical shape of the CTF rings in the power spectrum. All microscopes are equipped with 'stigmator' coils that allow compensation of the astigmatism. In the past, stigmation was performed manually by observing the pseudo real-time power spectrum produced by a digital camera from an amorphous carbon film area on the sample. Now stigmation can be performed automatically by software packages such as SerialEM (Schorb *et al* 2019) and EPU (Thermo Fisher Scientific, MA, USA), and this is the preferred method because it is more accurate. Astigmatism is not a resolution-limiting aberration because it does not affect the phases of the Fourier components of the sample. It is also easily detectable by observing the shape of the CTF rings in the power spectrum. Nonetheless, it is still good practice to correct the astigmatism, either manually or more accurately by automatic routines, before starting the data collection. This will improve the accuracy of the CTF fits during data processing and will have a positive effect on the rest of the workflow.

Coma is an antisymmetric aberration that is caused by angular deviation of the beam from the optical axis, referred to as 'beam-tilt'. It scrambles the phases of Fourier components of the sample and therefore could limit the resolution in reconstructions. The magnitude of the modulation increases as the third power of spatial frequency and could become the main limiting factor at resolutions better than ~3 Å. Coma can be corrected optically by software-driven beam-tilt alignment. The 'coma-free' alignment procedure takes about 5 min and involves collection of a series of images from an amorphous sample with the beam tilted in several directions, to build the so-called Zemlin tableau. The residual beam-tilt is then determined through numerical analysis of the CTFs of the tableau images. Most data acquisition software packages, such as SerialEM and EPU, include automated coma-free alignment routines. The alignment should be performed as one of the last steps before starting the data acquisition, after the illumination settings have been decided. The coma-free alignment should be repeated if the illumination settings, such as spot size or beam diameter, are changed. Modern data processing packages can measure and correct the coma (see section 4.8), but—as with astigmatism—it is highly advisable to minimize optical imperfections in the data.

The BIS acquisition technique uses beam-shift to move the illumination spot to each of the acquisition positions on the specimen. Ideally, shifting the beam should keep it parallel to the optical axis and therefore not cause coma aberration. However, in real optical systems beam-shift can cause some beam-tilt due to imperfect tuning of the condenser lens deflector system's 'beam-shift pivot points'. Additionally, with large offsets from the center of the lens, the local coma-free axis can be tilted relative to that at the center. Therefore, without additional illumination angle correction, images collected with BIS will have increasing amounts of coma with the increase in the distance from the central optical axis. To prevent such aberrations, up-to-date BIS data acquisition routines use beam-tilt compensation, also called 'aberration-free image shift' (AFIS). This approach requires additional calibration steps that involve performing coma-free measurements at BIS positions along the X- and Y-axes to determine the magnitude of the beam-tilt correction that must be applied during acquisition. These calibrations take longer to perform (~15 min by SerialEM) and are in principle valid only for the currently set spot size and beam diameter. The EPU software package uses a calibration that is performed by the service engineer and is not accessible by the user.

3.4.5 Use of phase plates to improve image contrast and the expected benefits

Phase plates are devices that generate phase contrast independent of defocus. They achieve this by introducing a phase shift between the unscattered, primary beam and the scattered beams at a diffraction plane inside the microscope. The optical principle was originally invented for the light microscope by Frits Zernike almost a century ago. In electron microscopy the development of phase plates was significantly delayed by practical challenges related to the charged-particle behavior of electrons. Many designs were tested in past decades, but despite some promising results none of them could be refined enough for widespread use. Just a few years ago, the Volta phase plate (VPP) became the first commercially viable device in terms of performance, robustness, and ease of use.

The VPP was developed in 2014, shortly after the onset of the 'cryo-EM revolution', and held promise for additional performance gains in single-particle analysis, in particular for small particles. Indeed, several groundbreaking results were produced with the VPP, such as the first high-resolution cryo-EM structures of the nucleosome (Chua *et al* 2016), hemoglobin (Khoshouei *et al* 2017), and a G-protein coupled receptor (GPCR) (Liang *et al* 2017). These exciting results created the subjective impression that the VPP is essential for determining the structure of small molecules (<200 kDa). Consequently, many groups adopted the VPP in their cryo-EM workflows with varying degrees of success. In the meantime the cryo-EM experimental technique and data processing methods continued to evolve and defocus-based phase-contrast results matching or exceeding those produced with the VPP started to appear. Recent defocus-based structures of hemoglobin (Herzik *et al* 2019) and streptavidin (Han *et al* 2020) clearly demonstrate that the VPP is not essential for cryo-EM of sub-100 kDa proteins. To determine if the VPP is beneficial for studies of small membrane proteins, we performed a direct comparison between VPP/no-VPP datasets (EMD-0993, EMPIAR-10359) using the same grid of a class B GPCR. The results (Danev *et al* 2020) indicate that the VPP in fact impairs the performance by producing ~0.3 Å worse resolution. Figure 3.8 shows representative images from the defocus and VPP GPCR datasets. The VPP significantly improved the contrast but ultimately produced a lower resolution reconstruction (VPP 3.0 Å versus defocus 2.7 Å).

The VPP uses the beam-induced change in electrical potential of a heavily irradiated spot on a thin carbon film to create the desired phase shift and this same carbon film is the source of its fundamental flaw—inherent information loss due to electron scattering by the film. Early measurements indicated that the loss is around 20% and we assumed that it is constant for all spatial frequencies. More recent experiments with non-biological test samples (Buijisse *et al* 2020) showed that the loss increases with spatial frequency and becomes severe below ~2.5 Å resolution. Theoretically, it may be possible to reduce the loss by using a thinner film, such as graphene, but for the VPP it would not be possible to resolve the issue completely.

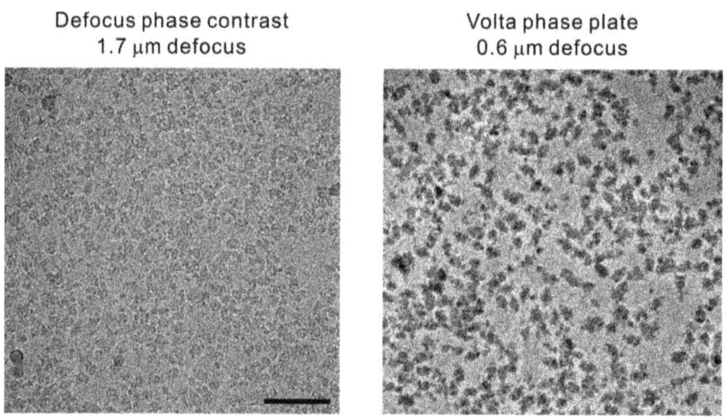

Figure 3.8. Representative images from a cryo-EM dataset from a class B G-protein coupled receptor. Approximately half of the dataset was collected with the conventional defocus approach (left) which generated images with low overall contrast but ultimately a higher resolution 3D reconstruction at 2.7 Å. The Volta phase plate (right) improved significantly the contrast of images and allowed data collection with lower defocus but produced a lower resolution 3D map at 3.0 Å. Scale bar = 50 nm.

Fortunately, there is a new phase plate development by the groups of Robert Glaeser and Holger Mueller (Schwartz *et al* 2019). They are working on a laser-based phase plate, which uses a high-intensity, standing light wave to manipulate the phase of the electron beam (figure 3.9). Because it does not use any material in the beam path, the laser phase plate is expected to solve the information-loss problem of the VPP as well as previous phase-plate designs. It will also have another big advantage by allowing real-time precise control of the phase shift and therefore will create opportunities for development of new data acquisition methods. The team has already demonstrated usable phase shifts with a prototype device, but there are several practical challenges that must be resolved before they could start cryo-EM application trials (Turnbaugh *et al* 2020). At this stage, it is difficult to predict when the laser phase plate will become commercially available. Depending on the success of planned experiments with the next prototype, it could be as early as 2–3 years from now.

The main, and most obvious, benefit from phase plates is the strong contrast enhancement in images produced by the greatly improved phase contrast at low spatial frequencies (figure 3.8). The expected advantages of using phase plates for single-particle analysis will stem from this low-frequency information boost. Phase plates are expected to improve the alignment accuracy for small molecules and extend the applicability of cryo-EM in the low molecular weight range, below 50 kDa. Furthermore, the additional low-frequency components should help in difficult cases of heterogeneous and/or flexible complexes that require extensive classification or motion analysis.

Phase plates are theoretically intended to be used in-focus, resulting in a unity CTF without zeros. Such images would have approximately twice the information content compared to defocus phase contrast. This would be a significant advantage,

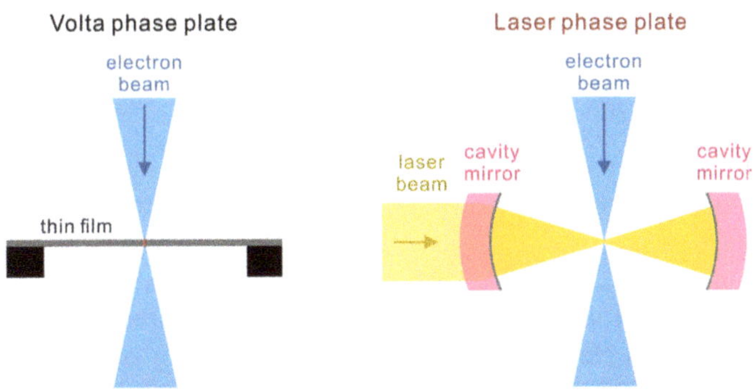

Figure 3.9. Illustrations of the principles of the Volta phase plate (left) and the laser phase plate (right). The Volta phase plate uses a thin film of amorphous carbon to generate a phase shift for the central beam of unscattered electrons. The thin film scatters some of the information carrying electrons and reduces the overall performance of the phase plate. The laser phase plate uses a resonant optical cavity formed by two high-precision mirrors to amplify the intensity of the source laser by tens of thousands of times. The focal point of the cavity, where the light field intensity is highest, is aligned on the crossover of the central electron beam and induces phase shift. The laser phase plate does not require placing of materials in the electron beam path and therefore is free from information loss.

but unfortunately the in-focus approach requires very accurate focusing, on the order of a few nanometers, which makes it incompatible with low-dose operation. The focusing and image acquisition areas are a few micrometers apart and the sample is never perfectly flat, leading to focusing errors of up to a couple of hundred nanometers. To overcome the focusing errors, phase-plate cryo-EM data acquisition must be performed with the minimal amount of defocus that would allow accurate CTF fitting for defocus determination during data processing. Experience with the VPP showed that the minimal defocus for accurate CTF fits was ~0.2 μm. However, to provide a margin for focusing errors and sample non-flatness, we typically used the VPP with 0.5 μm defocus (Danev *et al* 2017). The same low defocus approach would have to be used with the laser phase plate.

Phase plates are more demanding to use during experiments and require significantly more diligence and experience for data processing than does the defocus approach. At the same time, the performance and applicability range of the defocus method continues to evolve. The recent introduction in cryo-EM of a cold field emission gun that improves both the spatial and temporal coherence of the beam (Kato *et al* 2019) is expected to extend the applicability range of the defocus method toward smaller molecules. The laser phase plate is destined to become a permanent solution for the phase contrast deficit in electron microscopy, but the most important question from the cryo-EM perspective would be the cost/benefit of the device. The answer to the second part of this question will hopefully come soon, with the first cryo-EM experimental trials of the new device. Until the eventual introduction of the laser phase plate, we discourage the use of the current version of the VPP for single-particle analysis. It has not demonstrated any quantitative benefits in practice and its use may distract researchers from important sample-optimization steps by providing false hope for marginal samples.

3.5 Practical considerations: movie-mode data acquisition

Yifan Cheng
HHMI/Department of Biochemistry and Biophysics, UCSF, San Francisco, CA, USA

Direct electron detection cameras record images as movie stacks that contain many frames and enable correction of electron-beam-induced image motion (Bai *et al* 2013, Brilot *et al* 2012, Li *et al* 2013a). As these cameras are now commercially available and accessible to the broad structural-biology community engaging in single-particle cryo-EM studies, cryo-EM datasets are currently routinely acquired as movie stacks. Direct electron detection cameras typically operate in two different modes, single-electron counting or charge integration, regardless of which vendor manufactures the camera. Because single-electron counting produces better detective quantum efficiency (DQE) (figure 3.10) (Li *et al* 2013a), it is the more popular choice for high-resolution cryo-EM data acquisition.

While it is advantageous to use single-electron counting cameras for acquiring movie stacks for single-particle cryo-EM projects, the technical characteristics of such cameras impose certain constraints on how data acquisition should be strategized to fit the goals of a specific project, either to achieve the highest possible

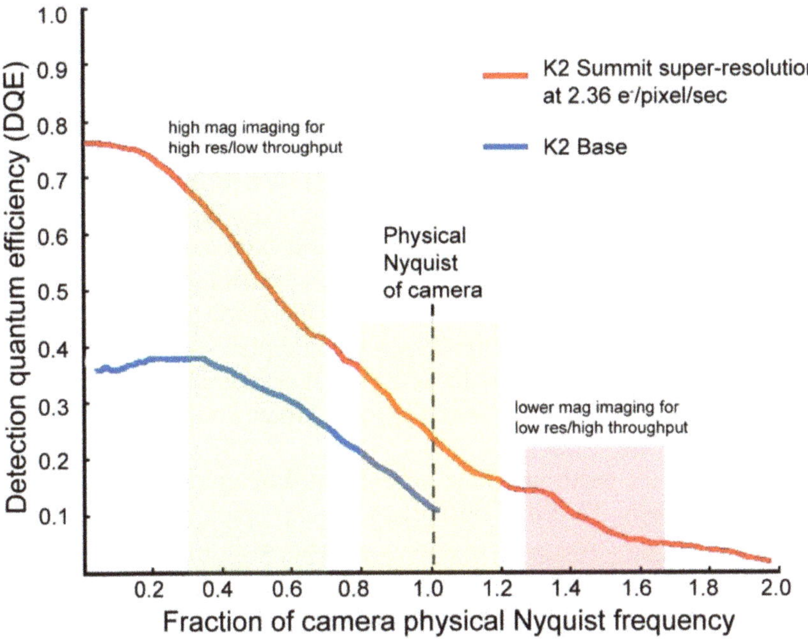

Figure 3.10. Performance of a Gatan K2 Summit single-electron counting camera. DQE curves are shown for the two options of operation, in the single-electron counting mode and in the charge-integration mode. Colored shadows correlate the choice of different magnification and pixel sizes with the DQE performance of the camera.

resolution, or to determine a structure as rapidly as possible at a modest resolution that is sufficient to build an atomic model, or both. Considering the usual constraints in availability of microscope time in almost all cryo-EM facilities, it is necessary to set the microscope differently to meet different goals. In the following, we will discuss a number of practical considerations that one needs to consider when strategizing single-particle cryo-EM dataset collection using movie mode.

3.5.1 Magnification and resolution

For a given type of camera, the magnification determines the pixel size (referred to the specimen) of the image being acquired, which in turn determines the Nyquist-resolution limit of the image. The Nyquist limit is the highest resolution that can be achieved by using the dataset, which is equal to 2× the pixel size. The DQE of the camera is a frequency-dependent curve that falls to ever decreasing values as one increases the resolution toward the Nyquist limit of the camera, as is shown in figure 3.10. For K2/K3 cameras from Gatan, there is also a difference between the Nyquist limit corresponding to the super-resolution pixel size and the Nyquist limit for the physical pixel size (Li *et al* 2013a). If the super-resolution mode is used, the resolution of the final reconstruction can reach beyond the physical Nyquist limit by 10%–20%, or even more.

While the DQE of a specific camera is independent of the pixel size for images being recorded with the camera, the choice of magnification/pixel size, in practice, places the targeted resolution of a reconstruction to within a certain range of DQE values of the camera. A higher magnification gives a smaller pixel size, which corresponds to a higher Nyquist-resolution limit of the image and it also puts the targeted resolution into a region with higher DQE, such as the green-shaded area in figure 3.10. The flip side of the coin is that micrographs recorded at a higher magnification cover smaller areas of the sample and each micrograph also contains fewer particles. Thus, a dataset containing a sufficient number of particles would require more micrographs, potentially lowering the throughput. On the opposite side, images collected at a lower magnification could place the desired resolution in the range of relatively lower DQE values (the red shaded area in figure 3.10). Images recorded at such a lower magnification cover a larger area and each micrograph contains more particles.

Considering that the needs are often different at different stages of a project, and in particular that one may need to optimize sample preparation as was described in section 2.5, a common strategy is to start a project with a lower magnification to generate a reasonable reconstruction in order to quickly assess the sample quality. This may then be followed by collecting datasets from optimized samples/grids, using a higher magnification to target for a higher resolution.

3.5.2 Dose rate

A major consideration of using a single-electron counting camera is to consider the proper dose rate, which strongly influences the camera performance and potentially the overall quality of the dataset. A major characteristic feature of a camera operating in single-electron counting mode is the coincidence loss, which occurs when more than one electron strikes the same pixel during the same time frame of a single counting cycle (Li *et al* 2013a, 2013b). When this happens, only one electron is counted, resulting in the loss of other electrons striking the same pixel in the same frame. The influences of coincidence loss on image have been characterized (Li *et al* 2013a, 2013b). It generates a nonlinear correspondence between the readout of the pixel intensity and the total dose of electrons striking the sample (figure 3.11(a)). In the Fourier transform of an image, coincidence losses generate a dark region in the center of the transform (figure 3.11(b)) that suppress the low-frequency amplitudes (figure 3.11(c)). Coincidence loss also reduces the DQE, presumably more in the low-resolution range to a much greater extent than near Nyquist resolution. Avoiding a significant coincidence loss is one of the primary considerations when using a single-electron counting camera for data acquisition.

Coincidence loss is determined by the internal frame rate of the camera. The Gatan K3 camera operates at 1500 frames per second (fps), significantly higher than the 400 fps of the previous generation of K2 cameras. For practical consideration by users, the coincidence loss is directly related to the dose rate on the camera being used for data collection. The manufacturer recommends a dose rate of ~8 e^{-}/pixel/second (eps) for the K2 camera and ~20 eps for the K3 camera. Higher dose rates

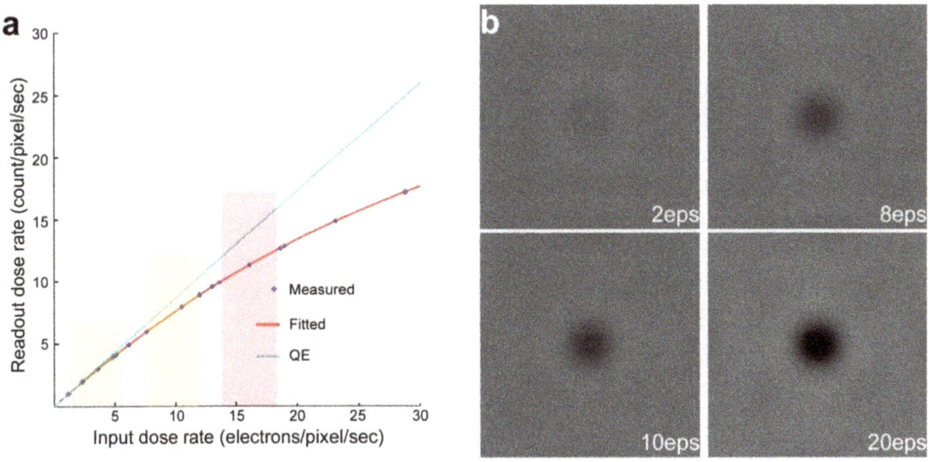

Figure 3.11. Examples of the effect of coincidence losses for different dose rates at the camera. (a) The detector conversion efficiency (DCE) of the camera shows a nonlinear relationship between the incident-electron dose rate and the measured response of the camera. (b) Fourier transforms of images recorded on a Gatan K2 camera, in counting mode, at different dose rates and without any specimen. In the absence of coincidence losses, these Fourier transforms would all have a uniform distribution of intensities, since the corresponding electron micrographs are all examples of pure shot noise. Note the dark region in the center of the FFT that corresponds to the suppression of low-resolution signal that occurs due to coincidence losses. A uniform distribution of the Fourier transform of the camera output is approached only at the lowest dose rate of incident electrons shown here.

than these lead to progressively higher coincidence loss, poorer DQE performance, and potentially relatively poorer image contrast. Using a lower camera dose rate reduces coincidence loss and its associated affects, but it also prolongs the exposure time and even reduces the SNR for a fixed amount of time per movie frame, which could potentially reduce the accuracy of motion correction. Furthermore, when the dose rate on the camera becomes too low, such as 1–2 electrons per second for the K2/K3 cameras, it will lead to insufficient events per frame and, according to the manufacturer, potentially generate false counts.

Note that the dose rate on the camera (in units of e^{-}/pixel/second) is correlated with, but different from, the dose rate at the specimen (in units of $e^{-}/Å^{2}$/second). The two values are related by the magnification and pixel size of images being recorded. The accumulated total dose on the specimen (in units of $e^{-}/Å^{2}$) is then determined from the dose rate on the specimen times the total exposure time. To achieve the desired total dose on the specimen, a lower dose rate on the camera leads to a longer exposure time, which can potentially influence the effectiveness of motion correction, as discussed below.

3.5.3 Strategies for motion correction

A major advance that has resulted from collecting cryo-EM datasets in movie mode is the ability to correct beam-induced image motion, so-called 'motion correction' (Campbell *et al* 2012, Li *et al* 2013a). Before summing all frames together, individual frames within a movie stack are aligned after determining the motions between these frames. The most obvious benefits resulting from motion correction are a significant improvement in visibility of Thon rings in Fourier transforms of cryo-EM micrographs (figures 3.12(a) and (b)), and the improvement of resolution in the final reconstruction (Bai *et al* 2013, Campbell *et al* 2012, Li *et al* 2013a). Currently, motion correction is often carried out by two different approaches and at two different stages of the structure determination pipeline.

The first approach is to correct image motion captured in micrographs, in which the motions between frames are determined by a cross-correlation based method. To mitigate the poor SNR of individual movie frames, which produces very noisy results obtained directly from cross correlation between two individual frames, least squares fitting of cross-correlations between all frames was used to obtain a smooth motion trajectory captured by the movies (Li *et al* 2013a). As it was recognized early on that beam-induced motion is not a rigid-body motion, but a distortion of the specimen area being illuminated by the electron beam (Brilot *et al* 2012), images are

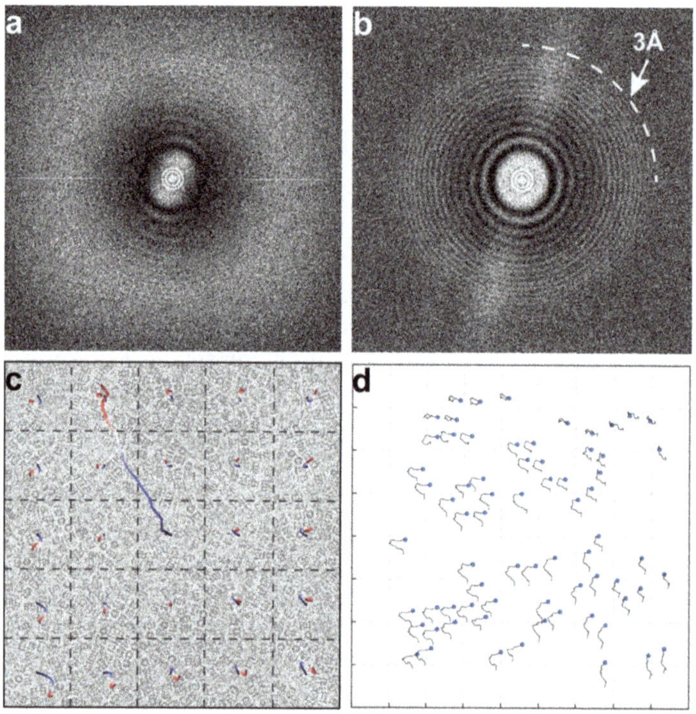

Figure 3.12. Motion correction of cryo-EM images recorded as a movie. (a) Fourier transform calculated from the sum of frames before motion correction. Beam-induced motion deteriorates the Thon rings in the direction perpendicular to the motion. (b) Fourier transform calculated from the sum of frames after motion correction. Motion correction restores the Thon rings to high resolution in all direction. (c) A micrograph is divided into multiple patches so that the non-uniform motion can be corrected across the field of the micrograph. (d) Trajectories of particle motion captured during the Bayesian polishing procedure (image adapted from Zivanov *et al* 2019, CC BY 4.0).

divided into smaller patches. While the motion within each patch is still determined initially as a rigid-body movement, a polynomial function is then generated by fitting the motion determined for each small patch in order to describe the motion trajectory of the entire field of view (figure 3.12(c)). This strategy is implemented in MotionCor2, a popular motion correction program (Zheng *et al* 2017). Because this approach can be applied to micrographs regardless of the type of sample being imaged, it is often used on-the-fly during cryo-EM data acquisition to produce motion-corrected micrographs immediately after acquisition.

In parallel, it is also possible to correct motion at the level of each individual particle during the refinement stage of 3D reconstruction (figure 3.12(d)). While there are a number of methods developed to do this, the most popular one that is currently used is the Bayesian polishing procedure implemented in RELION (Zivanov *et al* 2019). The key feature is that, at the time of carrying out this polishing procedure, a high-resolution reference map, accurate values of the orientational parameters, and the defocus values of particles are already known. Therefore, this procedure is often applied at the final stage of refining a reconstruction.

In practice, both procedures are used for high-resolution structure determination, one immediately after data acquisition, and the other at the end of image processing pipeline to further improve the resolution. Note that the computational correction of motion only compensates for the motion between adjacent frames, but it does not correct the motion that occurs within each frame. During the Bayesian polishing, the motion captured within the frame is characterized as part of a temperature factor (Zivanov *et al* 2019).

3.5.3.1 Output frame rate

While the internal frame rate of the K3 camera is much higher than that of K2 camera, 400 fps for a K2 camera and 1500 fps for a K3 camera, the total frame number per second that a camera outputs is a lot less, typically around 10–40 fps. In theory, a high output frame rate leads to a shorter time accumulation per frame and can potentially improve the accuracy of motion correction. However, because of the shorter accumulated exposure time per frame, the lower average intensity of individual frames leads to a poor SNR, which could negatively impact the accuracy of motion correction.

Certain computational approaches could partially mitigate the problem, such as using rolling averaging to extrapolate motion between frames that have a very poor SNR (Zheng *et al* 2017). One question concerning accurate motion correction is whether raw frames, 400 fps of K2 camera or 1500 fps of K3 camera, can be used directly for motion correction. The requirement to avoid coincidence losses results in such a sparse distribution of events per individual counting frame that they contain little of the low-frequency information regarding the shapes and locations of objects being imaged. Without a sufficient number of electrons to provide sufficient low-frequency information, motion cannot be corrected. Again, a rolling averaging or other computational tricks may eventually facilitate a more complete motion correction that might be achieved by using individual raw frames.

3.5.4 Total dose or exposure time

Concerning the radiation damage to biological specimens, the total electron dose used in collecting a typical cryo-EM dataset is set to ~20 $e^{-}/Å^2$ or more at the specimen. Radiation damage induced signal loss starts at high resolution and progressively develops toward the lower resolution range as the total accumulated dose increases, as is described in section 3.2. Since cryo-EM data are recorded as movie stacks (Li *et al* 2013a), there may still be useful low-resolution information in the later frames of a movie that could improve low-frequency image contrast. One is thus tempted to use a higher electron exposure.

The introduction of dose weighting, which down-weights the higher resolution Fourier components in the later frames by applying a progressively larger temperature factor, allows cryo-EM datasets to be collected with much higher total dose (Grant and Grigorieff 2015). The goal of dose weighting is to prevent further noise from accumulating at high resolution, once a point has been reached at which the movie frames no longer contain high-resolution signal. The benefit of doing so is to still enhance the low-resolution SNR, often loosely referred to as the 'image contrast', without reducing the high-resolution SNR.

3.5.5 File size of movie datasets

By using an optimal dose rate on the camera, the exposure time is dependent on the magnification/pixel size and the targeted value of the total accumulated dose on the specimen. An inconvenience of collecting a cryo-EM dataset with a large total dose is that fewer movies can be recorded in a given session. In addition, having to work with a larger number of frames per movie increases both the computation time needed for motion correction and the file size required for the back-up of raw data.

After motion correction, which is currently typically done on-the-fly, only two sums, a direct summation of all frames after motion correction and a dose-weighted sum, are output for further processing (Zheng *et al* 2017). Raw movie frames are typically backed up for long term storage, and thus the file size is not a major concern. In practice, it is becoming common to collect cryo-EM datasets with accumulated total doses of ~100 $e^{-}/Å^2$.

3.5.6 Summary

Two related practical questions that face any single-particle cryo-EM project are first what the target resolution is and then how fast one wants to achieve the goal, giving the usual constraint that an unlimited amount of microscope time will not be available to any given project. The practical considerations discussed above show that the data collection strategy always involves a balance between different, opposing factors. If the project aims to achieve the highest possible resolution, assuming the sample is ready for that goal, one may want to consider collecting the dataset using a higher magnification, in order to gain a better value of the camera DQE. One may also want to use a lower dose rate on the camera, so as to reduce coincidence losses and further boost the DQE, and one may want to collect images with a higher total dose on the specimen in order to enhance the low-resolution SNR (or the 'image contrast'). If, on the other hand, the throughput is of primary concern, the strategy will need to be adjusted following the considerations discussed above.

In practice, for many single-particle cryo-EM projects the requirements and goals at different stages are different. In the early stage, while optimization of sample preparation is still in progress, one may want to obtain a structure as quickly as possible in order to better assess the procedures used in sample preparation. At this stage, the strategy of data acquisition is best-focused on throughput. At a later stage, after sample optimization is accomplished, improving resolution to the highest possible level may become the focus. The data acquisition strategy should then be adjusted accordingly. While this brief section does not provide detailed suggestions that will fit all cases, discussion of these practical considerations can help readers to strategize the data acquisition properly, based on one's requirements.

References

Bai X C, Fernandez I S, McMullan G and Scheres S H 2013 Ribosome structures to near-atomic resolution from thirty thousand cryo-EM particles *eLife* **2** e00461

Baker L A and Rubinstein J L 2010 Radiation damage in electron cryomicroscopy *Method Enzym.* **481** 373–90

Baker L A, Smith E A, Bueler S A and Rubinstein J L 2010 The resolution dependence of optimal exposures in liquid nitrogen temperature electron cryomicroscopy of catalase crystals *J. Struct. Biol.* **169** 431–7

Bammes B E, Jakana J, Schmid M F and Chiu W 2010 Radiation damage effects at four specimen temperatures from 4 to 100 K *J. Struct. Biol.* **169** 331–41

Bartesaghi A *et al* 2018 Atomic resolution cryo-EM structure of β-galactosidase *Structure* **26** 848–56.e3

Brilot A F, Chen J Z, Cheng A, Pan J, Harrison S C, Potter C S, Carragher B, Henderson R and Grigorieff N 2012 Beam-induced motion of vitrified specimen on holey carbon film *J. Struct. Biol.* **177** 630–7

Buijisse B, Trompenaars P, Altin V, Daneve R and Glaeser R M 2020 Spectral DQE of the Volta phase plate *Ultramicroscopy* **218** 113079

Campbell M G *et al* 2012 Movies of ice-embedded particles enhance resolution in electron cryo-microscopy *Structure* **20** 1823–8

Cheng A *et al* 2016 Strategies for automated cryoEM data collection using direct detectors *Methods in Enzymology* vol 579 (Amsterdam: Elsevier) pp 87–102

Cheng A *et al* 2018 High resolution single particle cryo-electron microscopy using beam-image shift *J. Struct. Biol.* **204** 270–5

Chua E Y D, Vogirala V K, Inian O, Wong A S W, Nordenskiöld L, Plitzko J M, Danev R and Sandin S 2016 3.9 Å structure of the nucleosome core particle determined by phase-plate cryo-EM *Nucleic Acids Res.* **44** 8013–9

Cohen H A, Schmid M F and Chiu W 1984 Estimates of the validity of projection approximation for three-dimensional reconstructions at high resolution *Ultramicroscopy* **14** 219–26

Danev R, Belousoff M, Liang Y-L, Zhang X, Wootten D and Sexton P M 2020 Routine sub-2.5 Å cryo-EM structure determination of B-family G protein-coupled receptors *bioRxiv* 2020.08.21.260851

Danev R, Tegunov D and Baumeister W 2017 Using the Volta phase plate with defocus for cryo-EM single particle analysis *eLife* **6** e23006

Eades A 2006 Obtaining TEM images with a uniform deviation parameter *Ultramicroscopy* **106** 432–8

Fernandez I S, Bai X C, Hussain T, Kelley A C, Lorsch J R, Ramakrishnan V and Scheres S H 2013 Molecular architecture of a eukaryotic translational initiation complex *Science* **342** 1240585

Fujiyoshi Y 1998 The structural study of membrane proteins by electron crystallography *Adv. Biophys.* **35** 25–80

Glaeser R M 1971 Limitations to significant information in biological electron microscopy as a result of radiation damage *J. Ultrastruct. Res.* **36** 466–82

Glaeser R M *et al* 2011 Precise beam-tilt alignment and collimation are required to minimize the phase error associated with coma in high-resolution cryo-EM *J. Struct. Biol.* **174** 1–10

Glaeser R M, Downing K, DeRosier D, Chiu W and Frank J 2007 *Electron Crystallography of Biological Macromolecules* (Oxford: Oxford University Press)

Grant T and Grigorieff N 2015 Measuring the optimal exposure for single particle cryo-EM using a 2.6 Å reconstruction of rotavirus VP6 *eLife* **4** e06980

Han Y, Fan X, Wang H, Zhao F, Tully C G, Kong J, Yao N and Yan N 2020 High-yield monolayer graphene grids for near-atomic resolution cryoelectron microscopy *Proc. Natl Acad. Sci. USA* **117** 1009–14

Hayward S B and Glaeser R M 1979 Radiation damage of purple membrane at low temperature *Ultramicroscopy* **04** 201–10

Henderson R 1990 Cryo-protection of protein crystals against radiation damage in electron and x-ray diffraction *Proc. R. Soc.* B **241** 6–8

Henderson R 1995 The potential and limitations of neutrons, electrons and x-rays for atomic resolution microscopy of unstained biological molecules *Q. Rev. Biophys.* **28** 171–93

Henderson R 2018 From electron crystallography to single particle cryoEM (Nobel lecture) *Angew. Chemie Int. Ed.* **57** 2–24

Henderson R, Baldwin J M, Ceska T A, Zemlin F, Beckmann E and Downing K H 1990 Model for the structure of bacteriorhodopsin based on high-resolution electron cryo-microscopy *J. Mol. Biol.* **213** 899–929

Herzik M A, Wu M and Lander G C 2019 High-resolution structure determination of sub-100 kDa complexes using conventional cryo-EM *Nat. Commun.* **10** 1032

Iancu C V, Wright E R, Heymann J B and Jensen G J 2006 A comparison of liquid nitrogen and liquid helium as cryogens for electron cryotomography *J. Struct. Biol.* **153** 231–40

Kato T, Makino F, Nakane T, Terahara N, Kaneko T, Shimizu Y, Motoki S, Ishikawa I, Yonekura K and Namba K 2019 CryoTEM with a cold field emission gun that moves structural biology into a new stage *Microsc. Microanal.* **25** 998–9

Khoshouei M, Radjainia M, Baumeister W and Danev R 2017 Cryo-EM structure of haemoglobin at 3.2 Å determined with the Volta phase plate *Nat. Commun.* **8** 16099

Koster A J and de Ruijter W J 1992 Practical autoalignment of transmission electron microscopes *Ultramicroscopy* **40** 89–107

Leapman R D and Sun S 1995 Cryo-electron energy loss spectroscopy: observations on vitrified hydrated specimens and radiation damage *Ultramicroscopy* **59** 71–9

Li X, Mooney P, Zheng S, Booth C R, Braunfeld M B, Gubbens S, Agard D A and Cheng Y 2013a Electron counting and beam-induced motion correction enable near-atomic-resolution single-particle cryo-EM *Nat. Methods* **10** 584–90

Li X, Zheng S Q, Egami K, Agard D A and Cheng Y 2013b Influence of electron dose rate on electron counting images recorded with the K2 camera *J. Struct. Biol.* **184** 251–60
Liang Y-L *et al* 2017 Phase-plate cryo-EM structure of a class B GPCR–G-protein complex *Nature* **546** 118–23
Liao M, Cao E, Julius D and Cheng Y 2013 Structure of the TRPV1 ion channel determined by electron cryo-microscopy *Nature* **504** 107–12
Mastronarde D N 2005 Automated electron microscope tomography using robust prediction of specimen movements *J. Struct. Biol.* **152** 36–51
Meents S, Gutmann S, Wagner A and Schulze-Briese C 2010 Origin and temperature dependence of radiation damage in biological samples at cryogenic temperatures *Proc. Natl Acad. Sci. USA* **107** 1094–9
Miyazawa A, Fujiyoshi Y and Unwin N 2003 Structure and gating mechanism of the acetylcholine receptor pore *Nature* **423** 949–55
Peet M J, Henderson R and Russo C J 2019 The energy dependence of contrast and damage in electron cryomicroscopy of biological molecules *Ultramicroscopy* **203** 125–31
Pfeil-Gardiner O, Mills D J, Vonck J and Kuehlbrandt W 2019 A comparative study of single-particle cryo-EM with liquid-nitrogen and liquid-helium cooling *IUCrJ* **6** 1099–105
Ripstein Z A and Rubinstein J L 2016 Processing of cryo-EM movie data *Methods Enzymol.* **579** 103–24
Rubinstein J L and Brubaker M A 2015 Alignment of cryo-EM movies of individual particles by optimization of image translations *J. Struct. Biol.* **192** 1–11
Russo C J and Henderson R 2018a Ewald sphere correction using a single side-band image processing algorithm *Ultramicroscopy* **187** 26–33
Russo C J and Henderson R 2018b Charge accumulation in electron cryomicroscopy *Ultramicroscopy* **187** 43–9
Russo C J and Henderson R 2018c Microscopic charge fluctuations cause minimal contrast loss in cryoEM *Ultramicroscopy* **187** 56–63
Scheres S H 2014 Beam-induced motion correction for sub-megadalton cryo-EM particles *eLife* **3** 1–8
Schorb M, Haberbosch I, Hagen W J H, Schwab Y and Mastronarde D N 2019 Software tools for automated transmission electron microscopy *Nat. Methods* **16** 471–7
Schwartz O, Axelrod J J, Campbell S L, Turnbaugh C, Glaeser R M and Müller H 2019 Laser phase plate for transmission electron microscopy *Nat. Methods* **16** 1016–20
Stagg S M *et al* 2008 A test-bed for optimizing high-resolution single particle reconstructions *J. Struct. Biol.* **163** 29–39
Suloway C *et al* 2005 Automated molecular microscopy: the new Leginon system *J. Struct. Biol.* **151** 41–60
Tan Y Z *et al* 2018 Sub-2 Å Ewald curvature corrected structure of an AAV2 capsid variant *Nat. Commun.* **9** 1–11
Taylor K and Glaeser R M 1976 Electron microscopy of frozen hydrated biological specimens *J. Ultrastruct. Res.* **456** 448–56
Thorne R E 2020 Hypothesis for a mechanism of beam-induced motion in cryo-electron microscopy *IUCrJ* **7** 1–6
Turnbaugh C *et al* 2020 High-power near concentric Fabry–Perot cavity for phase contrast electron microscopy arXiv: 2012.08638

Unwin P N T and Henderson R 1975 Molecular structure determination by electron microscopy of unstained crystalline specimens *J. Mol. Biol.* **94** 425–40

Weis F and Hagen W J H 2020 Combining high throughput and high quality for cryo-electron microscopy data collection *Acta Crystallogr.* D **76** 724–8

Wolf M, DeRosier D J and Grigorieff N 2006 Ewald sphere correction for single-particle electron microscopy *Ultramicroscopy* **106** 376–82

Wu C *et al* 2019 High-quality, high-throughput cryo-electron microscopy data collection via beam tilt and astigmatism-free beam-image shift *J. Struct. Biol.* **208** 107396

Wu W, Thomas J A, Cheng N, Black L W and Steven A C 2012 Bubblegrams reveal the inner body of bacteriophage *φ*KZ *Science* **335** 182

Zhang X, Jin L, Fang Q, Hui W H and Zhou Z H 2010 3.3 Å cryo-EM structure of a nonenveloped virus reveals a priming mechanism for cell entry *Cell* **141** 472–82

Zheng S Q, Palovcak E, Armache J P, Verba K A, Cheng Y and Agard D A 2017 MotionCor2: anisotropic correction of beam-induced motion for improved cryo-electron microscopy *Nat. Methods* **14** 331–2

Zivanov J, Nakane T and Scheres S H W 2019 A Bayesian approach to beam-induced motion correction in cryo-EM single-particle analysis *IUCrJ* **6** 5–17

Chapter 4

Data processing

4.1 Overview

The primary goal in single-particle cryo-EM, as it is for other methods used in high-resolution structural biology, is to obtain a 3D density map that can be interpreted in terms of the atomic structure of the specimen. Just looking at cryo-EM images is not sufficient to achieve that goal, of course, since images only represent noisy, 2D views of what is a 3D object. Thus, data must be merged from many particles, not just to improve the signal-to-noise ratio but also to combine the full information that is available within each of the different views.

This chapter proceeds through the data-processing pipeline in seven stages, each of which has been improved through the contributions of many investigators. Section 4.2 begins by describing how particles are identified, so that their noisy images can be extracted for further analysis. These images contain substantial systematic error, however, due the intentional use of a large amount of defocus. As a result, the data contained in the raw images must be corrected for the effect of having a contrast transfer function (CTF), which varies from one image to the next. Section 4.3 describes how the CTF is estimated for each micrograph and in some cases for each particle.

The major techniques and software tools that are available to merge data and compute a 3D density map are then described in sections 4.4 and 4.5. In a first pass, which is described in section 4.4, it is assumed that the particles are structurally homogeneous, i.e. 'identical', at least to a given resolution. Understandably, particles contributing to a raw dataset are almost never structurally homogeneous. Fortunately, however, software tools have been developed that are often able to classify the extracted images of single particles into groups that are significantly more homogeneous than are the particles making up the initial dataset, a step that is covered in section 4.5.

Since successful recovery of a 3D density map requires merging data that correspond to many different views of the object, it is necessary that particles adopt a wide range of orientations within the specimen. At least half of the time, unfortunately, that does not

doi:10.1088/978-0-7503-3039-8ch4

happen. Since the problem of preferred orientation affects so many projects, section 4.6 makes a digression to describe what alternatives can be used to address the problem when it does occur. These include changing the way in which grids are prepared, or perhaps using images of tilted specimens.

Turning next to visualizing information contained in a 3D density map, it is almost always helpful to make a systematic correction for effects that produce a Gaussian decrease in the signal at high resolution. Section 4.7 explains that this decrease has both instrument-dependent and specimen-dependent contributions. This section then goes on to describe how to estimate the optimal size of the correction.

Finally, this chapter concludes with a section devoted to two issues that are becoming of increasing importance as the resolution of 3D reconstructions continues to advance. The first of these is corrections that must be made for the effect of higher order optical aberrations, of which correction for coma has recently become a routine step in data processing. The second issue is correction for curvature of the Ewald sphere, something that can be expected to grow in importance.

4.2 Automated extraction of particles

Dimitry Tegunov
Max Planck Institute for Biophysical Chemistry, 37077 Göttingen, Germany

4.2.1 From micrographs to particles

One of the central assumptions made in most single-particle algorithms is that there is, in fact, only a single particle in each image, located roughly at its center. However, typical micrographs contain tens or hundreds of particles in random locations, interspersed with contaminants that should be excluded from analysis. Before proceeding with any single-particle analysis (SPA) steps, particles of interest need to be located in each micrograph. A smaller, quadratic image region around each particle position is extracted, and the results are combined in a particle stack. This stack is used as input for most SPA programs.

Depending on the sample, identifying the right particles may not be easy (figure 4.1). Proteins can aggregate or dissociate, and non-spherical particles imaged at different angles will produce projections of very different size and shape. The grid-preparation process can also add contaminants that look like particles. While SPA can be used to make sense of this heterogeneity afterwards, starting with a clean selection of particles will benefit the analysis process. At the same time, it is important to remember that overly strict selection will introduce a bias that can affect the conclusions of the experiment. For instance, excluding all particles much smaller or bigger than average in an attempt to remove aggregates or dissociated subunits can also exclude entire projection directions or important biological states.

High-resolution maps can require averaging hundreds of thousands of structurally homogeneous particles. Due to compositional and conformational heterogeneity, some projects will start with many more—using 5%–10% of the initial particle set in the final maps is not unheard of (Plaschka *et al* 2015). Selecting so many particles manually is an extremely laborious task and accuracy will likely suffer from

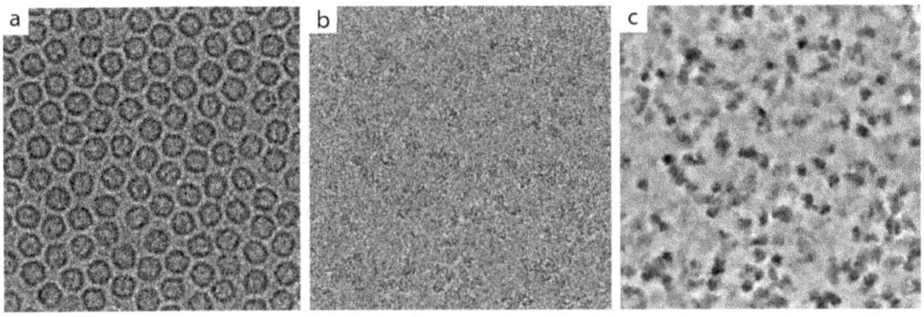

Figure 4.1. Examples of raw and denoised data. (a) Raw micrograph showing spherical apoferritin particles at high defocus. The particles are clearly visible and easy to select. Most 'real' samples are not as friendly as this one. (b) Raw micrograph of a SARS-CoV-2 polymerase sample at low defocus. The particles are barely visible and difficult to distinguish. (c) A denoised version of the micrograph shown in (b). While the particles are not as well-separated as in (a), they are easier to spot than in the raw data.

human fatigue after a few dozen micrographs. Automated-selection approaches have therefore been an area of active research since the early days of SPA (van Heel 1982, Nicholson and Glaeser 2001). Thanks to recent advances in machine learning (LeCun *et al* 1998, Krizhevsky *et al* 2012) (ML), the latest methods can reach human-like performance with minimal intervention in many cases (Tegunov and Cramer 2019, Bepler *et al* 2019, Wagner *et al* 2019). In the following sections, we will review and compare the most popular automated approaches for particle selection as well as manual selection.

4.2.2 Manual selection

Manual particle selection is rarely a fun activity. Clicking hundreds or even thousands of targets in very noisy micrographs is tedious and leaves the user wondering why such a straight-forward task has not been automated away. However, even though modern automated approaches can do most of the work, it is often a good idea to start a project by selecting a few micrographs manually. On the one hand, taking some time to become familiar with the raw data can save a lot of time later, or even decide the fate of some projects. Forcing yourself to go through the decision process of what is 'a particle' multiple times and assessing the sample quality will help with that. On the other hand, creating a reference set of accurately selected particles will allow to improve the accuracy of ML-based methods, and it is a necessary step for the *de novo* creation of reference images for template matching.

Conventional cryo-EM images are notoriously noisy, making it difficult to recognize particles in raw data by eye (figure 4.1(b)). A phase plate (Danev *et al* 2017, Schwartz *et al* 2019) (see section 3.4) can help immensely by increasing the image contrast in the lower spatial frequencies, but it may be absent from a microscope, or it may complicate the acquisition process. At the very least, a strong low-pass or median filter needs to be applied to conventional data to be able to see the particles. These filters suppress the fine-grained noise by blurring the images. This can be done automatically in the manual selection tools of packages like RELION (Scheres 2015), cisTEM (Grant *et al* 2018), cryoSPARC (Punjani *et al* 2017), or EMAN (Bell *et al* 2016).

However, a more powerful approach that has become available recently is image denoising. Using the noise2noise principle (Lehtinen *et al* 2018), a convolutional neural network (CNN) can be trained to denoise the data. Coupled with a Wiener filter to reverse the CTF (see section 4.3.2), the resulting images provide contrast and clarity on a par with phase plate data (figure 4.1(c)), while the latter can be made even clearer using the same approach. Macromolecules lighter than 50 kDa can be distinguished against the background even at low defocus and those located close to each other are easy to tell apart. Denoising can be applied automatically to all displayed images in Warp (Tegunov and Cramer 2019) and similar implementations are available as part of Topaz (Bepler *et al* 2019) and crYOLO (Wagner *et al* 2019) to prepare denoised images for display in other tools.

Once the particles have been made more visible, the user proceeds with clicking their centers one-by-one. It is not necessary to hit the center exactly, but getting it roughly correct will accelerate the SPA, where the translational search-radius and thus the computational complexity can be reduced. A less tedious approach to manual selection is to start from the other end—an automated approach which does not require additional knowledge of the data (e.g. an unbiased algorithm, a pre-trained 'general' ML model, or templates prepared from similar data), is used to select the micrographs very permissively. Most of the correct particles will be selected alongside some false positives that need to be removed. Removing particles is usually quicker than adding new ones since there is no need to hit the center. In any case, put on some music and make a cup of tea—this will be a long process.

4.2.3 Unbiased automated approaches

Introducing bias during particle selection is a real danger as it will affect all downstream processing and is difficult to fix later without going back to square one and re-selecting everything. Thus, a selection algorithm that does not let the user introduce bias would be ideal. In reality, avoiding bias completely is very difficult because at least a vague idea of what 'a particle' should look like must be incorporated into the algorithm's design, and a preconception certainly exists in every user's mind based on past experience.

Luckily, many proteins imaged with cryo-EM do have one thing in common: they are spherical to some extent and, given a strong enough low-pass filter, resemble 'blobs'. Selection algorithms such as the difference of Gaussians (DoG) (Voss *et al* 2009) or Laplacian of Gaussians (LoG) (Zivanov *et al* 2018) take advantage of that property by correlating the data against blob images of different sizes. When an image is correlated against a template image, the result is an image where each pixel's value describes the similarity between the image and the template image shifted such that its center is located at that pixel's position (figure 4.2(b) and (c)). Thus, the pixels with the highest similarity values will be in regions that resemble the template image—in this case a blob (figure 4.2(b)). The algorithms proceed with selecting regions resembling blobs within a size range specified by the user, while excluding blobs that are too big or too small. Because the blobs are rotationally

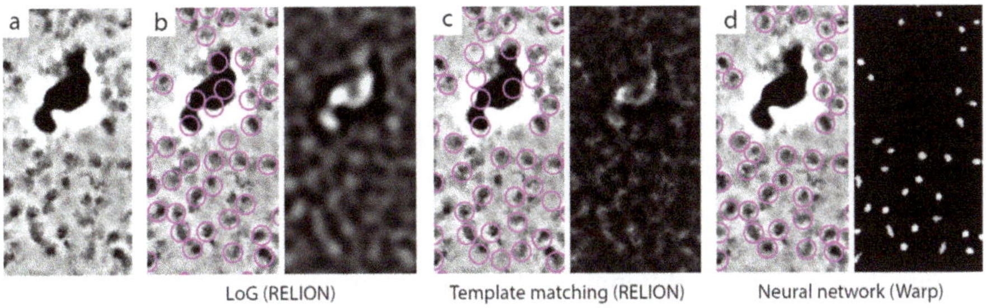

Figure 4.2. Selection results from different algorithms and the underlying correlation scores. (a) Denoised excerpt from the SARS-CoV-2 polymerase micrograph used for selection. (b) Selection result from RELION's unbiased LoG algorithm (left) and one of the blob-correlation images (right). Several particles were selected within the high-contrast artifact and empty ice. (c) Selection result from RELION's template-matching algorithm using nine templates showing the molecule of interest (left) and the maximum correlation scores over all templates and orientations (right). As with the LoG selector, high-contrast artifacts and empty ice were selected. (d) Selection result from Warp's convolutional neural net-based selector retrained using different micrographs from the same dataset (left) and probability scores for the 'particle' class. Problematic areas from (b) and (c) were avoided and all particles were selected.

invariant (i.e. they look the same regardless of the in-plane rotation) no additional rotation angles need to be tested and the algorithms execute very fast.

This approach is useful for quickly selecting close-to-spherical particles, but it may lose its utility for more irregular shapes such as filaments. Furthermore, because the blobs do not contain any fine features, the algorithm will also select similarly sized ice contaminants and it is more susceptible to image noise. Executing the algorithms on denoised images appears to improve the accuracy, but this workflow is not fully integrated in any package yet. The results of DoG or LoG selection usually require extensive cleaning through 2D classification (see section 4.2.5). Both algorithms require *a priori* knowledge of the approximate particle size and may exclude relevant views that deviate too far from that parameter.

4.2.3.1 Template matching

Using templates more specific to the data is a logical improvement upon matching against a blob. Micrographs are cross-correlated with several template images, representing various projection directions of one or more particle types. Since the templates are no longer rotationally invariant, correlation must be calculated for differently rotated versions of the same template. More advanced algorithms also apply the micrograph's CTF to the templates in order to model the data more closely. Every pixel in the final correlation image contains the maximum correlation value among all templates and their orientations at that pixel's position (figure 4.2(c)). Although the computational cost scales linearly with the number of templates and orientations, modern GPU-accelerated implementations such as those in RELION (Kimanius *et al* 2016), Gautomatch (Zhang 2016a), and cryoSPARC (Punjani *et al* 2017) can match micrographs against a set of templates quickly.

Templates are usually prepared *de novo* from the data they are used to select later. To do so, a few thousand particles are selected manually or through a template-free method, extracted into a particle stack, and subjected to 2D classification. The resulting 2D classes do not need to show high-resolution features because of the low resolution used later during matching, but they should reflect all projection directions present in the data. Selecting particles from micrographs with different defocus values can improve the template image quality, in particular if few particles are used. Variation in the ice thickness throughout a sample grid can affect the distribution of particle orientations, making certain orientations absent from parts of the data. Selecting from different grid squares rather than limiting the selection to a few adjacent holes can solve this problem. Another way to obtain templates for selection is to reuse 2D class averages from very similar samples, or to make 2D projections of the final 3D maps in multiple directions.

Template matching performed at high resolution has the highest potential for introducing bias. Even when executed on images containing pure noise, the resulting selection of 'noise particles' will contain enough random correlations with the original template to make a surprisingly faithful reproduction when averaged (Henderson 2013) (figure 4.3). Because of this, it is important to perform template matching at a resolution substantially lower than the one expected for the final maps. For instance, 15–20 Å is often a good resolution for template matching. However, if the 3D map refinement later gets stuck at 20 Å, it is necessary to go back to template matching and repeat it at a lower resolution. It would be difficult to trust the results otherwise.

Like the DoG and LoG selectors, template matching is susceptible to selecting high-contrast artifacts such as ice contaminants or hole edges (figure 4.2(b) and (c)). This may be surprising given the number and resolution of the positive examples provided to the algorithms. Humans can easily tell that ethane blobs or empty ice look nothing like the templates because they have a mental concept of each of these negative examples. However, a template-matching algorithm does not incorporate additional knowledge of the negative-example space and an ethane blob looks much more like a particle compared to empty ice at the noise level of a cryo-EM image and sometimes even compared to real particles. Setting the correlation-score threshold

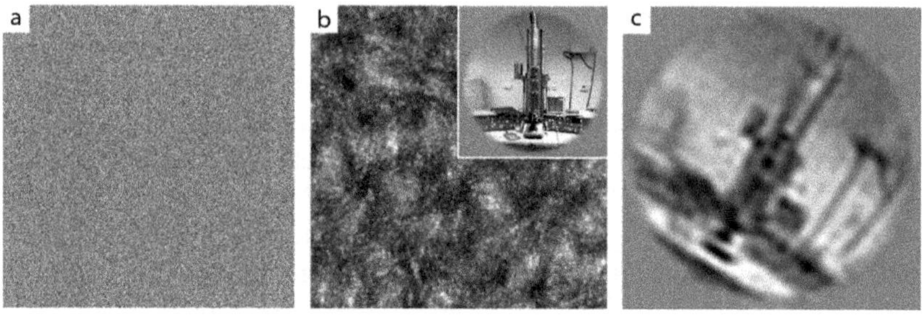

Figure 4.3. Dangers of template bias. (a) Image filled with completely random noise. (b) Correlation scores against the template, which was an image of a transmission electron microscope (insert, top-right). Any correlation peaks are purely random, as the original data did not contain any images of microscope. Nevertheless, the selection algorithm will choose the highest-correlation positions, even though their similarity to the template is merely fortuitous. (c) Class average from 10 000 positions selected from random noise using the high-resolution template. The microscope from the template is clearly visible—a result of the user having introduced template bias into their results.

forces the algorithm to select only positions with a high enough correlation. A low threshold will select more particles, but also include a high number of false positives. A high threshold will minimize the number of false positives, but will also miss some particles. Usually the parameter is set to maximize the number of particles, hoping that 2D and 3D classification will separate them from the numerous false positives later.

4.2.3.2 Convolutional neural networks and machine learning

Convolutional neural networks (LeCun *et al* 1998, 1989) have recently established themselves as the method of choice for a wide range of object-recognition problems in the field of computer vision (LeCun *et al* 2015). Such networks convolve the image data with a broad selection of small filters (imagine template matching, but with many more, much smaller templates, e.g. only 3 × 3 pixels) and apply a nonlinear function to the resulting correlation images. Each filter responds to some primitive visual feature. When multiple layers are connected by using the output of one layer as input for the next layer, concatenations of such filters can very efficiently model higher order features and their positive or negative influence on predictions of certain object classes (Kriegeskorte and Golan 2019, Olah *et al* 2018).

To learn the exact contents of all filters and their importance, the network is trained by presenting it with pairs of curated input and output data. For particle selection, such examples contain micrographs as input and particle locations as output. In one training step, the network is evaluated with one of the micrographs as input and the output is compared to the ground truth provided by the user. Any deviations are then back-propagated through the network to adjust its parameters a little. Unlike conventional template matching, the network must also explicitly learn to classify everything else as 'not a particle', gaining knowledge of the negative-example space. Following many training steps, the network converges to a state that keeps the output error to a minimum for the training data. Most importantly, if the training was successful, the network is also able to generalize its newly gained knowledge to correctly select new micrographs that were not part of the training set.

Modern implementations, such as those found in Warp, Topaz, or crYOLO, come pre-trained on a large number of different proteins and examples of contamination. Such 'general models' have a good idea of what a protein is and, more importantly, what it is not. Because the model's learning capacity is limited, it cannot memorize and reproduce any of the proteins seen during training in detail, thus avoiding the 'structure from noise' pitfall. Next to DoG-like selectors, general CNN models provide the least biased result. However, it is important to keep in mind that even a large and diverse set of images used in the pre-training of the general models does not guarantee that the models will work on every possible combination of particle shapes and sample conditions.

Like a human user looking at micrographs of an unusual sample for the first time, a CNN may need to look at a few examples to readjust its concept of what might be a particle in the new sample. In the currently available tools, it is best to start with the selection results of the general model, go manually through a few micrographs, and make corrections. During training, most tools augment the data by applying random rotation and minor distortions in every training step to make sure the model does not overfit even when only a few training examples are available. If the model does not choose well enough after the first training, it can help to select more micrographs with the new model and use the additional examples to train it further.

Extensively pre-trained CNN-based selectors currently offer the highest accuracy for most datasets (figure 4.2(d)) and the most intuitive approach to improving the performance, which is more akin to teaching a human than tuning numerical parameters. They have quickly become the most popular selection method, in particular for automated processing pipelines.

4.2.4 Particle extraction

Once their positions are known, particles need to be extracted as separate, square images. The size of the particle 'box' is an important parameter for several reasons. Once extracted, SPA algorithms cannot shift the particle images to bring in new content without re-extracting them, which is too laborious to be done routinely. Because some of the selected positions may not be at the particle's center, some margin must be left to avoid cutting off particles. Furthermore, modern SPA tools such as RELION (Scheres 2012b) or cryoSPARC (Punjani *et al* 2017) perform lossy interpolations in the frequency domain, which causes artifacts at the box periphery in the spatial domain. To make sure these artifacts do not impact the protein region, a rule of thumb is to make the box at least 50% larger than the particle. Finally, particles acquired at high defocus have their high-resolution signal delocalized far outside the apparent particle boundaries (Downing and Glaeser 2008). CTF correction during map refinement and reconstruction reverses this effect, but the delocalized signal cannot be brought back if it is not contained in the extracted box (figure 4.4).

All of this makes it tempting to make the box really big. Unfortunately, the computational complexity and memory footprint of all SPA algorithms scale with

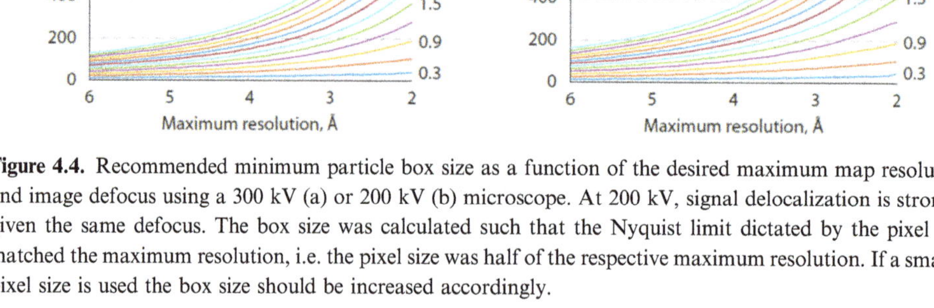

Figure 4.4. Recommended minimum particle box size as a function of the desired maximum map resolution and image defocus using a 300 kV (a) or 200 kV (b) microscope. At 200 kV, signal delocalization is stronger given the same defocus. The box size was calculated such that the Nyquist limit dictated by the pixel size matched the maximum resolution, i.e. the pixel size was half of the respective maximum resolution. If a smaller pixel size is used the box size should be increased accordingly.

the second power of the box size. It can help to ignore the CTF constraint in the beginning, in particular if the micrographs were acquired at low defocus (e.g. 1.5 μm or less), but keep it in mind if the map resolution hits a limit dictated by the signal delocalization.

4.2.5 Cleaning up the results through classification

Regardless of the type of algorithm used for selection, the results will likely contain a non-negligible number of particles that should be excluded from further analysis. If not excluded, they will decrease the resolution of the final 3D maps and can lead to visible artifacts. While the classification algorithms in SPA tools are engineered to find compositional and conformational differences between similar macromolecules, they can also be used to group and remove bad particles. The main requirement for this is that the images are at least somewhat similar between each other, and they are present in high enough amounts to facilitate the formation of dedicated classes. Otherwise the bad particles will be distributed among classes containing good particles.

2D classification refines hundreds of 2D image classes *ab initio* by starting with random shapes and iteratively adjusting particle positions and in-plane rotation. It can be more useful for cleaning up particle sets coming from unbiased selection or template matching, as those can contain a high percentage of non-protein contamination (figure 4.5(a)). Such high-contrast 'blobs' have no high-resolution similarity between each other, and benefit from a high number of classes that lets them form their own low-resolution, thinly populated classes. At the same time, underrepresented views of interesting macromolecules can also form low-resolution classes with

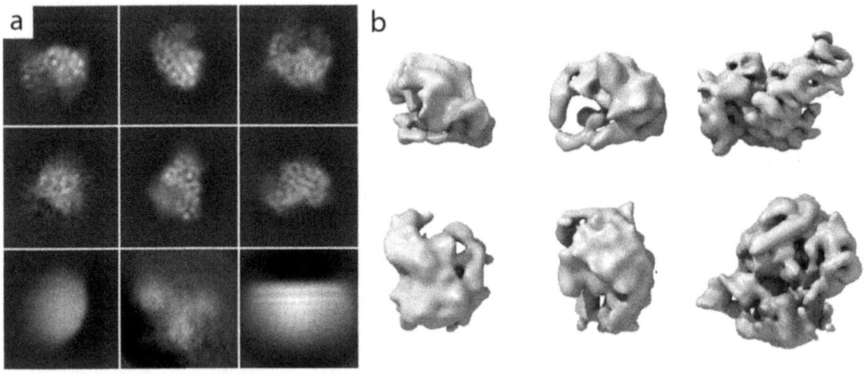

Figure 4.5. 2D and 3D classification for cleaning up selection results. (a) Good (top and middle rows) and bad (bottom row) 2D classes obtained using cryoSPARC from SARS-CoV-2 polymerase data selected with RELION's LoG picker. The bad classes show high-contrast, non-protein artifacts. Only particles from the good classes should be used for further analysis. 2D classification algorithms in RELION and cryoSPARC do not perform 'gold-standard' data separation, which can result in overestimated resolution and minor overfitting artifacts, seen here as 'streaks' radiating from the protein. These artifacts are harmless and will disappear later in 3D refinement. (b) Bad (left and middle columns) and good (right column) 3D classes obtained using cryoSPARC's *ab initio* map generation from SARS-CoV-2 polymerase data selected with Warp's convolutional neural net-based selector. No significant amounts of non-protein particles were selected. Thus, the bad classes show proteins, but not the molecule of interest. This separation is best done in 3D.

few particles. Removing them impairs the angular distribution, leading to anisotropic map resolution in 3D. Thus, 2D classification is best used to get rid of obvious non-protein particles, while leaving further cleaning to 3D classification.

3D classification, coupled with recent progress in *ab initio* 3D map generation (Punjani *et al* 2017), refines several (up to 10–20, depending on computational resources) 3D map classes by also adjusting the out-of-plane rotation of particles. This can be more useful for improving the results of CNN-based selectors, where most of the bad particles are dissociated-protein subunits rather than featureless contaminants (figure 4.5(b)). Such particles often average well in 3D and can be removed more reliably than in 2D. Furthermore, 3D classification initialized with low-resolution copies of the protein of interest can help remove particles belonging to the correct protein, but lacking high-resolution features, e.g. due to denaturation.

4.3 CTF estimation and image correction (restoration)

Benjamin A Himes and Nikolaus Grigorieff
Janelia Research Campus, Howard Hughes Medical Institute, Ashburn, VA, USA

Images recorded in the electron microscope have contrast that is affected by lens aberrations and imaging defocus (see section 1.2). These parameters may be manipulated by the microscope operator to enhance the contrast, in turn enabling 3D reconstruction of the object being imaged. Fortunately, lens aberrations and defocus do not lead to significant information loss thanks to the high degree of coherence of the electron beam. The relationship between lens aberrations and the contrast in the image is defined by the contrast transfer function (CTF). To calculate a 3D reconstruction, CTF effects will have to be accounted for. The more accurately the CTF is known, the higher the potential resolution of the reconstruction.

The CTF was introduced in section 1.2 as a function producing sinusoidal modulations of the elastic $\boldsymbol{F}(s_x, s_y)$ structure factors and the amplitude damping factor $\mathscr{F}\{\mu\}$. The general assumptions underlying the theory presented in section 1.2 and here are as follows:

i. The scattering is sufficiently weak that only interactions with the unscattered, incident beam must be considered, and further interactions of the specimen with already-scattered electrons can be ignored. This is known as kinematic scattering and leads to linear image formation.
ii. The Fourier transform of the specimen potential is assumed to have Hermitian (Friedel) symmetry. This is only strictly true for a pure phase object, and it is approximately correct when the amplitude contrast is small.
iii. A small amount of amplitude contrast, assumed to be 7%–10% for frozen-hydrated specimens, which only arises when nonlinear terms are considered, may be incorporated ad hoc in order to better match experimental data.

In the following description we will also ignore the frequency dependence of the amplitude term, as well as the variable amplitude losses with atom type. Given these simplifications, it is common practice to further assert that $\mathscr{F}\{\mu\}$ is proportional to $\boldsymbol{F}(s_x, s_y)$ so that the amplitude-contrast ratio '*w*' can be written as

$$\mathscr{F}\{\mu\}(s_x, s_y) = w/\sqrt{1 - w^2}\boldsymbol{F}(s_x, s_y). \tag{4.1}$$

The approximation that the amplitude contrast is a constant does not limit the resolution in most cryo-EM experiments since amplitude contrast constitutes only a small fraction of the total contrast. Additionally, any errors due to this assumption can be partially compensated by adjusting the phase aberration function, $\gamma(s)$, which we discuss in the following section. Given equation (4.1) we write for the CTF (Wade 1992)

$$\mathrm{CTF}(s) = -\sqrt{1 - w^2}\,\sin\gamma(s) - w\cos\gamma(s). \tag{4.2}$$

The CTF is the Fourier transform of the objective lens point spread function, which causes delocalization of the signal in real space. This can be observed in figure 4.6

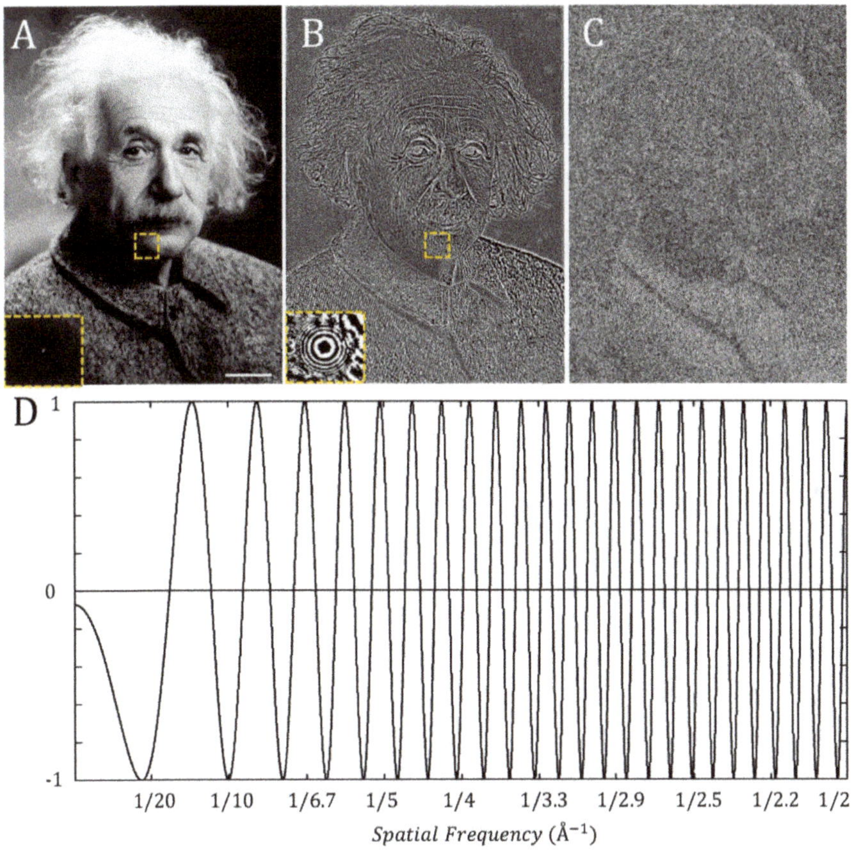

Figure 4.6. (A) Image of Albert Einstein (from Wikimedia Commons) with pixel size scaled such that his head is roughly the diameter of a ribosome. A particularly bright pixel is highlighted in the dashed orange box. Scale bar = 50 Å. (B) Image after application of a CTF that corresponds to a defocus of 1 μm. Information in the bright pixel is now delocalized by the point spread function, which displays alternating zones of positive and negative contrast, i.e. positive and negative deviations from the average intensity value. (C) Image at Scherzer defocus (~0.07 μm), showing reduced low-frequency contrast. (D) One-dimensional plot of the CTF at 1 μm underfocus. This image of Einstein has been included for illustrative purposes only; it has not been included for any promotional purposes, or to indicate any link between this publication and the Einstein estate.

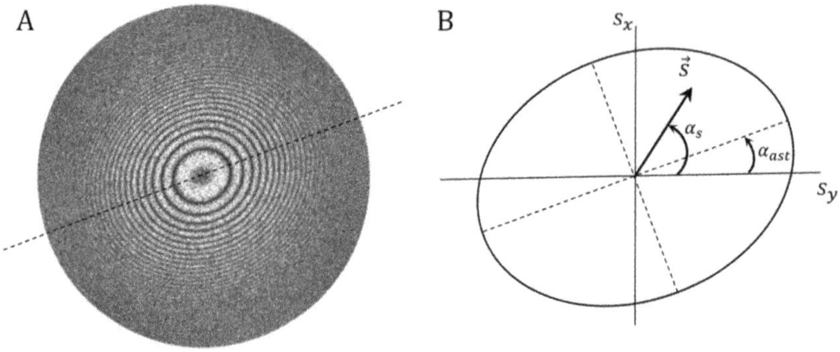

Figure 4.7. (A) 2D Thon ring pattern showing the sinusoidal oscillations that characterize the CTF, in this case with residual astigmatism resulting in an elliptical distortion. (B) Schematic showing the parameters describing the distortion due to astigmatism (equation (4.4)).

where the strong point feature in the panel (A) call-out box is shown, in panel (B), to be spread over many angstroms after application of the CTF. In addition to causing delocalization, the CTF also acts as a filter defined by zones of contrast reversal oscillating between −1 and 1. This means that some spatial frequencies in an image, usually measured in Å^{-1}, appear with unaltered contrast, others with inverted contrast, and some not at all when they are near a zero crossing of the CTF (figure 4.6(D)).

Furthermore, experimentally observed contrast transfer is characterized by a slowly varying attenuation toward higher spatial frequencies (larger values of s), commonly referred to as an envelope (sections 1.2 and 4.7). The attenuation can be the result of partial beam coherence as well as other systematic errors that will be discussed in section 4.8.

For the purpose of this section, we will only consider the combined effects of spherical aberration and defocus, including the presence of objective lens astigmatism, which leads to a dependence of the defocus on the two-dimensional (2D) Fourier coordinates s_x and s_y. We can rewrite $\gamma(s)$ as

$$\gamma(s_x, s_y) = 2\pi\left[\frac{C_s}{4}\lambda^3 s^4 - \frac{\Delta Z(s_x, s_y)}{2}\lambda s^2\right], \tag{4.3}$$

where the astigmatism is parameterized according to the notion use in figure 4.7 and folded into the regular defocus term by

$$\Delta Z(s_x, s_y) = \frac{1}{2}[\Delta Z_1 + \Delta Z_2 + \Delta\Delta Z \cos(2[\alpha_s - \alpha_{\text{ast}}])] \tag{4.4}$$

with $\alpha_s = \tan^{-1} s_y/s_x$. In equations (4.3) and (4.4) ΔZ is the defocus at Fourier coordinates s_x and s_y, ΔZ_1 and ΔZ_2 are the maximum and minimum defocus values generated by the astigmatism, respectively, $\Delta\Delta Z = \Delta Z_1 - \Delta Z_2$, and λ the wavelength of the electrons. The spherical aberration constant[1] C_s is determined by the

[1] C_s can vary slightly with a change in the objective lens current, which changes the magnetic field inside the lens, however, this is negligible in practice.

design of the objective lens and, hence, the user can adjust the CTF primarily through changing the defocus and the wavelength. Most cryo-EM images are recorded at underfocus, which is achieved by weakening the objective lens, i.e. reducing its current. The imaged plane at focus under these conditions lies further from the objective lens and closer to the electron source in the microscope.

In equation (4.2) we see that the amplitude contrast term, $-w\cos\gamma$, is always negative for low spatial frequencies, which means imaging with overfocus would create an additional zone of weak contrast at very low spatial frequency ($s = \sqrt{2\Delta Z/C_s\lambda^2}$). To avoid this, cryo-EM data are commonly collected with an underfocused objective lens. By examining equation (4.3) we also see that underfocus can serve to partially cancel the effects of spherical aberration. A special focal condition that maximizes the width of the band of spatial frequencies prior to the first zero crossing of the CTF is called the Scherzer defocus (Scherzer 1949), $\Delta Z = -\sqrt{C_s\lambda}$. The effect of using Scherzer defocus for a weak phase object is illustrated in figure 4.6(C). While useful in material science applications, this imaging condition weakens low spatial frequency features that are important for particle alignment in cryo-EM. On the other hand, larger defocus enhances low spatial-frequency contrast and therefore helps in recognizing and aligning particles in an image. However, it also has the undesirable effect of reducing the spatial frequency of the first CTF zero, increasing the number of phase reversals in a given frequency interval, and leading to increased delocalization of the signal in real space due to the point spread function (see above). The latter is discussed further in section 4.3.2. In a single-particle experiment, it is therefore necessary to find a defocus, usually between 1 and 3 μm at 300 kV, that generates sufficient contrast while limiting detrimental effects.

4.3.1 CTF estimation

A full treatment of the effects of the CTF usually proceeds in two stages: CTF estimation and CTF correction. CTF estimation often makes use of the sinusoidal modulations predicted by equation (4.2), which were experimentally verified by Thon (1966, 1971) following theoretical work by Hanszen *et al* (Hanszen 1967, Hanszen and Morgenstern 1965). The sinusoidal modulations, sometimes referred to as Thon rings, form a characteristic pattern of rings or ellipses observed in computer-generated power spectra. They can be used to determine the defocus and astigmatism to within about 100 Å (Mindell and Grigorieff 2003), permitting 3D reconstruction at about 2 Å resolution (Jensen 2001). Once a reconstruction has been determined, the defocus parameters can be further refined and other, less significant errors can be measured and corrected (section 4.8). A full correction is therefore an iterative process that starts with an initial estimation of defocus and astigmatism from the electron micrographs themselves, without reference to a 3D reconstruction.

Figure 4.8(A) shows a typical power spectrum calculated from a cryo-EM image, highlighting Thon rings. These rings appear on a smoothly varying background (seen much more easily in a radial average of the 2D power spectrum), which decreases toward high spatial frequencies and which prevents the oscillations from

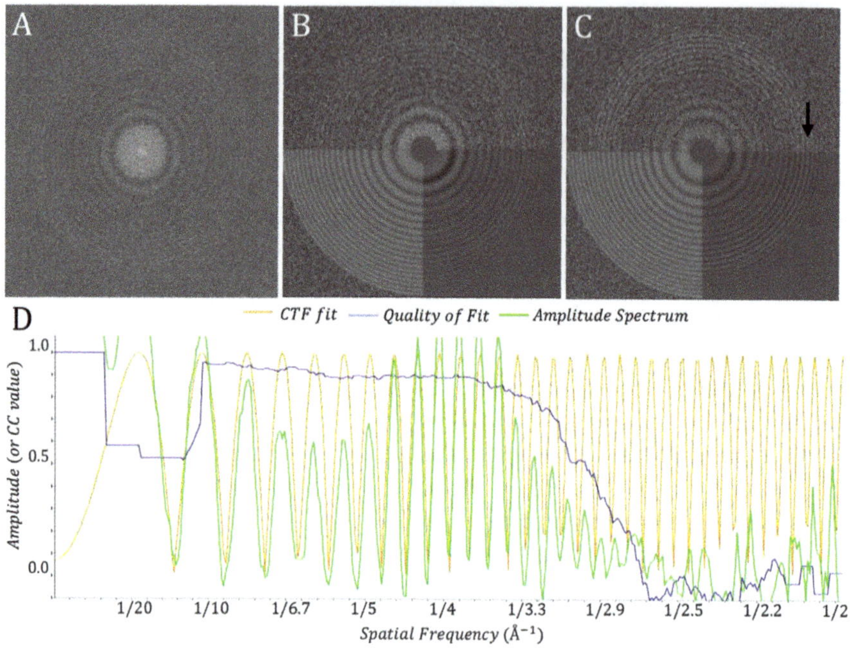

Figure 4.8. (A) Power spectrum calculated from an image of beta galactosidase (Bartesaghi *et al* 2015) after averaging the aligned movie frames (38 frames per move at 1.2 electrons/Å^2/frame). (B) Power spectrum after background subtraction and fitted with a calculated CTF (inset) (Rohou and Grigorieff 2015). (C) Power spectrum calculated as a sum of three-frame averages. The Thon rings are more clearly visible than in (A), particularly around the water ring at 3.7 Å resolution, marked by the black arrow. (D) 1D plot generated by CTFFIND4 (Rohou and Grigorieff 2015), showing the amplitude spectrum (green), the model fitted to the data (yellow), and a correlation-based score function (blue), which can be used to assess the spatial-frequency range that was fitted successfully (quality of fit > 0.5).

reaching zero. This background is primarily due to the noise associated with the detection of a given number of electrons (shot and detector noise), as well as contributions made by the usually ignored nonlinear terms to the image intensity, including the inelastically scattered electrons. After subtraction of the background term, which may be accomplished by a variety of approaches (Ludtke *et al* 1999, Penczek *et al* 2014, Sander *et al* 2003, Zhang 2016b, Rohou and Grigorieff 2015), the calculated CTF is compared to the observed power spectrum $A_d(s_x, s_y)$ between spatial frequencies $s_{\min}$ and $s_{\max}$ (figure 4.8(B)) by computing their cross correlation, given as

$$\mathrm{CC} = \frac{\sum\limits_{s_{\min} < |s| \leqslant s_{\max}} A_d(s_x, s_y) \cdot |\mathrm{CTF}(s_x, s_y)|}{\sqrt{\sum\limits_{s_{\min} < |s| \leqslant s_{\max}} A_d(s_x, s_y)^2 \cdot \sum\limits_{s_{\min} < |s| \leqslant s_{\max}} |\mathrm{CTF}(s_x, s_y)|^2}}. \tag{4.5}$$

A naïve approach to finding the best fit between the data and the model would involve an exhaustive search of all three parameters shown in figure 4.7(B). Since this approach is computationally expensive, a 'divide-and-conquer' approach is adopted in many algorithms. For example, in the presence of moderate astigmatism an average defocus can be determined first by an exhaustive search of one parameter, followed by a local search to refine all three parameters including astigmatism (Zhang 2016b, Rohou and Grigorieff 2015). Some algorithms also estimate the astigmatism angle, α_{ast}, (equation (4.4)) by mirroring the power spectrum along the x- or y-axis and determining the rotation angle that aligns the mirrored version with the original in a one-parameter search.

To obtain an accurate fit and limit the effect of systematic errors and noise, the low spatial-frequency limit s_{min} is usually set to a value between 1/40 and 1/50 $Å^{-1}$. This will exclude frequencies at which the contrast in cryo-EM images is affected by residual inelastically scattered electrons, a term that is not modeled correctly by equation (4.2). The optimal value for the high-frequency limit s_{max} depends on the strength of the Thon rings and background noise and it is usually set between 1/3 and 1/5 $Å^{-1}$.

Apart from the envelopes that occur at higher resolution (section 1.2), the amplitude of the Thon rings is also affected by a limited depth of field (DeRosier 2000) in the case of thicker samples (section 4.8). Furthermore, Thon rings are attenuated by sample drift occurring during image acquisition, which leads to blurring in the image and loss of high-resolution signal in the direction of the drift. Thick samples (1000 Å and more) will also reduce the visibility of Thon rings due to increased background and loss of electrons to inelastic scattering. For this reason, the strength and visibility of Thon rings often serves as a proxy to the overall quality of the data, an indicator that the data have the potential to yield a high-resolution reconstruction. While this criterion is often useful, weaker Thon rings may also simply be the result of fewer particles in the field of view, thus limiting the overall signal in the image that could otherwise be suitable for high-resolution reconstruction. The strength and visibility of the Thon rings can be quantified by cross-correlation with a calculated CTF (equation (4.5), figure 4.8(C)) (Rohou and Grigorieff 2015).

When micrographs are recorded as movies (section 3.5), blurring due to sample drift can be reduced by aligning the movie frames to each other to restore high-resolution contrast. With perfectly aligned frames, the contrast in the image, as well as the strength of the Thon rings, is maximized. However, movie alignment comes with its own errors and limitations, and local movement cannot always be fully corrected. It is therefore sometimes advantageous to calculate the power spectrum directly from the movie frames, or from sub-averages of multiple frames. In this case, the Thon ring pattern is the average of all calculated power spectra and, because it is calculated from multiple shorter time intervals, it is less affected by sample drift. For Thon rings generated by vitrified ice, an optimal interval for averaging is given by the time it takes to accumulate about 3–4 electrons/$Å^2$ (at 300 keV, figure 4.8(D)). At that point, the water molecules will have moved on average by about 1.5–2 Å, which is still small enough to maximize the intensity of the Thon rings at a resolution of about 3–4 Å (McMullan *et al* 2015).

Finally, most samples are tilted to some degree with respect to the optical axis of the microscope. This is done deliberately in a tomographic series, or it may be done to overcome limitations of preferred particle orientation (section 4.6, Tan *et al* 2017).

Even when sample tilt is not introduced intentionally, it is often present as a result of residual stage tilt or local sample undulations (Booy and Pawley 1993, Vonck 2000). Sample tilt leads to a variable defocus across the recorded image; to obtain a more accurate defocus estimate for each location, sample tilt axis and angle have to be determined in addition to the average defocus. This can be done, for example, by modeling the defocus variation across the image along a tilted plane (Mindell and Grigorieff 2003, Su 2019). More complex sample geometries may including tilt axis direction, tilt angle and other geometry descriptors as search parameters to achieve the best fit between locally calculated power spectra and corresponding CTF patterns (Tegunov and Cramer 2019).

The need for sample-tilt estimation depends on the degree of tilt. Defocus variation in images of nominally untilted samples can also be addressed on a per-particle basis by performing a local refinement of the defocus parameters against locally calculated power spectra (Zhang 2016b). When a 3D reference reconstruction is available, per-particle defocus values can be estimated by maximizing the correlation (or another similarity measure) between a particle image and a CTF-treated matching projection (Grigorieff 2007, Punjani *et al* 2017, Grant *et al* 2018, Zivanov *et al* 2018). This approach can also accommodate different particle heights in the ice layer of the sample (Noble *et al* 2018a). In this case, the signal available for per-particle CTF estimation is generated only by one particle and is therefore noisier than the signal in a Thon ring pattern calculated from a local patch or the entire micrograph. This increased level of noise imposes a lower molecular-mass limit on the particle of about 300–400 kDa. It also requires images with strong signal at spatial frequencies of at least 1/3–1/4 Å^{-1} and a good reference reconstruction with corresponding resolution, below which errors in the estimation may be larger than the potential gain in defocus accuracy from per-particle CTF estimation.

4.3.2 Image correction

After determining the defocus and astigmatism values for each particle image, as well as particle orientations (Euler angles) and 2D coordinates within each image, a 3D reconstruction can be calculated (section 4.4). As discussed above, one of the features of the CTF affecting cryo-EM images are zones of weak or zero contrast. It is therefore impossible to fully restore the signal spectrum from a single image by simply dividing by the CTF. A partial correction that will not restore image amplitudes consists of restoring the phases of the spatial frequencies that were inverted by the CTF (phase flipping, van Heel *et al* 2000). To calculate a fully corrected reconstruction of the object, data from many images have to be merged (section 4.4), i.e. CTF correction and 3D restoration are accomplished in a single step. CTF correction can be illustrated by the simpler case of calculating 2D class averages (section 2.2), which lacks the 3D reconstruction step. A 2D class average is generated from a set of aligned cryo-EM images x_i: $i \in M$ corresponding to the same particle view. It is convenient to refer to the Fourier transform of these images $X_i(s_x, s_y)$, which may be defined by the structure factor $\boldsymbol{F}(s_x, s_y)$, corrupted by wave aberrations defined by the CTF and two additional noise terms (section 1.2):

$$X_i(s_x, s_y) = \mathrm{CTF}_i(s_x, s_y) \cdot [\boldsymbol{F}(s_x, s_y) + \boldsymbol{N}_s(s_x, s_y)] + \boldsymbol{N}_i(s_x, s_y). \tag{4.6}$$

The noise terms $\boldsymbol{N}_i(s_x, s_y)$ represent different realizations of the 'shot' and detector noise in the particle images. In equation (4.6) 'structural' noise $\boldsymbol{N}_s(s_x, s_y)$ added by the embedding medium (ice), which is also affected by the CTF can also be considered but is often ignored. For images represented by equation (4.6) with $\boldsymbol{N}_s = 0$, Saxton (1978) derived a Wiener filter $\Omega(s_x, s_y)$ for TEM; optimal in the sense that it minimizes the sum of squared differences between the CTF-corrected average, $A_\Omega(s_x, s_y)$, and the underlying structure factor, $\boldsymbol{F}(s_x, s_y)$. Defining the ratio of particle signal power (before CTF aberrations) and average noise power $\sum_{i=1}^{M} |\boldsymbol{N}_i(s_x, s_y)|^2/M$ as

$$\mathrm{SNR}_{\boldsymbol{F}}(s_x, s_y) = \frac{M\,|\boldsymbol{F}(s_x, s_y)|^2}{\sum_{i=1}^{M} |\boldsymbol{N}_i(s_x, s_y)|^2}, \tag{4.7}$$

the Wiener filter is given by

$$\Omega(s_x, s_y) = \frac{1}{1 + 1/\mathrm{SNR}_{\boldsymbol{F}}(s_x, s_y) \sum_{i=1}^{M} \mathrm{CTF}_i^2(s_x, s_y)} \tag{4.8}$$

and the CTF-corrected, Wiener-filtered average is

$$\begin{aligned} A_\Omega(s_x, s_y) &= \frac{\sum_{i=1}^{M} \mathrm{CTF}_i(s_x, s_y) X_i(s_x, s_y)}{\sum_{i=1}^{M} \mathrm{CTF}_i^2(s_x, s_y)} \Omega(s_x, s_y) \\ &= \frac{\sum_{i=1}^{M} \mathrm{CTF}_i(s_x, s_y) X_i(s_x, s_y)}{1/\mathrm{SNR}_{\boldsymbol{F}}(s_x, s_y) + \sum_{i=1}^{M} \mathrm{CTF}_i^2(s_x, s_y)}. \end{aligned} \tag{4.9}$$

$1/\mathrm{SNR}_{\boldsymbol{F}}(s_x, s_y)$ was approximated by a constant in many early cryo-EM software packages, however, more rigorous statistical approaches (Scheres 2012a, Sindelar and Grigorieff 2012) now determine it as a function of the data. It can be seen that this term's magnitude relative to the sum of squared CTF values (the second term in the denominator of equation (4.9)) is smaller the higher the ratio of structure factor to noise (equation (4.7)) and the larger the dataset (large M). In the limit of very large datasets, $\Omega(s_x, s_y) = 1$, and the CTF-corrected average will simply be a sum of CTF-multiplied images, divided by the sum of squared CTF values. If the CTF-corrected average, $A_\Omega(s_x, s_y)$, is estimated using a maximum likelihood approach,

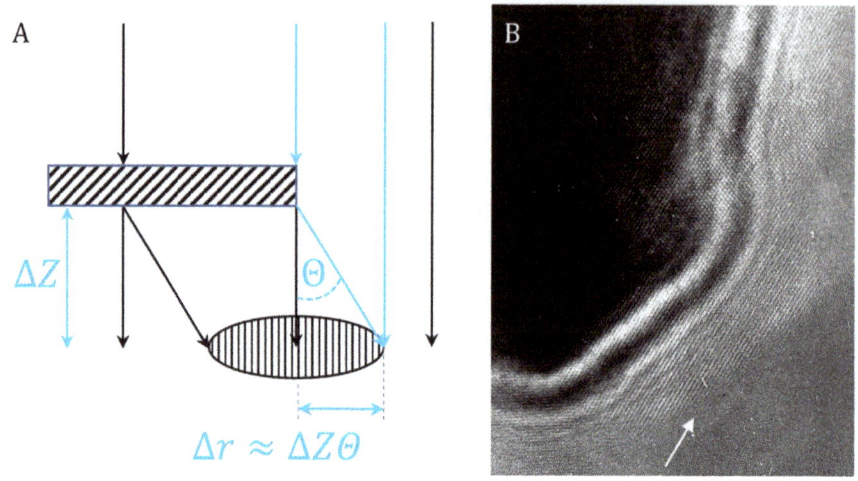

Figure 4.9. (A) Diagram explaining the displacement of signal from the edge of a particle. Phase contrast from the edge of the particle and corresponding to a spatial frequency $1/d$, where d is the resolution, will be displaced by $\Delta r = \Delta Z\Theta$, where ΔZ is the defocus and the scattering angle Θ is given by $\Theta \approx \lambda/d$, with λ the wavelength of the electrons ($\lambda \approx 1/50$ Å at 300 keV). (B) Image of the edge of an InGaAs semiconductor crystal showing 3 Å lattice fringes extending beyond the crystal edge due to an image defocus of about 1 μm.

the terms in the sums in equation (4.9) are replaced by their probability-weighted estimates (Scheres 2012a, section 4.4).

Another result of the CTF-related degradation of images is the displacement (delocalization) of signal in real space, away from its location in a fully corrected image (see above). The amount of displacement, visible as Fresnel fringes in the images, is dependent on the spatial frequency of the signal, as well as the amount of image defocus. To fully recover the signal in a CTF-corrected average at a given resolution, d, it is therefore important that the window size used to extract particles from the micrographs is sufficiently large to include the fringes corresponding to the resolution d, as well as avoiding aliasing of the CTF oscillations in the Fourier transform (Penczek *et al* 2014). As a rule of thumb, the displacement of signal from the edge of a particle is given by the product of scattering angle, λ/d, and the defocus ΔZ (figure 4.9). To include these fringes for a particle with diameter D_P, the size of the window, D_W, should be at least

$$D_W = D_P + 2\frac{\lambda}{d}\Delta Z. \tag{4.10}$$

This means, for example, that 2 Å signal in an image of a particle of 200 Å diameter, recorded at 1 μm defocus and 300 kV requires a particle box size of about 400 Å.

4.3.3 Magnification distortion

An additional source of image distortion that can limit the attainable resolution of a 3D reconstruction, and which therefore requires correction, is anisotropy in the image magnification (Grant and Grigorieff 2015, Zhao *et al* 2015). This 'magnification distortion', which usually affects only certain magnifications, is due to sub-optimal presets applied to the stigmators in the projector system of the microscope that require a service engineer and/or software update to rectify. The distortion leads to particles being stretched in one direction and compressed in the orthogonal direction compared to its average dimensions (figure 4.10(A)). The effect of magnification distortion on the final reconstruction depends upon the amount of stretching and compression, which can be up to a few percent, as well as upon the average particle diameter. For example, for a virus capsid with a 700 Å diameter, a 2% distortion (difference between most stretched and most compressed dimension) would lead to a displacement of the particle boundary from the particle center of 2% × 350 Å = 7 Å, or about 3.5 Å from the

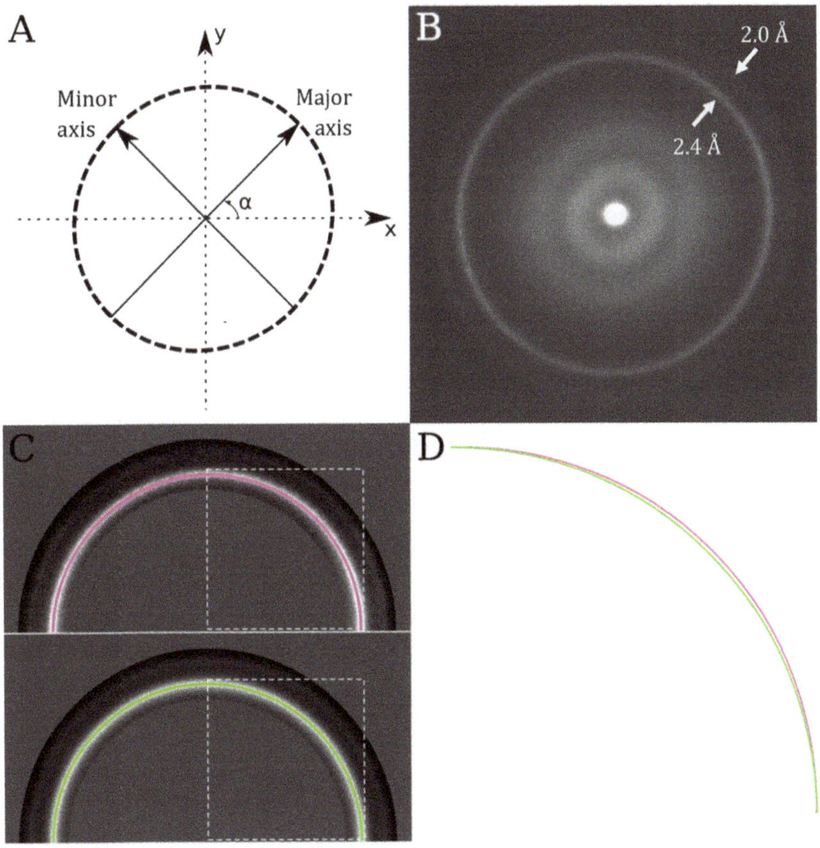

Figure 4.10. (A) Similar to axial astigmatism, magnification distortion can be parameterized by an ellipse; in this case an angle (alpha) and two scale factors along the major and minor axes. (B) Sum of the amplitude spectra from ten images of polycrystalline gold. Rings corresponding to spatial frequencies of at 2.4 and 2.0 Å are visible. (C) Top-half of the image shown in A, with the gold rings masked out and a path tracing the ~2.4 Å gold ring. Bottom-half of the rotational average of the image shown in (A), also with the gold rings masked out and a path tracing the ~2.4 Å gold ring. The dashed white box illustrates the area, which is shown zoomed in and overlaid in panel (C). (D) Overlay of the section of the paths traced in (B) surrounded by the dashed white box. A mismatch indicating a magnification distortion of about 2% is visible. Reproduced with permission from Grant and Grigorieff (2015). Copyright 2015 Elsevier.

average boundary position. An average 3.5 Å shift would eliminate signal beyond 7 Å resolution at the particle periphery in 2D class averages and 3D reconstructions, and severely attenuate signal at lower resolution.

The amount and direction of magnification distortion can be measured directly from the distortions observed in the particles (Yu *et al* 2016, Zivanov *et al* 2018), or from an image of a polycrystalline gold sample (figure 4.10(A)) (Grant and Grigorieff 2015). Since the distortion present in a given instrument does not change significantly over several months, it can be measured every few months, and images can be computationally corrected using the measured parameters by interpolation and resampling of the micrographs before proceeding with other steps in the image processing pipeline (Grant and Grigorieff 2015). The observed magnification distortion in instruments installed since the problem was recognized in 2015 has become less severe and correction may therefore not be required on newer instruments.

4.3.4 Concluding remarks

The steady improvement of algorithms has led to streamlined image processing of cryo-EM data, and many of the correction steps discussed here are now routine and fully automated. It also means that the demands on microscope hardware and operator skills in aligning instruments have been lowered. Most image distortions and misalignments can now be detected and corrected for by modern processing packages. Nevertheless, further improvement may come from making corrections for more complicated imaging errors, such as residual off-axis coma (Glaeser *et al* 2011) and for a small, defocus-dependent change in magnification when the electron beam is not completely parallel. Another source of error comes from inelastic scattering, which is currently not included in the correction of images, except for the ad hoc removal of background in power spectra (see section 4.3.1 above). The contribution of inelastically scattered electrons can be further reduced when an energy-filter is available. Additional correctors in the electron microscope, such as a chromatic aberration corrector, may convert some of the inelastically scattered electrons into electrons that contribute useful phase contrast to the image. This may then make additional algorithms necessary to accommodate this additional contrast.

4.4 Merging data from structurally homogeneous subsets

Basil J Greber

California Institute of Quantitative Biosciences (QB3), University of California, Berkeley, CA 94720, USA

Molecular Biophysics and Integrative Bio-Imaging Division, Lawrence Berkeley National Laboratory, Berkeley, CA 94720, USA

Present address: Institute of Cancer Research, Division of Structural Biology, Chester Beatty Laboratories, London SW3 6JB, UK

In this subsection, we will cover different aspects relevant to the three-dimensional (3D) reconstruction of a single cryo-EM map, preferably at high resolution, from a structurally homogeneous set of particle images. The methods employed to obtain homogeneous particle subsets from unclassified, heterogeneous cryo-EM datasets are explained in section 4.5. Here, we will start our discussion by estimating how many particle images are needed to obtain a 3D reconstruction, considering both the coverage of Fourier space required to obtain an isotropic reconstruction and the accumulation of sufficient signal to arrive at a desired resolution. Subsequently, we will explore the basics of image alignment and the algorithms used to obtain 3D reconstructions from aligned particle images. Finally, we will briefly touch on some of the methods that allow us to determine entirely unknown structures *ab initio*, in the absence of a previously known reference volume.

4.4.1 How many particle images are needed for a 3D reconstruction?

4.4.1.1 Geometrical considerations

As discussed in section 1.2, the central-section theorem states that the Fourier transform of a 2D projection of a 3D object is a central section through the 3D Fourier transform of said object. This principle is of great importance in cryo-EM because it provides us with one way to reconstruct a 3D volume from a set of 2D projection images (see also section 4.4.2.2), which in our case are particle images extracted from single-particle cryo-EM data. Recovering the information to fully sample the 3D Fourier transform at a given resolution requires views from a number of projection directions that is proportional to the diameter of the object, as indicated in equation (1.7) of section 1.2 (Crowther *et al* 1970). This geometrical consideration provides a lower bound for the number of orientations we need to sample for a successful cryo-EM experiment. For an asymmetric object, such as a ribosome, approximately 30 views are required to achieve 30 Å resolution (De Rosier and Klug 1968) and proportionally larger numbers are needed for higher resolutions, for example, ten-fold more for 3 Å resolution for the same object.

Ideally, the projections obtained in a cryo-EM dataset show the object from all sides, but we should emphasize that even for an asymmetric object this is not strictly required. In principle, a tomographic series, corresponding to a 180° rotation around one axis for an asymmetric object, is sufficient to provide the information needed to reconstruct a 3D volume of an object lacking symmetry (De Rosier and Klug 1968, Crowther *et al* 1970). Projections that are 180° apart are equivalent, i.e. they are the same except for the fact that they are mirror images of each other. This means that even particles that exhibit some degree of preferential orientation on the cryo-EM grid, due to their shape or because certain portions of the molecule like to interact with the specimen support film or the air–water interface, may produce high-resolution structures without further intervention (figure 4.11). Nevertheless, even for particles where a tomographic series of views can be obtained and Fourier space can be sampled sufficiently with a limited set of projection directions, the availability of additional views may facilitate the recovery of a correct structure and may be particularly beneficial in the case of *ab initio* structure determination in the absence of a high-quality reference (e.g. Punjani *et al* 2017).

It is quite common, however, that particles adopt highly preferred orientations on the grid, an issue that is addressed in depth in section 4.6. To summarize briefly, in cases of more severe orientation bias, it may not be possible to adequately sample

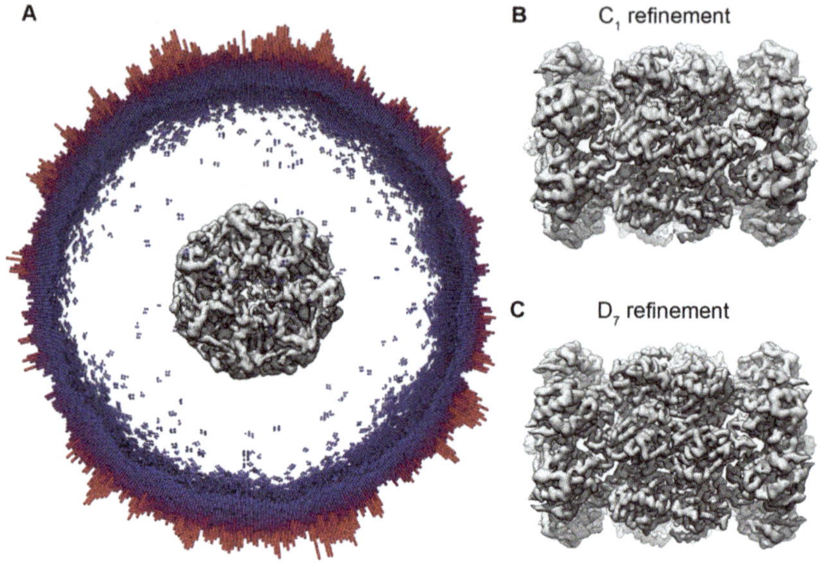

Figure 4.11. Full coverage of Fourier space in spite of the absence of certain views. (A) A cryo-EM dataset of proteasome particles (Campbell *et al* 2015) was processed without imposing symmetry after first removing the top views, leaving only side views. The cryo-EM density (gray) is shown at the center. The projection histogram is shown as spikes colored blue (few projections) to red (many projections) and forms a ring round the reconstruction, while almost no projections from directions away from this ring are present. The projections here cover 360°, but 180° would be sufficient in principle even for an object lacking symmetry altogether. (B, C) Map of the asymmetric proteasome reconstruction (B) compared to the map of the reconstruction with D_7 symmetry imposed (C) computed using the particle images whose orientation distribution is shown in (A). Both maps exhibit isotropic resolution. The proteasome cryo-EM dataset (Campbell *et al* 2015) used to prepare this figure was retrieved from the EMPIAR database (Iudin *et al* 2016), with accession code EMPIAR-10025. Computations were performed using RELION 3 (Zivanov *et al* 2018) and the figure was created using UCSF Chimera (Pettersen *et al* 2004).

Fourier space to obtain a high-quality structure without adjustments to the preparation of the sample or the data collection strategy. Examples of sample-preparation methods that can improve the orientation distribution include the addition of detergents (Chen *et al* 2019), the use of support films or affinity grids (Han *et al* 2012), or the use of advanced sample-preparation robots that reduce the time particles have to interact with the air–water interface (Noble *et al* 2018b). In some cases, the orientation bias even varies from square to square or from grid to grid, possibly due to differences in ice thickness. A proven and reliable method to address preferential orientation by adjusting the data collection strategy is tilting of the specimen (Tan *et al* 2017). To achieve high resolution, this approach requires gold grids instead of the more traditional copper grids with carbon support films to mitigate beam-induced specimen motion (Russo and Passmore 2016).

Symmetrical molecular complexes (including both point-group symmetrical objects and helical filaments) contain more than one asymmetric unit per particle. Therefore, each individual particle projection may present asymmetric units in different, symmetry-related views. However, this is not guaranteed for all particle orientations. For example, top views of ring-shaped multimeric complexes (e.g. the view of the proteasome in figure 4.11(A)) present all asymmetric units under the same view, related only by an in-plane rotation. Nevertheless, in most cases symmetry helps increase the number of different views that can be obtained from a single particle (or from a set of particles adopting a preferred orientation). Consequently, more limited angular sampling is sufficient to obtain a 3D reconstruction. For helical or icosahedral objects, even 1–2 views may allow recovery of a low-resolution 3D map (De Rosier and Klug 1968). Indeed, early cryo-EM work on viruses produced 3D maps that, albeit at low resolution, provided valuable biological insight from only a few dozen viral particles (Stewart *et al* 1991, Vogel *et al* 1986).

4.4.1.2 How many perfect molecular images do we need to obtain a high-quality reconstruction?

The considerations outlined above tell us about the number of orientations that we need to sample, but they do not, on their own, inform us on how many particles images we need to collect in order to obtain a reconstruction at a given resolution. This is because, in addition to adequately sampling Fourier space, we also need to accumulate enough signal to achieve a signal-to-noise ratio that allows us to distinguish real molecular features from noise. The criteria by which we judge whether we have achieved the necessary signal-to-noise ratio at a certain resolution are explained in more detail in section 5.2. For this section, we will content ourselves with the observation that due to the radiation sensitivity of biological specimens and the need to use low electron doses to preserve high-resolution structural information (Glaeser 2016, Taylor and Glaeser 1976), the signal contained in each individual particle is low, and we do need to average 'many' projection images (see section 3.2). The following paragraphs will try to define what constitutes sufficiently 'many' images.

An important and illuminating question to consider is how many projection images we would need in order to reach the desired resolution under the assumption that these images are ideal. This type of analysis explores the physical boundaries of what cryo-EM can achieve, and it proved to be instrumental in highlighting the potential of the method and in guiding its development toward its current state. Predictions of the number of particle images required to achieve resolutions that would allow building of reliable atomic models (approximately 3 Å, see section 4.7.5) turned out to be particularly insightful because they demonstrated a need to improve the quality of the molecular images and invest in the development of better electron detection technology than the CCD detectors and film that were widely used in the first decade of this century (Henderson and McMullan 2013).

As early as 1995, Richard Henderson proposed that it should be possible to compute single-particle cryo-EM maps at 3 Å resolution from as few as 12 600 ideal particle images (Henderson 1995). This number is independent of molecular weight or particle diameter: larger particle size contributes to improved alignment quality, thereby lowering the number of images needed, but conversely, retrieving all the information needed to fill the 3D Fourier transform requires more views for a larger particle (see equation (1.7) in section 1.2 and section 4.4.1.1). These two effects cancel each other out. Remarkably, this study also predicted a lower size limit of approx. 40 kDa for high-resolution single-particle cryo-EM, far below what was feasible in practice at that time, when most cryo-EM studies involved megadalton-sized complexes. The estimated particle number for crossing the 3 Å threshold was subsequently revised to 1400 particles (Glaeser 1999) and was later even further reduced to 600 particles after it was shown that previous resolution criteria had been too stringent (Rosenthal and Henderson 2003). In the presence of symmetry, these numbers represent the number of asymmetric units needed and thus the number of required projection images is reduced even further in that case.

To fully appreciate the significance of even the most conservative of these estimates, it is important to remember that until the introduction of direct electron detector technology, the only single-particle cryo-EM map approaching 3 Å resolution was that of aquareovirus (Zhang *et al* 2010), a landmark structure that was computed from 18 464 icosahedral particles encompassing nearly 1 200 000 asymmetric units—almost 100× more than predicted by Henderson in 1995 (figure 4.12(A)). Today, comparable numbers of particles allow reconstructions of density maps of viral particles to better than 2 Å resolution (Tan *et al* 2018), at which point holes in aromatic side chains become visible (figure 4.12(B)).

4.4.1.3 How close to these ideal numbers has cryo-EM brought us when using ideal specimens?

Following the introduction of direct electron detectors, cryo-EM has progressed rapidly toward achieving its true potential (McMullan *et al* 2014, Bai *et al* 2015a, Glaeser 2019). While this subsection does not aim to provide an exhaustive review of the state of the field, we want to highlight a number of remarkable achievements that illustrate the recent progress toward matching the theoretical predictions.

In recent years, apo-ferritin, an octahedral iron-storage complex with 24 asymmetric units per particle, has been used as a benchmark for optimization of image acquisition and data processing (Russo and Passmore 2014, Zivanov *et al* 2018, Wu *et al* 2020, Danev *et al* 2019, Kato *et al* 2019, Nakane *et al* 2020, Yip *et al* 2020). The highest-resolution structures, at 1.2–1.7 Å resolution, have been computed from hundreds of thousands to millions of particles, corresponding to millions or tens of millions of asymmetric units. However, computations using ever-smaller numbers of particles reaching approximately 3 Å resolution from roughly 50–100 particles or 1200–2400 asymmetric units (Yip *et al* 2020, Kato *et al* 2019) have shown how few particle images are required with this ideal complex (figure 4.12(C)). Although these data have achieved what Henderson (1995) and Glaeser (1999) predicted, further technical innovation is needed to fully realize the most recent prediction (Rosenthal and Henderson 2003).

Another striking example of recent progress toward fulfilling the promises of the theoretical predictions is the structure determination of ever-smaller molecules by single-particle cryo-EM methods. These structures include hemoglobin (Khoshouei *et al* 2017, Herzik *et al* 2019), alcohol dehydrogenase (Herzik *et al* 2019), the SARS-CoV-2 ORF3a membrane channel (Kern *et al* 2020), streptavidin (Han *et al* 2019, Fan *et al* 2019), and the SAM-IV riboswitch RNA (Zhang *et al* 2019), all of them at

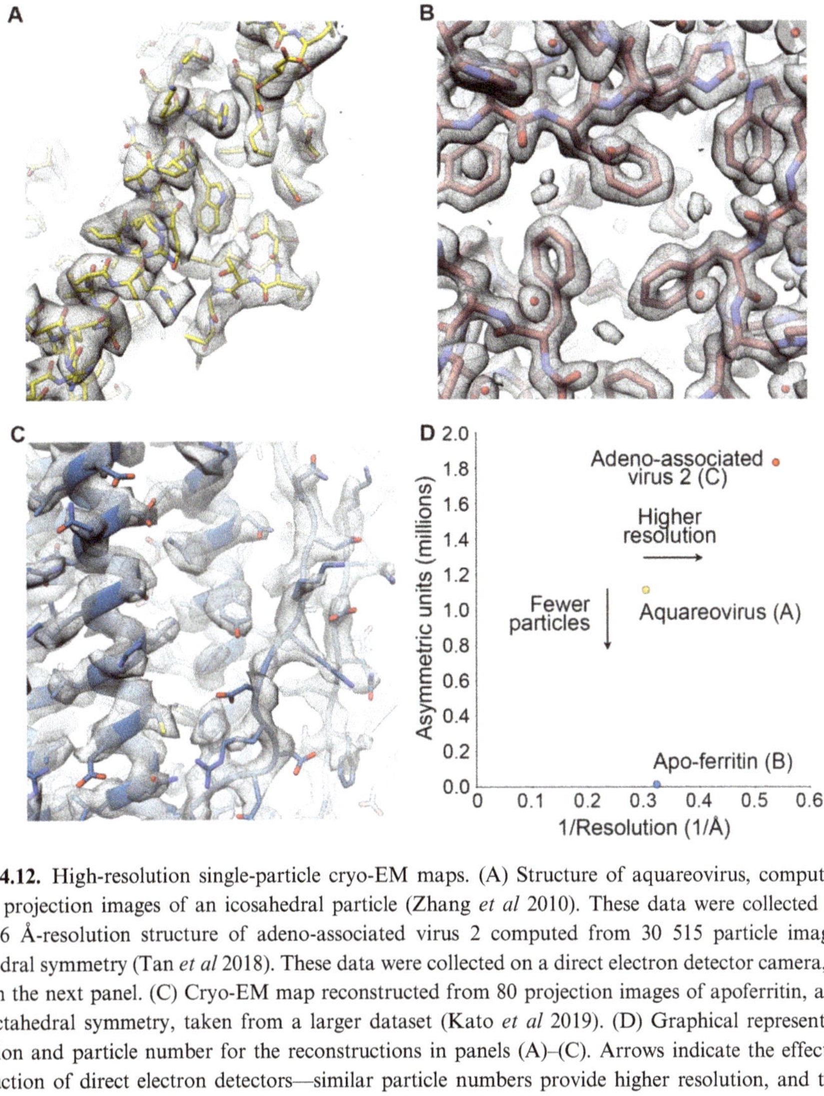

Figure 4.12. High-resolution single-particle cryo-EM maps. (A) Structure of aquareovirus, computed from 18 464 projection images of an icosahedral particle (Zhang *et al* 2010). These data were collected on film. (B) 1.86 Å-resolution structure of adeno-associated virus 2 computed from 30 515 particle images with icosahedral symmetry (Tan *et al* 2018). These data were collected on a direct electron detector camera, as were those in the next panel. (C) Cryo-EM map reconstructed from 80 projection images of apoferritin, a particle with octahedral symmetry, taken from a larger dataset (Kato *et al* 2019). (D) Graphical representation of resolution and particle number for the reconstructions in panels (A)–(C). Arrows indicate the effects of the introduction of direct electron detectors—similar particle numbers provide higher resolution, and the same resolution can be achieved using fewer particles. Panels (A)–(C) were prepared using UCSF Chimera (Pettersen *et al* 2004). Panels (A) and (B) use publicly available cryo-EM maps and atomic coordinates (EMD-5160, EMD-9012, PDB 3IYL, PDB 6E9D). Panel (C) shows a map computed from a small subset of the publicly available cryo-EM dataset EMPIAR-10248 (Kato *et al* 2019) fitted with the model PDB 6S61.

better than 4 Å resolution and for particles smaller than 100 kDa (table 4.1). In addition to the symmetrized reconstruction at 2.9 Å, the alcohol dehydrogenase structure could also be resolved at 3.1 Å from the remarkably low number of approximately 12 000 particles without symmetry imposed (Herzik *et al* 2019). It is important to note that the low molecular weights of these complexes approach previous predictions for the lower bound of high-resolution structure determination by cryo-EM.

Table 4.1. Selected cryo-EM reconstructions. These are examples representing the current frontiers of the method in terms of absolute resolution, small number of particles used, or low molecular weight of the studied molecule. The metrics that stand out are printed in bold.

Particle (accession code)	Molecular weight (MDa)	Symmetry (asymmetric units/particle)	Particle number (asymmetric units)	Resolution (Å)	Comments	Reference
Adeno-associated virus 2 (EMD-9012)	3.9	Icosahedral (60)	30 515 (1 830 900)	**1.86**		(Tan *et al* 2018)
Apoferritin (EMD-11638)	0.48	Octahedral (24)	363 126 (8 715 024)	**1.22**	Cold-FEG	(Nakane *et al* 2020)
Apoferritin (EMD-11103)	0.48	Octahedral (24)	1090 676 (26 176 224)	**1.25**	Monochromator, C_S corrector	(Yip *et al* 2020)
Alcohol dehydrogenase (EMD-0406)	**0.08**	C_2 (2)	11 672 (23 344)	2.9		(Herzik *et al* 2019)
		C_1 (1)	**11 672** (11 672)	3.1		
Hemoglobin (EMD-0407)	**0.06**	C_2 (2)	24 308 (48 616)	2.8		(Herzik *et al* 2019)
ORF3a (EMD-22898)	**0.06**	C_2	91 298 (182 436)	2.1	Contains lipid nanodisc	(Kern *et al* 2020)
Hemoglobin (EMD-3488)	**0.06**	C_2 (2)	175 300 (350 600)	3.2	Volta phase plate	(Khoshouei *et al* 2017)
Streptavidin (EMD-20907)	**0.05**	D_2 (4)	11 402 (45 608)	2.6		(Han *et al* 2019)
Streptavidin (EMD-0689)	**0.05**	D_2 (4)	45 686 (182 744)	3.2	Volta phase plate	(Fan *et al* 2019)
SAM-IV riboswitch (EMD-20755)	**0.04**	$\mathbf{C_1}$ (1)	796 923 (796 923)	3.7	RNA particle	(Zhang *et al* 2019)

Given the rapid pace of technological development in cryo-EM and the application of new technology such as the instrumentation that enabled to approximately 1.2 Å resolution reconstructions of apo-ferritin (Yip *et al* 2020, Nakane *et al* 2020) to more samples, ever higher resolutions from fewer and fewer particles are very likely to be achieved at a rapid pace.

4.4.1.4 How many particles will I need using my specimen?

The examples cited in table 4.1 mostly represent ideal cases of comparably stable, homogeneous, and otherwise well-behaved complexes that can be prepared in very thin ice. Other specimens may require larger datasets for high-resolution structure determination because of conformational or compositional heterogeneity, issues with stability—e.g. because they do not tolerate freezing in very thin ice or require specimen support films that increase background noise—or because their orientation distribution under-samples some of the views required. Such specimen-dependent effects, along with any contributions from beam-induced motion and the microscope optics toward dampening of high-resolution information, can be expressed as a B factor, in analogy to temperature factors used to describe the decay of high-resolution information in x-ray crystallography (Liao and Frank 2010). The B factor can be used to estimate how much the resolution will improve when the dataset size is increased and can thus guide the experimental strategy toward high-resolution structure determination.

Common ways to analyze the relationship between particle number and resolution and to deduce a B factor are the Henderson–Rosenthal plot (Zivanov *et al* 2018, Rosenthal and Henderson 2003) or the ResLog plot (Stagg *et al* 2014). The Henderson–Rosenthal plot involves graphing of log(N), the logarithm of the particle number N of data subsets of different sizes, against $1/d^2$, the inverse square of the resolution of 3D maps obtained from these subsets. Figure 4.13 shows such a plot for two molecular complexes: a bacterial microcompartment shell, which is a giant icosahedral particle, and human transcription factor IIH, a moderately flexible complex of intermediate size. This plot allows the estimation of the B factor on the basis of the slope of a linear fit of the data (Rosenthal and Henderson 2003, Liao and Frank 2010), and extrapolations to larger dataset sizes can be made if needed (LeBarron *et al* 2008). Using this approach it is possible to use an initial dataset to estimate whether high-resolution structure determination at a desired resolution is possible within the constraints of practical data collection and processing workflows. For the examples shown in figure 4.13, achieving 2.5 Å resolution with either complex would require more than 10 000 000 particles of the same quality in the final reconstruction, which is clearly not feasible in practice. In such cases improvements of the biological specimen, the cryo-EM grid material or grid preparation, the microscope, the detector, the acquisition strategy, or the processing methodology are required to extract more high-frequency signal from each individual particle.

The ResLog plot (Stagg *et al* 2014) is conceptually similar to the Henderson–Rosenthal plot, but uses $1/d$ rather than $1/d^2$ on the resolution axis. The analysis of how this plot changes in response to the perturbation of particle alignments led to the proposal that the slope reflects the contribution of each additional particle to the reconstruction (i.e. data quality, with better data quality being analogous to a lower

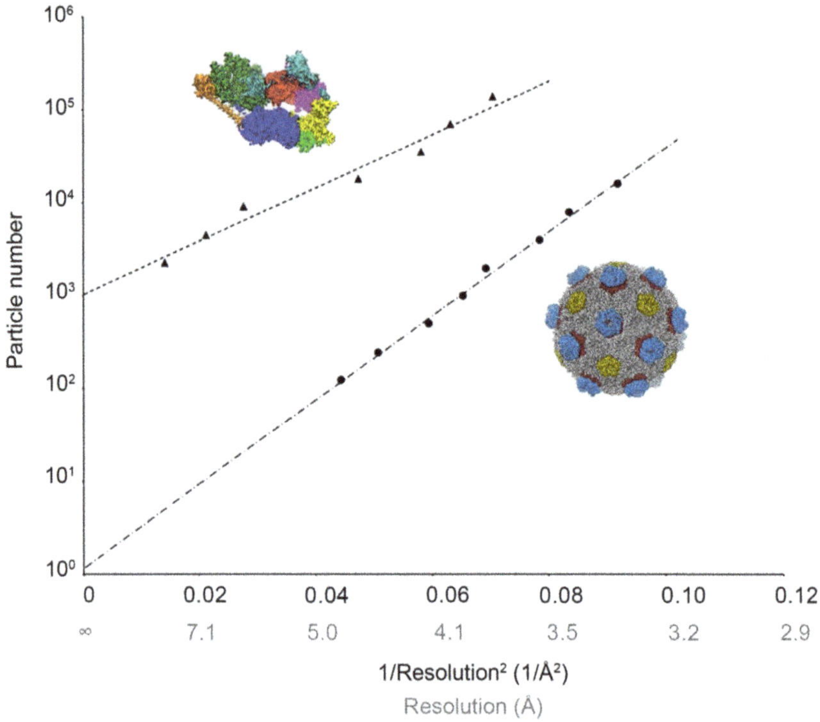

Figure 4.13. Analysis of the relationship between particle number and resolution. Cryo-EM datasets of transcription factor IIH and a bacterial microcompartment shell (Greber *et al* 2019a, Greber *et al* 2019b) were split into subsets of different size and these subsets were refined and post-processed in RELION 3 (Zivanov *et al* 2018) to determine the resolution of each reconstruction. The logarithms of the particle numbers were plotted against the resolutions achieved, yielding a linear relationship that can be used to assess particle alignability, data quality, and the number of images required to reach a given resolution with a larger dataset.

B factor as discussed above), while the *y*-axis-intercept informs on the quality of the alignment and classification of the particle dataset (Stagg *et al* 2014). Note that in the above example (figure 4.13), the bacterial microcompartment shell, which aligns exceedingly well due to its 6.5 MDa-mass and high symmetry, has a very small *y*-axis-intercept, but a somewhat higher slope. The higher slope, or higher B factor, might originate from the fact that the data were collected using a microscope equipped with a side-entry holder, which is more prone to drift than the autogrid loaders in contemporary high-end instruments and may dampen the high-resolution information. It is also worth noting that reconstructions with very few particles may start deviating from the linear trend observed (not shown in the graph).

4.4.2 Obtaining a 3D reconstruction

After considering how many particles we need to obtain high-quality reconstructions, we will now discuss how such a particle dataset can be converted into a 3D reconstruction. The required computational procedures—image alignment, orientation assignment, and 3D reconstruction—are implemented in a large number of different software packages. Discussion of all of these software packages or providing an exhaustive description of all the algorithms used in each of them is beyond the scope of this section. Instead, we provide an overview of the basic concepts and will mention specific algorithms or packages as examples.

4.4.2.1 Particle image alignment and Euler angle assignment

The particle images that are initially extracted from cryo-electron micrographs (see section 4.2) show the molecules in unknown orientations. Before a 3D reconstruction can be obtained, the relative projection angles as well as any required in-plane transformations (rotation and *x*/*y*-coordinate shifts) need to be determined for all the particle images. Most current software packages employ an approach that uses iterative refinement of the orientation parameters based on comparison with a reference structure, either in the form of cross-correlation-based projection matching (figure 4.14) or by maximum likelihood methods (see below).

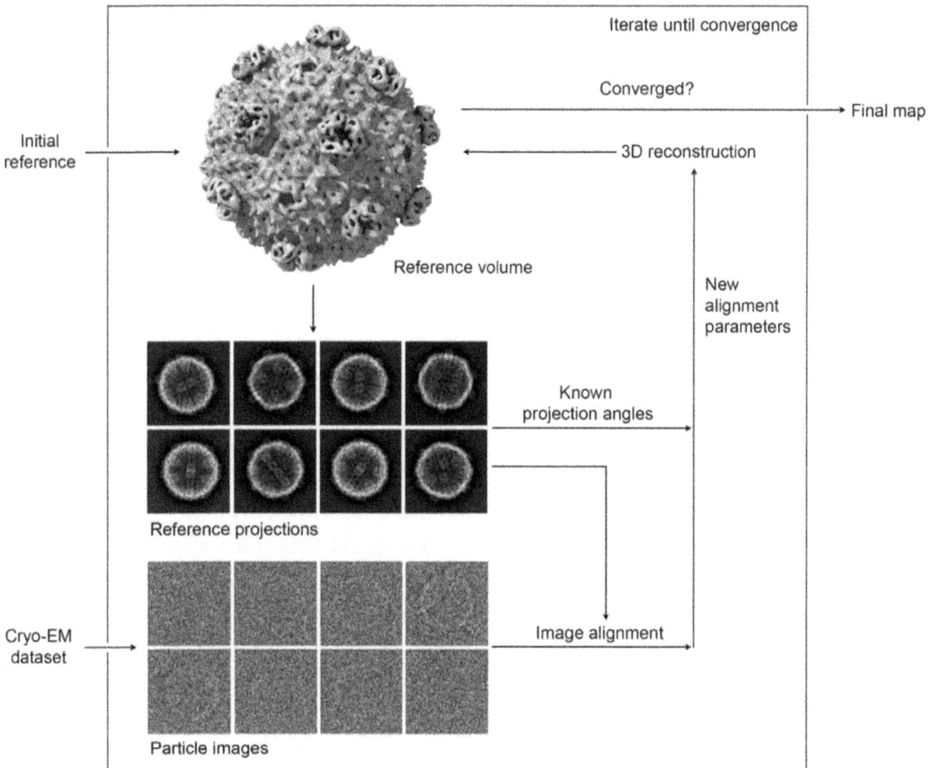

Figure 4.14. Iterative particle alignment and 3D reconstruction scheme in cryo-EM. Cryo-EM map refinement uses a particle dataset and an initial reference as input. The reference is projected using known projection angles, and these projections are compared to the particle images, yielding alignment parameters for the images. The map and alignment parameters are then iteratively improved until convergence is reached and the final cryo-EM map is produced as the output, along with the orientation parameters that were applied to the particles to create the map. Particles images and a cryo-EM map of a bacterial microcompartment shell (Greber *et al* 2019a) were used in this example; the image of the cryo-EM map was created using UCSF ChimeraX (Goddard *et al* 2018).

Projection matching (Penczek *et al* 1994) has been widely adopted and is implemented in many commonly used software packages, including SPIDER, EMAN, and FREALIGN (Frank *et al* 1996, Ludtke *et al* 1999, Grigorieff 2016). During projection-matching-based alignment, particle images are compared to reference projections generated from a known 3D density map. The reference might be a previously determined structure, an *ab initio* reconstruction, or the result of a previous iteration of the refinement. Each particle image is then assigned to the most similar projection, according to the values of the cross-correlation function. Because the relative orientations of the reference projections are known, this information, combined with the *x*/*y*-coordinate shifts and in-plane rotation angle needed to center and orient the particle on the reference projection, can be used to produce the reconstruction of a new 3D volume. This new volume then serves as a new reference structure for the next round of iterative alignment, until convergence is reached.

Variations of this general iterative refinement procedure exist. In EMAN a 2D class-averaging step is inserted after initial projection matching and before 3D reconstruction. This 2D classification step uses the particles that were assigned to one projection at the projection matching stage, but the resulting 2D class averages are allowed to deviate substantially from that projection, which accelerates convergence and reduces model bias. The eventual 3D reconstruction is performed with the resulting class averages, not with the individual particle images (Ludtke *et al* 1999).

It is important to note that this iterative refinement process is prone to noise bias or over-fitting due to the low signal-to-noise ratio of the cryo-EM particles. This can lead to inflated resolution estimates or inaccurate results if the noise starts to strongly influence image alignment. Although these issues most severely affected data collected on film or CCD cameras, with the latter featuring particularly low signal-to-noise ratios (Ruskin *et al* 2013), even data from the latest direct detector cameras with their enhanced signal-to-noise characteristics and improved alignability are not immune to these effects. In the worst case, a cryo-EM map can be obtained from what may be mostly random noise (Henderson 2013, Subramaniam 2013, van Heel 2013). Although such extreme cases are rare, the scheme used to determine the alignment parameters needs to include safeguards that prevent over-fitting even with high-quality data. These include gold-standard refinement of two fully independent particle half-sets (Scheres and Chen 2012), which is now the most commonly used approach against over-fitting and implemented in many software packages, and frequency-limited refinement, as used in FREALIGN and CISTEM (Grant *et al* 2018, Grigorieff 2016). While these safeguards can mitigate inflated resolution estimates as measured by self-consistency criteria (section 5.2), verification of the overall correctness of the structure is still required (see section 4.2.3 on initial models).

More recently, maximum likelihood methods (Sigworth *et al* 2010, Scheres *et al* 2005) have been widely adopted for both particle alignment and classification. Maximum likelihood refinement aims to find the most likely 3D reconstruction given the cryo-EM dataset, i.e. it maximizes the probability that the observed particle images would be generated, given the cryo-EM reconstruction produced during the refinement. Pioneered by the XMIPP package (Scheres *et al* 2005, Sorzano *et al* 2004b) and popularized by RELION (Scheres 2012a, 2012b), variants of these methods are now implemented in a large number of additional software packages, including CRYOSPARC, CISTEM, and SPHIRE (Grigorieff 2016, Grant *et al* 2018, Punjani *et al* 2017, Lyumkis *et al* 2013, Moriya *et al* 2017). In contrast to image alignment by maximum cross correlation-based projection-matching methods, where each particle image is assigned a single orientation, maximum likelihood methods consider all possible orientations of the particle and assign weights to each direction according to their probability (Scheres 2012a, Scheres *et al* 2005). These probability distributions are broad if the assignments are very uncertain, for example, at the beginning of a refinement run. An initially broad probability distribution mitigates the effect of model bias and prevents the orientation search from being trapped in local minima during the first iterations of refinement or when the data are weak, and it can become sharp as refinement progresses if the assignments are very reliable, e.g. in the case of very high-quality data and high-resolution reconstructions.

A notable exception to the commonly used reference projection-based schemes for angle assignment in single-particle cryo-EM is the method of angular reconstitution (van Heel 1987), as implemented, e.g. in the IMAGIC package (van Heel *et al* 1996). This approach uses the principle that the respective Fourier transforms for any two 2D projections of the same 3D object share one common line in reciprocal space. As detailed in van Heel (1987) this principle can also be re-stated in real space: two 2D projections from one 3D object will always share one common 1D projection. When applying this approach in practice, 1D projections from the 2D images are computed across all angles in real space and stored in a so-called sinogram. Comparing the sinograms of two 2D images allows identification of their shared 1D projection. Three of these common lines (in reciprocal space) or shared 1D projections (in real space) define the orientation of 2D class averages—which must be used instead of original projection images in order to improve the signal-to-noise ratio—relative to each other. Orientations determined in this way then allow, as before, reconstruction of a 3D volume (van Heel 1987). Therefore, this method can be used entirely independently of a reference. Alternatively, it can use a set of reference projections, termed an anchor set, generated from a previously determined 3D map to guide the angle assignment process (van Heel 1987). A recent paper describing the structure determination of a worm hemoglobin explains how this approach can be applied in practice in great detail (Afanasyev *et al* 2017).

4.4.2.2 Algorithms for 3D reconstruction

Once a set of aligned projection images is in hand, the task is to find the 3D structure that generated this set of 2D projections. A detailed treatment of the theoretical aspects of this topic exceeds the scope of this section, but this has been covered extensively in the literature, including in an excellent review (Penczek 2010).

Cryo-EM poses a set of unique challenges to 3D reconstruction algorithms (Penczek 2010). The problem of 3D reconstruction from cryo-EM datasets with finite numbers of particles is not invertible, i.e. the solution cannot be uniquely recovered and needs to be approximated. In addition to this inherent difficulty there are also practical challenges—a suitable algorithm needs to be able to handle thousands to millions of particle projections, which might themselves be possibly quite large in terms of the number of picture elements per image. As mentioned previously, these images may have been generated for projection angles that may show uneven distributions. And finally, the reconstruction needs to be done in a computationally efficient way, such that the computation time and hardware required stay within practical limits, even during iterative refinement schemes or 3D classifications with a large number of classes where 3D reconstructions have to be computed many times. A number of different algorithms that satisfy these criteria have been developed.

The commonly used 3D reconstruction algorithms can be subdivided into two main classes (Penczek 2010):

(i) Algebraic methods in which the reconstructed object is retrieved by solving a set of linear equations that relate the volume to be reconstructed and the observed particle projections. In simplified terms, the output volume is iteratively adjusted such that its projections better match the observed output projections. The most prominent representatives of this class of algorithms in cryo-EM are the algebraic reconstruction technique (ART), implemented in XMIPP as ART with blobs (Marabini *et al* 1998, Marabini *et al* 1997), and the sequential iterative reconstruction technique (SIRT) (Gilbert 1972), implemented among several other algorithms in SPIDER (Frank *et al* 1996). Because both of these algorithms are iterative, they tend to require more computation time than non-iterative methods.

One of the major challenges in single-particle cryo-EM 3D reconstruction is that the orientation distribution of the input particles cannot be easily controlled. Even in cases where the available views are sufficient to reconstruct the 3D volume (section 4.4.1.1), the distribution of views is rarely, if ever, uniform. Therefore, overabundant views are often present and can possibly induce artifacts such as elongation of the reconstruction in one direction. However, it has been shown that if the free parameters are correctly chosen, both ART and SIRT—as well as the Fourier space methods discussed below—can produce correct reconstructions even in the presence of overrepresented views (Sorzano *et al* 2001).

(ii) Fourier transform methods that rely on the central-section theorem. These operate in Fourier space and can be further subdivided into filtered back projection and direct Fourier inversion methods. The advantage of the transform methods is that they are comparably fast and at the same time provide excellent reconstructions. These algorithms have been implemented in many cryo-EM packages, including SPIDER, EMAN, IMAGIC, and RELION (Frank *et al* 1996, Ludtke *et al* 1999, van Heel *et al* 1996, Scheres 2012b).

Filtered back projection methods apply a radially increasing weigh to data in Fourier space, to compensate for the decreasing density with which the data are sampled at increasingly higher resolution, followed by back projection in real space. Direct Fourier inversion methods operate solely in Fourier space and treat the Fourier components of the 2D projection images as samples of the 3D Fourier transform of the object that is being reconstructed. These methods encompass the most accurate currently available algorithms (Penczek 2010). In particular, the gridding-based direct Fourier inversion algorithms implemented in SPIDER and RELION have been shown to produce near-perfect reconstructions on noise-free projections (Scheres 2012b, Penczek *et al* 2004), although the RELION implementation appears to perform slightly better when using noisy projections (Scheres 2012b).

Correction of the image-distorting effects of the CTF (section 4.3) can be achieved in a number of ways, but implementation typically differs between algebraic and Fourier transform methods. In the context of algebraic methods one possible solution is to incorporate the CTF (or its inverse Fourier transform, the point spread function) as a term in the mathematical model that produces the equations that are being solved, resulting in retrieval of a CTF-corrected structure (e.g. Zhu *et al* 1997). Alternatively, the projection data themselves can be treated by phase flipping and then subjected to a method called iterative data refinement (Sorzano *et al* 2004a) to further reduce the effects of the CTF. In contrast, as discussed in Penczek (2010), the Fourier transform methods implement a Wiener filter (see section 4.3.2) before or during the 3D reconstruction process, or when merging volumes computed from data acquired at different defocus values.

4.4.2.3 How to obtain an initial 3D reconstruction

The map refinement strategies outlined above assume that an initial reconstruction exists, which is used to start the particle orientation search. This subsection will briefly introduce some of the approaches that allow us to create such an initial map if none is currently available. Different software packages provide a number of strategies for initial reconstruction generation that are based on several fundamentally different principles. These include (i) specimen tilting methods, (ii) common lines-based methods, and (iii) random initialization methods. It is worth noting that the tilting methods can be used to determine the handedness of the molecule under study, while common lines and random initialization do not provide this information. A description of a number of these methods as well as examples of their application to the structure determination of the 50S subunit of the bacterial ribosome and a comparison of the results can be found in Voss *et al* (2010).

The sections below will discuss approaches that are used for *ab initio* 3D reconstruction of single particles exhibiting no or only low symmetry. For highly symmetrical particles or filaments with helical symmetry, different approaches exist. While convergence is usually quite robust for icosahedral or other highly symmetrical complexes, correct determination of the initial model and helical symmetry (Diaz *et al* 2010, Egelman 2007) are crucial in the case of filaments because errors at this stage can lead to erroneous results (e.g. Egelman 2014). However, a detailed treatment of these methods is beyond the scope of this section.

4.4.2.3.1 Methods that involve tilting of the specimen

Tilting methods used for initial 3D reconstruction of single particles approach the problem of the unknown relative orientations of particle projections by introducing known angles between two experimental projections of the same particle. This is achieved by recording a first image and then tilting the specimen before recording a second image (figure 4.15(A)). Tomographic reconstruction (Frank 2008) even uses a set of multiple exposures separated by small stage tilts to arrive at a series of projections with known Euler angle differences, permitting 3D reconstruction of single objects. One drawback of this method is that coverage of Fourier space is incomplete because the stage cannot be tilted to 90°, resulting in an artifact termed the missing wedge. The random conical tilt (RCT) method (Radermacher *et al* 1987) reduces this wedge of missing projections to a missing cone. RCT uses one un-tilted

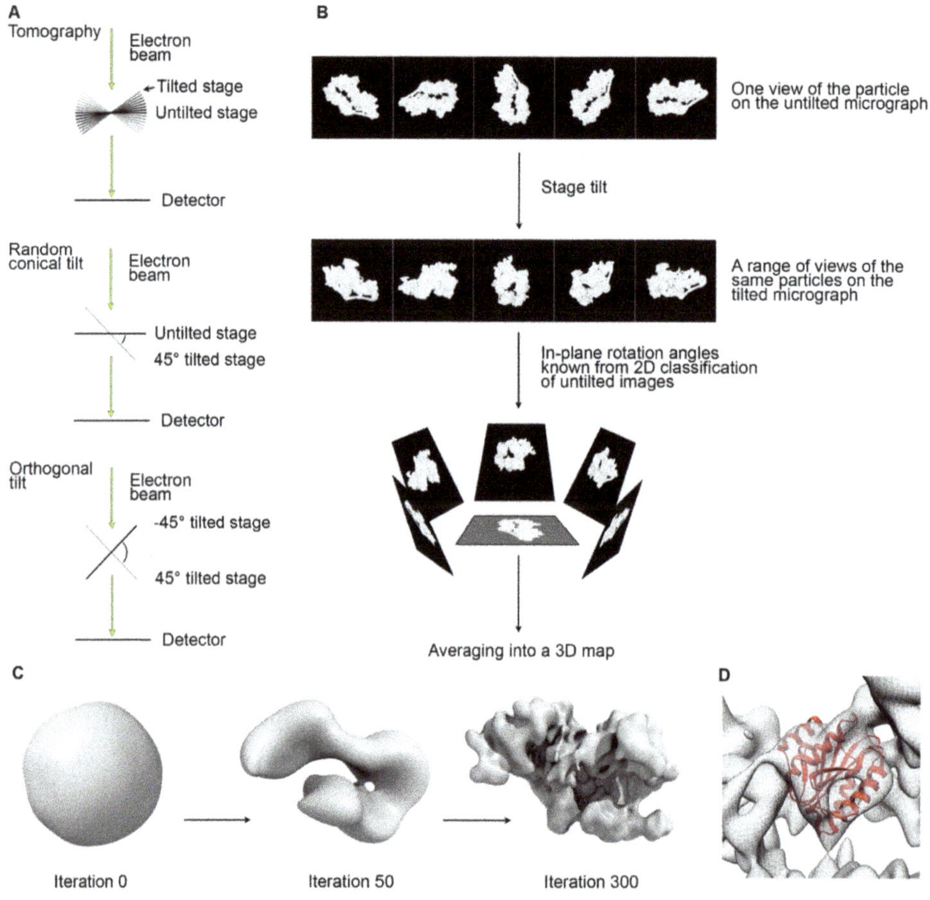

Figure 4.15. *Ab initio* 3D reconstruction methods. (A) Stage tilt geometries for tomographic, random conical tilt, and orthogonal tilt reconstruction methods. (B) Schematic of how stage tilt can produce a range of views from a single un-tilted view. (C) *Ab initio* structure determination using random initial assignment and stochastic gradient descent (as implemented in Zivanov *et al* 2018) using a pre-classified dataset of transcription factor IIH (Greber *et al* 2019b) as test dataset. The starting volume is roughly spherical but converges to the true structure eventually. (D) Docking of a previously solved structure is one way to validate an *ab initio* reconstruction. Here, the x-ray crystal structure of the TFIIH subunit p44 (PDB ID 4WFQ, Kim *et al* 2015) was fitted into the density. Map representations were created using UCSF ChimeraX (Goddard *et al* 2018) from EM databank entry EMD-0452.

and one tilted exposure (typically at 45°) of the same field of view including many particles that are randomly oriented in-plane. This creates two views for each particle that are related by the stage tilt angle (figure 4.15(A) and (B)). The particles in the un-tilted exposures can be subjected to 2D alignment and classification, yielding 2D classes that contain particles that have been imaged under the same projection angle, but exhibit different in-plane rotations. These 2D-classified particles from the un-tilted exposure can then be correlated with the very same particles in the tilted exposure. Because the difference in the projection angle between these two views of the same particles has been introduced by stage tilt and is thus known, and because particles assigned to a given un-tilted 2D class are known to be seen from the same projection angle, this procedure defines a set of tilted particles that form a cone and can then be merged into a 3D reconstruction (figure 4.15(B)).

The orthogonal tilt reconstruction (OTR) method (Leschziner and Nogales 2006) is based on the same principle but uses opposite tilts (typically +45° and −45°) for the two paired exposures to eliminate the missing cone in the resulting reconstructions. Because the stage is tilted for both exposures, the identification of identical views, required for the 2D class-averaging step, requires that the particles assume different orientations on the grid. Therefore, the OTR method relies on the presence of a range of views on the grid and will, by default, produce isotropic reconstructions. On the other hand, RCT can also be applied to samples that exhibit highly preferred orientation, although the resulting reconstruction will suffer from some anisotropy. These distinct properties are highly relevant to the choice of the data collection and 3D reconstruction method: if the sample does suffer from preferential orientation, RCT is the geometrical method of choice. However, when this is not the case, and particularly if multiple states exist and the generation of multiple references is required, it can be a significant advantage to have isotropic reconstructions that can be used for the initial model. It is worth noting that the results of both RCT and OTR are typically limited to low resolution, either because negatively stained samples are used, or because of the high electron exposure associated with taking two images of the specimen (tilted and un-tilted), degrades the high-resolution signal even when using unstained, frozen-hydrated specimens.

4.4.2.3.2 Common lines-based methods

The basic principle on which common lines-based methods rest was introduced above (section 4.4.2.1), but their application to obtain initial 3D reconstructions in practice depends on the software package and strategy used. For example, structure determination of worm hemoglobin (Afanasyev *et al* 2017) by angular reconstitution in IMAGIC (van Heel 1987, van Heel *et al* 1996) used an approach that initially assigns random starting angles to a small group of 2D class averages. The Euler angles assigned to the projections are then optimized one-by-one for each projection by using the remaining projections of the group as an anchor set, followed by the addition of further 2D class averages (Afanasyev *et al* 2017). Variants of common lines-based methods are available in other software packages as well: a Fourier space variant of the approach has been implemented in EMAN (Ludtke *et al* 1999) and SPIDER employs a simultaneous optimization variant of the method that enables angular assignment to more than three projection images at once (Penczek *et al* 1996). In all cases, 2D class averages are used during the reconstruction process to increase the signal-to-noise ratio in the projections that are being compared.

4.4.2.3.3 Random initialization

Random initialization methods start by assigning random starting angles to projections and aim to recover the true structure by iteratively aligning the projection images to the resulting reconstruction, which initially only loosely resembles the true structure, or not at all (figure 4.15(C)). Such methods have been implemented in a number of software programs, including FREALIGN and CISTEM (Grant *et al* 2018, Grigorieff 2016), which use their standard image alignment algorithm to start the reconstruction process from particles with random angular assignment, and PRIME (Elmlund *et al* 2013), which initializes a stochastic hill climbing optimization from a random starting configuration. While these concepts and a number of implementations existed before the development of direct electron detectors, the reliability of the resulting structures has greatly increased due to the dramatically improved quality of the data in the direct detector era. Recent years saw the introduction of stochastic gradient descent variants of the method, the application of which to cryo-EM was pioneered in the CRYOSPARC package (Punjani *et al* 2017) and which were also implemented in RELION (Zivanov *et al* 2018). Stochastic gradient descent uses random initialization to start the orientation search and then performs a very large number of fast iterations, which are greatly accelerated by only using small subsets of particle images in each iteration (Punjani *et al* 2017).

4.4.2.3.4 The importance of validation

With all of these methods it is critical to verify that the solution that has been retrieved is correct in order to avoid potentially catastrophic errors (Murray *et al* 2013). Map validation in general, and particularly at high resolution, is covered in detail in chapter 5. In the context of the topic of initial 3D reconstruction it is important to note that resolution assessment by Fourier shell correlation, even when using gold-standard or frequency-limited refinements to mitigate noise bias (Scheres and Chen 2012, Grigorieff 2016), is a measure of self-consistency rather than correctness of the map as such. Therefore, additional methods are required to verify correct overall structure determination. Here, we are specifically interested in these latter methods, i.e. ensuring that the initial model is valid. This can be done in many ways, which typically rely on reproducibility, consistency within the data, or agreement with external data. Examples of such procedures include (i) running a large number of independent reconstructions using the same method, possibly in the context of more elaborate protocols that improve convergence (e.g. Penczek and Asturias 2014), (ii) comparing the results from different methods to check for consistency, (iii) verifying that the resulting reconstructions has the expected biomolecular features (e.g. protein or nucleic acid secondary structure), (iv) fitting of known sub-structures (e.g. from x-ray crystallographic or NMR experiments), or (v) tilt-pair validation (Henderson *et al* 2011, Rosenthal and Henderson 2003). The tilt-pair validation method verifies that aligning particles from micrographs with a known tilt angle difference—the tilt pairs—to a common reference yield that same difference in their assigned Euler angles and thus gives a direct quantitative assessment of the quality of the reconstruction.

After an initial model has been obtained it can be used as a starting reference in any of the iterative refinement approaches outlined in section 4.4.2.1 and, one hopes, refined to higher resolution.

Acknowledgments

We thank all the authors who deposited their cryo-EM datasets in EMPIAR. The microcompartment shell data used in figures 4.13 and 4.14 were obtained in collaboration with Markus Sutter and Cheryl Kerfeld. This work was supported by NIGMS grants R01-GM63072, P01-GM063210, and R35-GM127018 to Eva Nogales.

4.5 3D classification of structurally heterogeneous particles

Xiao-chen Bai
Department of Biophysics, University of Texas Southwestern Medical Center, Dallas, TX, USA

4.5.1 Introduction

Nearly all macromolecular protein complexes are intrinsically dynamic during cellular processes. As a result the protein samples extracted from cells normally contain many different conformational or compositional states and this fact poses a great challenge for single-particle cryo-EM structural determination (Bai *et al* 2015b, Scheres 2016, Serna 2019). If the particles of different conformational states in a cryo-EM dataset are not classified into a number of structurally homogeneous subsets, the reconstruction of heterogeneous particles into a cryo-EM map will display blurred structural features and anisotropic resolution (Orlova and Saibil 2004). The regions with remarkably high structural flexibility may even become invisible in the cryo-EM map, leading to an incomplete structure which can significantly hinder structural interpretations.

In addition to utilizing biochemical approaches to trap the macromolecular complex in a relatively static state, many different 3D classification methods have been developed to tackle the problem of structural heterogeneity in cryo-EM. Historically, supervised 3D classification was the first approach to be developed (Gao *et al* 2004, Heymann *et al* 2004, Valle *et al* 2002). This method requires the user to provide several known 3D structures, which are used as references, and the particles in the dataset are classified based on the degree of their similarity to one or another of those input references. This approach, however, has a serious limitation, as it requires a large amount of prior knowledge about the structural variability in the protein sample, which is often not available to the user.

In contrast to the supervised 3D classification method, an unsupervised, maximum likelihood based 3D classification approach (ML3D), which does not require any prior information of the structural heterogeneity present in the data, was developed (Sigworth

1998, Sigworth *et al* 2010, Lyumkis *et al* 2013, Scheres *et al* 2008, 2007). The unsupervised ML3D is achieved by applying duplicate, identical 3D references in the first iteration of classification. In other words, the user only needs to provide a single, low-resolution map as well as a predicted number of classes. In this way, the particles are randomly assigned into different classes at the first iteration, resulting in similar, but not identical 3D reconstructions for different classes. Subsequently, the structural differences between classes will be amplified gradually until convergence through many iterations of simultaneous refinement of multiple references. Notably, distinct to the conventional projection-matching method that assigns the discrete angular parameter and class belonging into each particle, probability-weighted integrations over all possible angular and class assignments are calculated in the ML method. It has been demonstrated for many cases that ML-based 3D classification methodology is robust to noisy cryo-EM data and normally yields less biased results.

Furthermore, a Bayesian statistics-based framework was introduced to the single-particle image analysis by Sjors Scheres, by which parameters for 3D classifications are optimized further, without user intervention, to reduce the noise level in the reconstructions. Implementation of this kind of regularized maximum likelihood approach in the program called RELION has been proven to be a powerful tool for 3D classification that has been widely used to tackle the structural heterogeneity problem for many different types of samples (Scheres 2012a, 2012b).

Except for the regularized-ML-based 3D classification approach as in RELION, there were alternative 3D classifications algorithms invented, based on multivariate statistical analysis (MSA) (Elad *et al* 2008), 3D variance calculation (Zhang *et al* 2008), and non-statistical multi-particle refinement (Spahn and Penczek 2009). Additionally, a novel manifold-embedding based classification method has been proposed recently for mapping continuous structural changes in the dynamic protein macro-machineries (Frank and Ourmazd 2016).

4.5.2 Global 3D classification

As mentioned above, in the unsupervised 3D classification approach the user is required to provide the number of possible structures in the data, which is difficult to predict in most cases. Nevertheless, even if the number of desired classes specified by the user is smaller than the actual number of structures present in the data, at least a few major conformational states which exhibit the largest structural differences can usually be identified and classified.

When using ML3D one should start the first round of 3D classification with exhaustive angular searches, as the relative orientations of the particles are unknown (figure 4.16). In order to reduce the computational costs, exhaustive angular searches in 3D classification are typically performed using a relatively coarse angular sampling, such as 7.5 degrees, a sampling rate which is commonly used in RELION. Due to the coarse sampling rate, the classification approach with exhaustive angular search is primarily used to distinguish 'junk' from 'good' particles, or to separate particles with large conformational or compositional differences, rather than detecting small structural differences. Nevertheless, it is still necessary to perform such a 'rough'

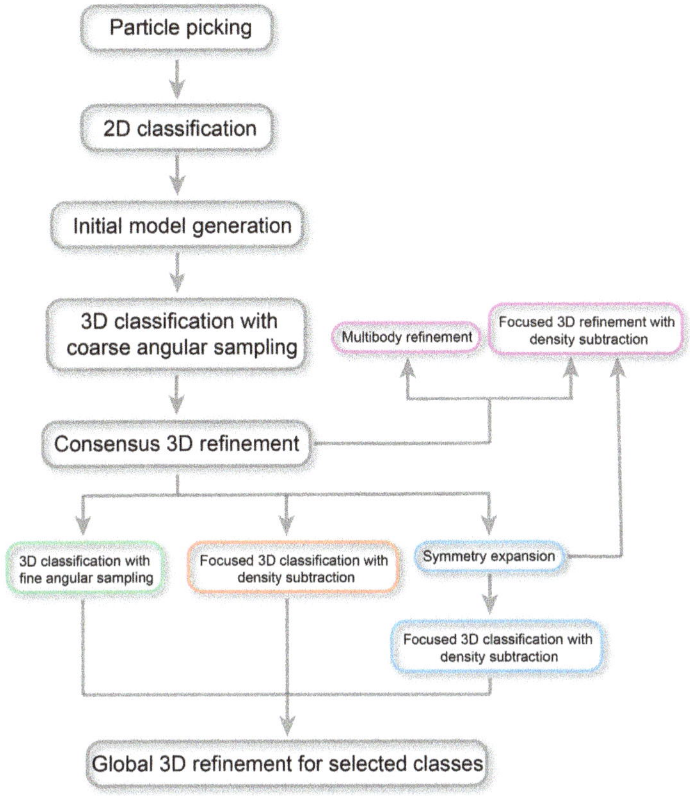

Figure 4.16. General workflow for dealing with structurally heterogeneous cryo-EM data. Different strategies for dealing with heterogeneous particles are highlighted with colored boxes, which are described in the text.

classification procedure in the initial stage of data processing, to efficiently remove junk particles and provide more homogeneous particle subsets for the later steps of data processing. For instance, during the sample preparation of mitochondrial ribosomes, these are always contaminated with cytoplasmic ribosomes that are affiliated with the outer surface of mitochondrial membranes. In this case, the ML-based 3D classification procedure was used for *in silico* protein purification, where mitochondrial ribosome particles were effectively separated from cytoplasmic ribosomes particles (Amunts *et al* 2014) (figure 4.17).

After a subset of particles is selected by initial 'rough' 3D classification, a subsequent 3D auto-refinement approach is commonly used to improve the accuracy of particle alignment further (figure 4.16). As a result, a 3D reconstruction with higher overall resolution can be obtained. However, this map is normally not the final reconstruction used for structural interpretation, as some degree of structural heterogeneity—particularly in a few local regions—may remain present in the structure. As mentioned above, small-scale local conformational changes or subtle compositional differences may not be able to be distinguished by employing 3D classification with coarse angular-sampling searches.

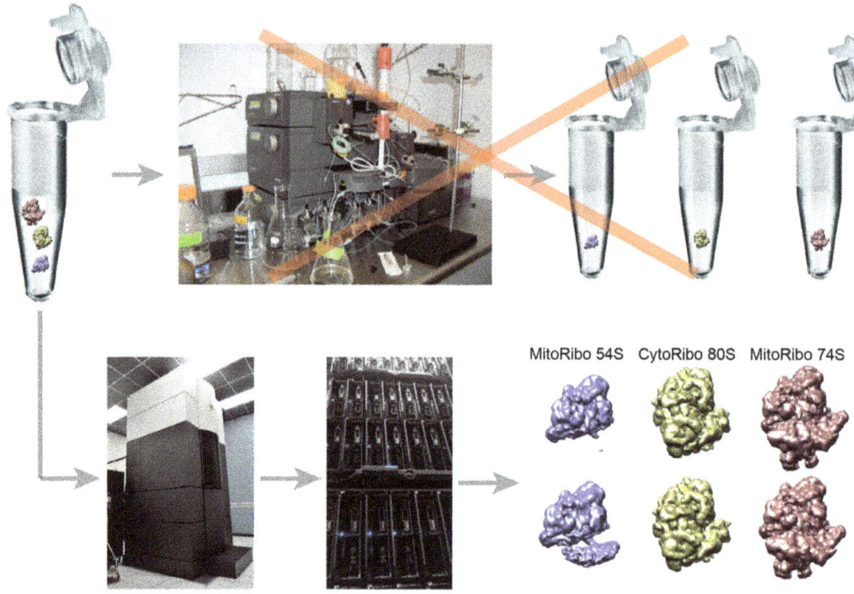

Figure 4.17. *In silico* protein purification. During the sample preparation, mitochondrial ribosomes were contaminated with cytoplasmic ribosomal particles. Instead of employing conventional biochemical approaches to purify mitochondrial ribosomes, the ML-based 3D classification procedure was used for *in silico* separation of mitochondrial ribosome particles from cytoplasmic ribosome particles.

To allow for the separation of structures with smaller differences, one needs to perform a 3D classification with finer angular samplings (the green box in figure 4.16). Needless to say, this approach requires a correspondingly higher computational capacity. To reduce the computational cost, when performing 3D classification with finer angular sampling, local angular searches are only performed around the most probable orientations derived from the previous consensus 3D refinement (figure 4.16). The range of local angular search is determined by the user, mainly according to the estimated level of structural heterogeneity in the data as well as the alignment accuracy calculated from the previous consensus refinement.

In the case of the 80S ribosome in complex with a small initiation factor eIF5B, the initial reconstruction of this complex showed blurred and incomplete density for eIF5B due to the low binding occupancy of eIF5B to 80S ribosomes. By using a cascaded 3D classification approach together with finer angular sampling and a local search, a small subset of particles, in which eIF5B is associated with 80S ribosome in a rigid conformation, could be identified and thus a cryo-EM map at 6.6 Å resolution could be calculated from this small subset, showing prominent density for the entire eIF5B (Fernandez *et al* 2013).

4.5.3 Masked 3D classification

Despite its great success there are important limitations on what can be achieved with the global 3D classification method. Particularly for complexes that contain small continuous flexing domains, separation of the particles into homogeneous subsets is difficult. As a result masked (or focused) 3D classification has been introduced as a powerful method to deal with structural heterogeneities that only exist within local regions of a large protein complex. By applying a 3D soft mask on a specific part of the 3D references, the classification will be focused on the region within the mask, while structural variabilities for the regions outside the mask are completely ignored (blue box in figure 4.16). This stage can also be used to separate fully assembled complexes from partially assembled complexes for which one or more subunits are missing (Abeyrathne *et al* 2016).

For those protein complexes that have relatively small flexible parts, accurately aligning the particle's images to the masked reference, which only contains a small portion of the protein mass, will be very difficult. In such cases, during the masked 3D classification, one normally fixes the orientation parameters of each particle to be the same as determined in the previous consensus refinement, and then performs only the class assignments. This type of masked 3D classification without angular searches can be used to effectively identify very small local conformational changes or even detect continuous motions (Fernandez-Leiro *et al* 2017, Nguyen *et al* 2016).

In masked classification, one needs to compare projections of a masked reference which contains only part of the protein complex to the entire experimental images of particles. This leads to a so-called inconsistencies problem, where the projections of the references and experimental images contain different amounts of protein mass but nevertheless have to be aligned together. It is easy to imagine that, if the mask applied in the 3D classification is very small, this type of inconsistencies problem will largely affect the orientation assignment and classification results. It has been demonstrated that the inconsistency in particle alignment could be reduced by an approach called 'signal subtraction', where the signal outside the mask is subtracted from the experimental particle images (Bai *et al* 2015b, Ilca *et al* 2015, Zhou *et al* 2015b). In this specific approach, one needs to first design a mask that includes only the flexible region of a protein complex. Then another mask, complementary to the designed mask and which only comprises rigid regions of the complex is applied to the reconstruction from the consensus 3D refinement. Projections of this masked reconstruction, using the angles determined from the consensus refinement, are then subtracted from all experimental particle images in the dataset (the orange box in figure 4.18). This procedure will generate a new dataset consisting of modified experimental particle images that only contain the regions of interest in the protein complex, which could be used for further masked 3D classification (the final gray box in figure 4.18). Similar to the conventional masked classification, if the flexible part of the complex is small, the combination of masked classification with signal subtraction can be performed with the orientation of each particle kept fixed at the values obtained from the consensus refinement.

Utilization of the signal subtraction approach in the case of human γ-secretase enabled distinguishing three states of the transmembrane domain of this complex that mainly result from the conformational differences of two transmembrane helices (figure 4.18). Moreover, in one of these conformations, an additional transmembrane helix, corresponding to a mass of 3 kDa, is visible, which could be attributed to the co-

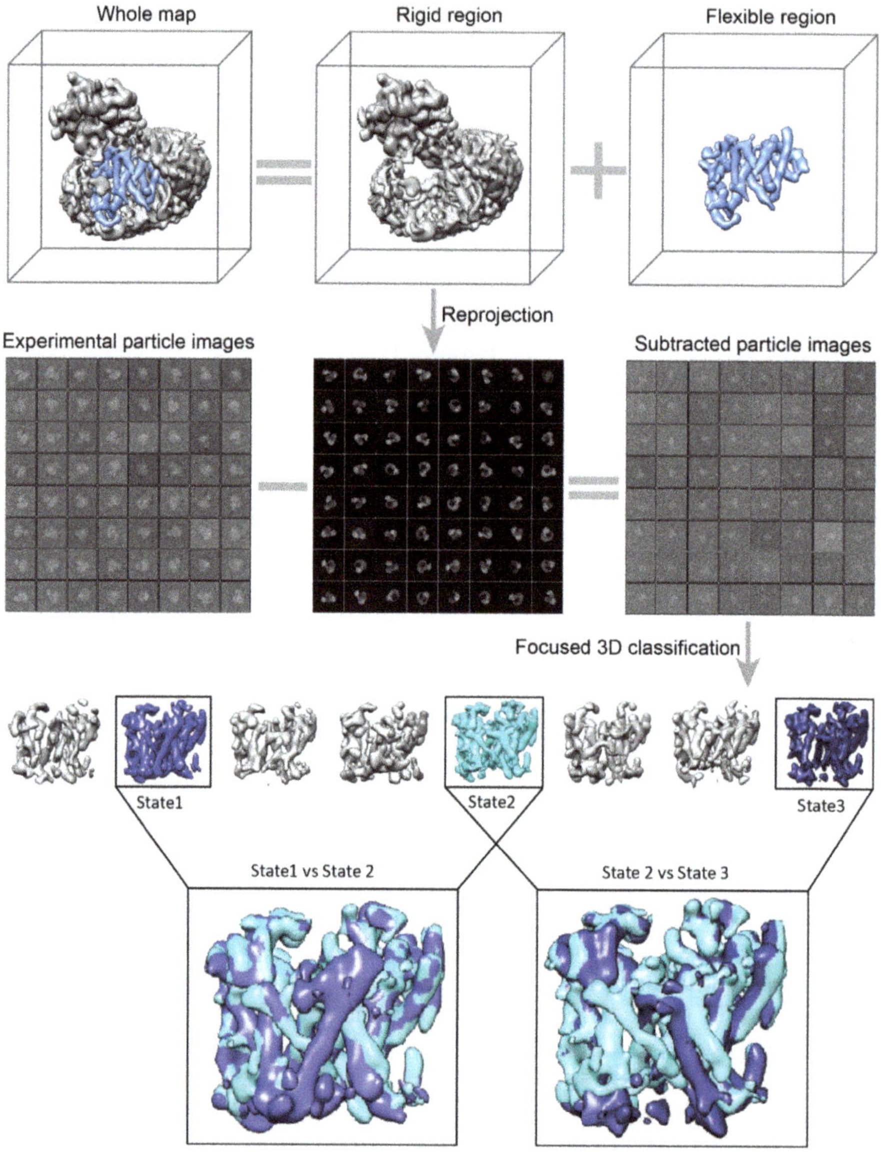

Figure 4.18. Particle signal subtraction. The rigid part of the complex one would like to ignore in masked classification is shown in gray. The part of the complex on which one would like to focus classification is shown in blue (top row of panels). During the particle signal subtraction, one subtracts the 2D projections of the rigid part of a 3D volume from the experimental particle. This results in modified experimental particle images that only contain the flexible part of the complex (middle row of panels). The masked 3D classification on the modified particle set can be used to effectively identify small local conformational changes (bottom panels).

purified substrate of γ-secretase (Bai *et al* 2015b) (figure 4.18). This result demonstrates that masked classification with signal subtraction could be potentially used in classifying particle images which only exhibit very small conformational or compositional differences.

4.5.4 3D classification of particles with pseudo-symmetry

In many protein complexes that have some sort of symmetry, the symmetry is only true for the stable part of the complex, whereas it is broken for the flexible region. An approach called 'symmetry expansion' was developed to deal with such pseudo-symmetric problems (Morais *et al* 2003, Zhang *et al* 2007, Zhou *et al* 2015a) (the blue box in figure 4.16). In this approach the dataset is artificially expanded according to the overall symmetric group (e.g. n-fold rotational symmetry) by replicating and rotating each particle by n times. In this way all the asymmetric subunits within one protein complex are overlapped with each other in the expanded dataset. As a result, the expanded dataset becomes n times larger than the original one and the subsequent 3D classification on the expanded dataset could be focused on any one of the n asymmetric subunits (figure 4.19). The masked 3D classification

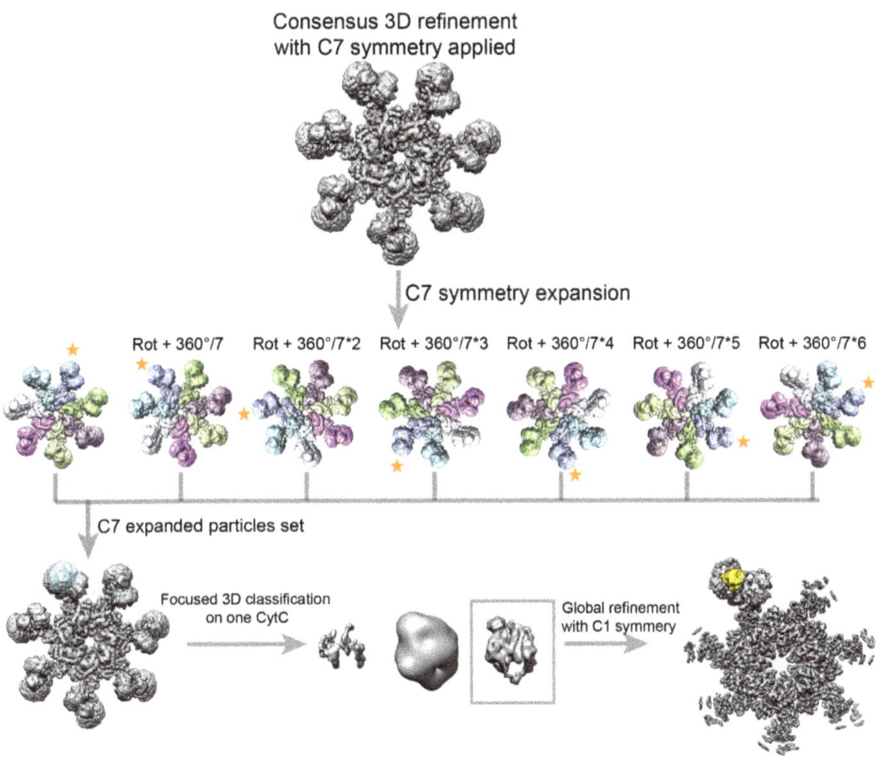

Figure 4.19. Symmetry expansion. A consensus refinement with imposed C7 symmetry resulted in a map in which the central disc was well resolved, but the resolution for the CytCs bound at spoke regions was poor (top panel). In order to deal with structural heterogeneities of CytCs, the dataset was expanded seven-fold by replicating each particle and adding $360°/7 \times n$ ($n = 1, 2, 3, \ldots 7$) to the Euler angle 'Rot' of every particle. In such way, all seven CytCs of every particle are placed in the same position (middle panel). Masked classification for a single CytC resulted in a subset of all CytCs that adopts a relatively stable conformation. Subsequent masked 3D refinement with C1 symmetry, using a soft mask around the entire central disc plus one single spoke, led to a cryo-EM map with one of seven CytCs resolved at significantly better resolution (bottom panel).

approach with symmetry expansion can still be coupled with signal subtraction as well as with local/skipped angular search to increase the accuracy of class assignment (Roh *et al* 2017).

The case of human apoptosome complex is an example in which seven identical subunits assemble as a ring-like structure with C7 symmetry. However, since the seven protruding spokes undergo conformational changes independently, the symmetry of this complex is partially broken. To solve the pseudo-symmetry problem, after refining the map with C7 symmetry imposed, the dataset was expanded by seven times by modifying the Euler angle 'Rot' of each replicated particle with RELION command 'relion_particle_symmetry_expand'. In this way, the seven spokes in each particle are placed in a similar position. Masked classification on one single spoke lead to identifying a subset of particles showing a relatively stable conformation of the spoke (figure 4.19) (Zhou *et al* 2015a).

4.5.5 Dealing with continuous motions

Masked 3D classification is commonly used to separate particles with discrete conformational states. Closely related to masked 3D classification, masked 3D refinement has been developed for dealing with continuous rigid-body motions between different subunits of a large protein complex (figure 4.16). In this approach, a 3D soft mask that only comprises part of the complex, which undergoes rigid-body movement with respect to the rest of the complex, is again applied to the 3D reference during the 3D refinement. Consequently, the 3D refinement will be focused predominantly on the regions inside the mask whereas the regions outside the mask become very ineffective for angular determination. As a result of masked 3D refinement, each particle is realigned to the masked region of a 3D reference so that the resulting cryo-EM density within the mask is normally improved to a large extent. It worth noting that, to provide enough signal for accurate particle alignment in the masked 3D refinement, the protein mass inside the mask must be 100 kDa or more. For those protein complexes having flexible parts smaller than 100 kDa, masked 3D classification would be an alternative solution. In addition, masked 3D refinement suffers from a similar inconsistencies problem as described in section 4.5.3. Therefore, the masked 3D refinement approach can also be combined with signal subtraction.

In the case of *Plasmodium falciparum* 80S ribosome, large, continuous ratchet-like motions were observed between large and small ribosomal subunits. Due to the inter-subunit motions the reconstruction obtained by global refinement showed well-resolved density for the large subunit but relatively poor density for the small subunit. By applying a 3D soft mask around the small subunit during the 3D refinement, the influence of the large subunit on the alignment of the small subunit was largely reduced, which resulted in a much improved reconstruction for the small subunit (Wong *et al* 2014).

A new masked 3D refinement approach, called 'multibody refinement', was developed recently to automatically refine multiple flexible regions of a protein complex without user intervention (Nakane *et al* 2018) (the purple box in figure 4.16).

In this approach, users need to first prepare soft masks for multiple individual bodies in a protein complex, which might undergo independent, rigid-body motions. Subsequently, different regions are refined simultaneously by using masked 3D refinement with signal subtraction. During the multibody refinement more accurate angular assignments for each body, obtained after one given iteration, are used for a better signal subtraction in the next iteration. In the case of the spliceosomal tri-snRNP complex, the cryo-EM densities for four independent regions were significantly improved after multibody refinement (Nguyen *et al* 2016, Nakane *et al* 2018, Plaschka *et al* 2017).

4.5.6 Conclusion

Cryo-EM has been developed as a cutting-edge technology for high-resolution structural determination of protein complexes. Cryo-EM has shown great advantages in dealing with structurally heterogeneous data, as compared to other techniques, such as x-ray crystallography. Nevertheless, there are clear limitations on what can be achieved with the current particle-sorting methods in cryo-EM, which hinders high-resolution 3D reconstruction for a wide range of protein samples. Therefore, there is a clear need for developing an even better image-classification algorithm to deal with those biological samples that exhibit a large degree of structural heterogeneity, or have a small flexible domain that undergoes continuous motions.

4.6 Preferred orientation: how to recognize and deal with adverse effects

Dmitry Lyumkis
The Salk Institute for Biological Studies, 10010 North Torrey Pines Road, La Jolla, CA 92037, USA
Department of Integrative Structural and Computational Biology, The Scripps Research Institute, La Jolla, CA 92037, USA

Cryo-electron microscopy (cryo-EM) has become the go-to method for determining structures of many large macromolecules and macromolecular assemblies (Fernandez-Leiro and Scheres 2016). Technological advances are continuing to propel the field, but numerous challenges remain (Nogales 2016, Vinothkumar and Henderson 2016, Lyumkis 2019). One of the major bottlenecks involves specimen vitrification for imaging under cryogenic conditions. Most current grid vitrification techniques generate a thin aqueous film of vitreous ice containing embedded protein 'particles'. There are two air–water interfaces (AWIs) characterizing a frozen-hydrated holey grid, one at the top and one at the bottom. These interfaces attract macromolecular objects and lead to macromolecules adopting 'preferred orientations', which means that there are several preferential views of the object within the dataset with respect to the electron beam. This in turn has a wide range of deleterious effects, including compromising the integrity of specimen, the quality and quantity of data necessary to obtain a reconstruction, and the fidelity of the reconstructed object. Here I will discuss how and why proteins interact with the AWI, the consequential preferred particle orientations in cryo-EM, and its effects on the map, with suggestions for overcoming the problems.

4.6.1 Protein interaction with the air–water interface

The AWI plays an important role in cryo-EM. For a typical imaging experiment, several microliters of biological sample are applied onto a grid, blotted away with filter paper, and plunge frozen into liquid ethane. Although various steps have been automated, the general workflow has not changed much since the early pioneering experiments (Dubochet and McDowall 1981, McDowall *et al* 1983) (figure 4.20). This procedure generates a thin aqueous film of vitreous ice (with depth on the order of tens of nanometers) characterized by two AWIs, one above and one below (reviewed in Dubochet *et al* 1988). Based on theoretical calculations, particles will interact with the AWI through diffusion ~10–1000 times within the timeframe between blotting and plunging (Naydenova and Russo 2017), and some of these interactions may result in particle adherence. The problems caused by those unwanted interactions and some ways to overcome them when they occur are introduced in section 2.5. Here we elaborate further on the issue with special emphasis on the preferential orientation of particles that often happens.

From research performed outside the cryo-EM community, we now know that interactions of proteins with the AWI can be modeled through a three-step process. First, there is a diffusion-mediated initial adsorption of the protein at the interface, followed by an accompanying conformational change and/or denaturation/unfolding of the adsorbed protein, and lastly the formation of a protein network (a viscoelastic film) with specific chemical and mechanical properties (Razumovsky and Damodaran 1999, Martin *et al* 2002). Recent experimental evidence showed that, despite the common assumption that particles are randomly dispersed within the vitreous ice layer, the reality is the reverse. In a survey of ~50 different cryo-EM

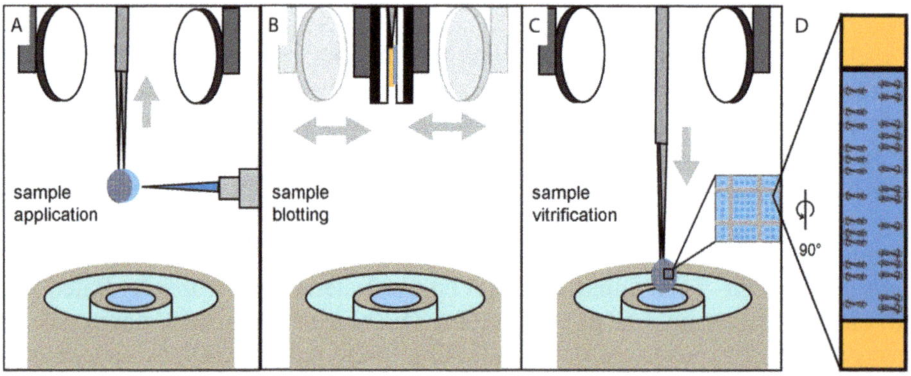

Figure 4.20. Sample vitrification using conventional two-sided blotting strategies. (A) The sample is applied to one side of the grid. (B) Tweezers holding the grid are moved upward into position for blotting and two pieces of filter paper blot the sample, wicking away excess liquid. (C) The sample is then plunged into liquid ethane for vitrification. (D) Cross-section of a single hole within a vitrified cryo-EM grid. This example is shown with particles adsorbed at the AWI.

samples vitrified using conventional approaches, Noble *et al* showed that >90% of the samples are characterized by particles predominantly adhering to one of two AWIs (Noble *et al* 2018a). In fact, the particles embedded between the two interfaces are few and far between. Specimen adherence to the AWI is ubiquitous and, as discussed later, may have significant detrimental consequences.

The AWI has been studied outside the cryo-EM field and the properties that are relevant to electron microscopists are briefly introduced below. There is now an extensive literature about the AWI and its effects on proteins (Barnes and Gentle 2011, Déjardin 2006). Early experiments demonstrated that protein molecules flowing over a glass rod diffuse to the AWI and unfold/denature in the process (Trurnit 1960). Proteins form a new phase at the AWI, which in turn causes more molecules to adhere. Under appropriate conditions (flow rate, rod thickness, sample concentration, etc) the majority of protein molecules will denature and form a viscoelastic film at the AWI. The rate of adsorption (Tripp *et al* 1995) and unfolding appear to be the most important factors that influence the surface properties of the interface (Martin *et al* 2002). The more stable a protein is in solution, the more robust is the resulting film at the AWI. Proteins that are less stable tend to rapidly adsorb at an interface and have higher tendencies to unfold (Razumovsky and Damodaran 1999, Martin *et al* 2005). This helps to explain why there are dramatic sample-dependent differences among specimens frozen in an otherwise similar manner on cryo-EM grids (Carragher *et al* 2019). Antiparallel β-sheets are often encountered at the AWI (both in their native and non-native conformations, i.e. induced by adsorption) and lead to subsequent protein adherence (Lad *et al* 2006, Martin *et al* 2003, Kudryashova *et al* 2003). Samples with high β-sheet content may be particularly susceptible to adsorption. Although numerous factors play a role, the viscoelastic film that results from adsorption to the AWI is likely driven by hydrophobic interactions (Norde 2011). There can be sequential adsorption of proteins onto a multi-layered interface and each layer will have distinct properties (Ganzevles *et al* 2008). In cryo-EM such sequentially adsorbed protein films would be expected to lead to increased background noise within the data. The local protein concentrations can get as high as 150–250 mg/ml (Kudryashova *et al* 2003), indicating that the AWI serves as an artificial concentrator for macromolecular samples. The protein network is unlikely to be uniform (Gunning *et al* 2004), implying that local patches of adsorbed protein may be mixed in with free monodisperse particles; if the latter remain intact, they may still contribute to a high-resolution map. Together, the adsorption properties of proteins at an AWI arise from combinations of molecular characteristics, including size, hydrophobicity, chemical reactivity, structural stability, dynamics, electrostatics, temperature, buffer conditions, the presence/absence of surfactants, among others, and are not determined by a single property (De Jongh *et al* 2004).

4.6.2 Preferred orientation and its effects in cryo-EM

In the last several years the cryo-EM community began appreciating the widespread influence of the AWI in sample vitrification, its effects on sample distribution, denaturation, aggregation, data acquisition, and cryo-EM reconstructions. The distinct phenomena will be described below.

Protein adherence to the AWI may affect the integrity of the sample. On the one hand, interactions may be localized to a small stretch of the sample (e.g. an exposed β-sheet region), with negligible effects on the overall structure. On the other hand adherence to the AWI may induce complete denaturation or aggregation along the interface. There are various scenarios in between (Taylor and Glaeser 2008) as summarized in figure 4.21. A recent study with the yeast fatty acid synthase showed that a substantial part of the complex was denatured in the high-resolution reconstruction (D'Imprima *et al* 2019). Moreover, ~90% of the particles in the study were affected, indicating that most of the acquired data were not useful for the purposes of obtaining a high-resolution map. Importantly, such effects were not observed in negative stain reconstructions. Denaturation of one particle can induce denaturation and/or aggregation in the vicinity, leading to a cascading effect. There are many

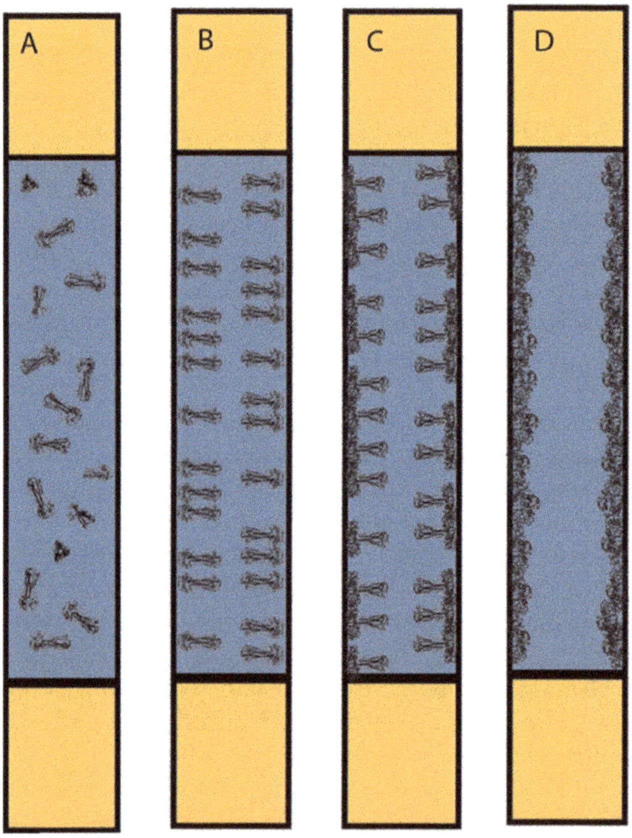

Figure 4.21. Scenarios of protein interaction at the air–water interface. (A) The ideal behavior of particles residing in random orientations and locations within the ice layer, away from the AWIs. (B–C) The behavior of particles adhered to the AWI in one orientation (B) without and (C) with local partial denaturation. (D) Complete sample denaturation at the AWI.

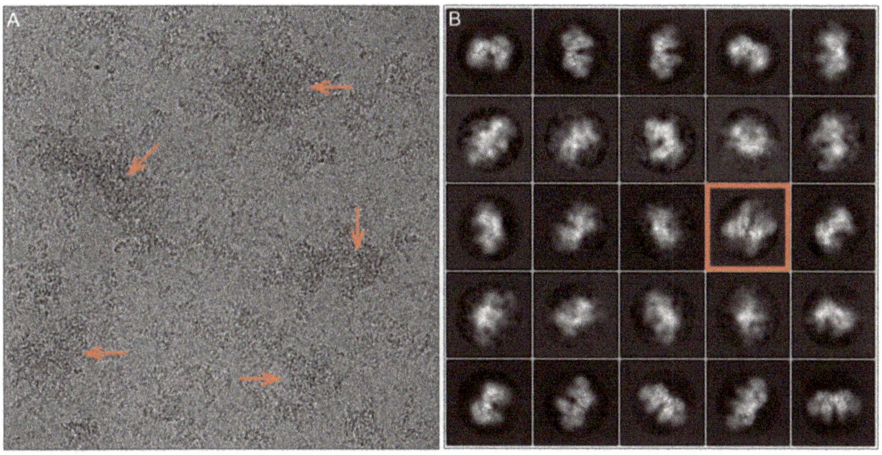

Figure 4.22. Qualitative indicators of particle adherence to the AWI. (A) Patchy areas of particle density of HIV intasomes from a sample that is similar to that described in Passos *et al* (2020). These areas are indicative of local aggregation and/or denaturation through adherence to the AWI and were not observed in negative stain or in size-exclusion chromatographic profiles. (B) Class averages computed from non-aggregated areas containing individual HIV intasome particles. The selected class represents the 'top-like' view of the two-fold rotationally symmetric object, but corresponds to only 2% of the total particle population, indicating that the sample is preferentially adhered along its other interfaces.

examples of images containing patches of protein particles surrounded by areas of empty ice, as exemplified by an HIV intasome sample (figure 4.22(A)) that is similar to that described in Passos *et al* (2020). This is likely the result of localized adherence and/or aggregation induced by protein unfolding at the interface (Martin *et al* 2002). The interface in such cases is a deterrent, but not necessarily an outright barrier, since it is still possible to obtain good class averages (figure 4.22(B)) and a high-resolution reconstruction within regions that appear more monodisperse. In other instances the experimentalist may observe complete denaturation of the complex within a field of view of cryo-EM grids (Carragher *et al* 2019). In such cases structure determination is impossible and it is necessary to adjust vitrification conditions.

At the level of the reconstructed map the most readily apparent consequence of preferred orientation is a loss of resolution along one direction. To understand why the decrement is specific to a single direction it helps to review the geometry of the imaging experiment. Consider the simplest scenario where there is only one preferred orientation. The scattered electrons form a projection of the object along the Z-direction (with respect to the beam path). If there are no other views of the object, then the XY-direction will be well-resolved, but the information along the Z-direction will not be well localized. In the reconstruction, this appears as an apparent stretching or blurring of the map (or, equivalently, a loss of resolution) along the Z-direction. In most cases when there is a single, strongly preferred orientation (and in the absence of symmetry), the map will have poor resolution along Z and this may also compromise the interpretability of the features along orthogonal planes. In the absence of methods to ameliorate the effects (discussed later), a single preferred orientation can mislead or even impede proper structure interpretation. Conversely, when there are many different preferred orientations from distinct directions, the map should be mostly isotropic. Indeed, there may be numerous intermediate scenarios, where the overall effect is less clear.

Preferred orientation results in a non-uniform distribution of projections, which in addition to affecting directional resolution also attenuates global resolution. This effect has only recently been described (Baldwin and Lyumkis 2020) and requires further explanation. Conventionally, we measure global resolution in cryo-EM by splitting the data into two halves, independently refining the subsets for reconstructing half-maps and calculating the correlation between these two as a function of spatial frequency (Scheres and Chen 2012). This produces the Fourier shell correlation (FSC) curve (Harauz and van Heel 1986). A cutoff, either at a fixed (Rosenthal and Henderson 2003) or variable threshold (van Heel and Schatz 2005, Rohou 2020), is then used to define the nominal resolution value. The FSC depends on both signal and noise. Recently, we showed that, in the simplest of possible theoretical treatments, the effect of the sampling on the FSC can be approximately decoupled from the other aspects affecting FSC (Baldwin and Lyumkis 2021). We characterize 'sampling' simply as the number of times that a representative point in Fourier space has been measured. In the presence of non-uniform sampling caused by preferred orientation the noise variance in 3D has greater fluctuations, which leads to a net increase in variance. An increase in the noise variance leads to a corresponding attenuation of the FSC curve, the extent of which can be predicted by a geometrical factor that we call the sampling compensation factor (SCF); therefore, it is necessary to compensate for the geometry of the sampling to correctly estimate the FSC (Baldwin and Lyumkis 2020). Since most samples prepared using current vitrification techniques suffer from at least some non-uniformity in sampling caused by preferred orientation, the attenuation of resolution seems to be a widespread phenomenon. The observation additionally implies that most currently obtained cryo-EM reconstructions may benefit by improving the sampling distribution, as will be described later. The extent of attenuation in real cryo-EM data will vary and is the subject of ongoing investigation.

4.6.3 Quantifying preferred orientation and its effects on cryo-EM reconstructions

The degree of preferred orientation and its effects on cryo-EM reconstructions can be quantified experimentally. It is necessary to distinguish assessing four distinct phenomena: (i) the extent of particle denaturation, (ii) the degree of preferential orientation, (iii) the effects of the non-uniform orientation distribution on directional resolution, and (iv) the attenuating effects on global resolution. Although the combined effects of these phenomena will vary, the reconstructed object is always negatively affected when the orientation distribution is not uniform (figure 4.23(A) and (B)).

There is currently no direct way to quantitatively assess how adherence to the AWI affects particle integrity. However, several indirect measurements can provide some idea of the extent of the problem. Most notably, cryo-electron tomographic

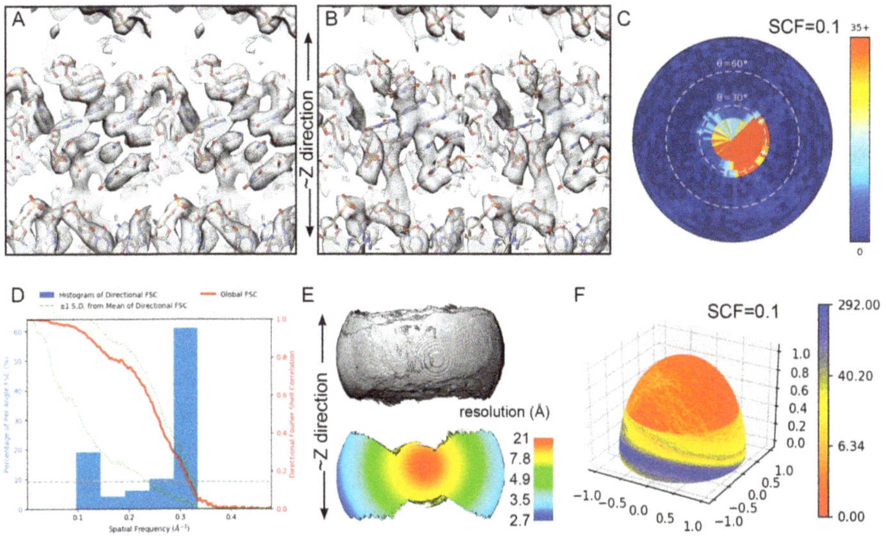

Figure 4.23. Quantitative assessments of the effects of preferred orientation. Stereo views of a reconstruction from (A) a uniform orientation distribution or (B) a set of views comprising tilt (theta) angles up to 30°, i.e. characterized by a 60° missing cone. The deleterious effect on the reconstruction is clearly apparent in (B). (C) Euler angle orientation distribution plot. (D) Histogram of directional resolution values, shown alongside the global resolution curve for the reconstruction, indicating ~3.2 Å globally, but a large spread of directional resolution values. (E) 3D FSC shown as (top) an isosurface or (bottom) an orthoslice and thresholded at 0.143. Notice that the ~*Z*-resolution is very poor. (F) surface sampling plots, introduced recently (Baldwin and Lyumkis 2020), showing largely missing information along the top of Fourier space, consistent with a ~60° missing cone characterizing the reconstruction. In (C) and (F), the SCF value of 0.1 for the orientation distribution was calculated as described (Baldwin and Lyumkis 2021).

reconstructions can provide a direct readout of particle location (Noble *et al* 2018a). If most particles reside within the AWI, this implies that the particle is adhered to the interface and its structural integrity may potentially (but not necessarily) be compromised. In many published structures, the number of particles selected from micrographs far outweighs the number of 'good' particles that produce a high-resolution reconstruction. Sometimes this value is less than 10% of the particles contributing to the high-resolution map (D'Imprima *et al* 2019, Zhang *et al* 2018, Herzik *et al* 2019). The presence of damaged/denatured particles is one major cause of the phenomenon (D'Imprima *et al* 2019), but there are other possible causes, including false choices, thick ice, errors in CTF estimation, etc. D'Imprima *et al* demonstrated substantial denaturation of the fatty acid synthase molecule at the AWI, which affected >90% of the selected particles (D'Imprima *et al* 2019). Since denaturation will affect computational processing, at least some amount of classification usually benefits cryo-EM refinement. Particles that fall into uninterpretable classes during 2D, or in particular 3D classification, are often referred to as 'bad' particles. Denaturation should affect smaller particles more since they have a proportionally higher mass residing at the AWI and thus less useable signal for Euler angle assignment (Herzik *et al* 2019). Through combined use of cryo-ET, 3D classification, and careful tracking of particle assignment, one can obtain indirect insight into the extent of particle denaturation. Given the obvious caveats, such data must be interpreted with caution.

A quantitative understanding of the effects of non-uniform sampling on a reconstructed object begins with determining the Euler angles corresponding to the set of particles comprising the data. It is necessary to know how these preferential views relate to the reconstructed object. Once the Euler angles are assigned, the user can inspect a plot of the angular distribution, typically phi versus theta angles (figure 4.23(C)). Such plots describe how the particles corresponding to a reconstructed object relate to one another in 3D space. Several approaches also provide a quantitative readout of the uniformity of the orientation distribution (Naydenova and Russo 2017, Baldwin and Lyumkis 2020), for example, as indicated by the SCF value, introduced in section 4.6.2, above. If there are localized patches in the display of angular distributions this indicates that the specimen is preferentially adhered to the AWI. The number of patches should approximately correspond to the number of distinct class averages (after accounting for in-plane alignment). Importantly, such plots are necessarily relative to the reconstructed object and are accurate *only if the reconstruction is correct*. If the reconstruction is incorrect so is the orientation distribution and such plots are meaningless. Even when the reconstruction is correct there is probably some error associated with the orientation distribution. In almost all experimental cases some false positives will be sprinkled throughout Euler space. Therefore, while such plots are informative, they should not be taken at face value.

A variation of conventional FSC analysis affords a quantitative assessment of directional resolution for a reconstructed cryo-EM map. Whereas the global FSC is evaluated on shells (Harauz and van Heel 1986), restricting the evaluation to a cone provides a consistency measure along a particular viewing direction. Repeating this calculation for different directions generates a complete measure of directional resolution across Euler space (Dang *et al* 2017, Dudkina *et al* 2011, Diebolder *et al* 2015, Tan *et al* 2017). Combining these 1D conical FSC curves, obtained from different angular directions, generates a real 3D array that we term the 3D FSC (Dang *et al* 2017, Tan *et al* 2017) (figure 4.23(D) and (E)). The 3D FSC provides a quantitative assessment of directional resolution for the reconstructed object. It can be visualized or displayed as an isosurface at a nominal threshold (currently only fixed thresholds are implemented) or as an orthoslice through the volume. The latter helps to visualize the directional differences with spatial-frequency readouts. A standalone tool for calculating 3D FSCs is available at https://github.com/LyumkisLab/. To facilitate its use by cryo-EM practitioners, there is also a web server, available at 3dfsc.salk.edu.

Recently we developed several tools that help to understand the effects of sampling on cryo-EM data (Baldwin and Lyumkis 2021). We introduced surface sampling plots as a complement to the conventional Euler distribution plots (figure 4.23(F)). The plot provides a quantitative measure of Fourier space coverage and highlights regions of sampling modulation. Regions that are over/under-sampled are expected to have a direct relationship to increased/worsened resolution in the appropriate direction. In this way, the plot complements the results from the 3D FSC to provide an intuitive understanding of the degree of orientation anisotropy, which is not always apparent from Euler distribution plots. Furthermore, the degree of orientation non-uniformity will impact the SCF value, which is also reported. The lower the SCF value, the more the FSC has been attenuated due to fluctuation in the noise variance. The plot also highlights if there are zeros in the Fourier transform, which are indicative of completely missing data. Ideally, such scenarios should be avoided and alterations to the data collection scheme are warranted, as will be suggested below. These tools can also be obtained from https://github.com/LyumkisLab/.

It is worth noting that having a completely uniform orientation distribution is not a requirement to obtain good experimental reconstructions, although it helps. Even in the presence of preferred orientation, it is possible to reconstruct a nearly isotropic map. Symmetry can help to this end. If there are multiple symmetry units, then the projection is distributed around Fourier space in accordance with the point-group symmetry. This can dramatically improve the sampling, and the corresponding geometrical SCF factor, in particular for tetrahedral, octahedral, and icosahedral point groups. However, the manner in which the symmetry axis relates to the axis of preferred orientation is important and cannot be neglected, as discussed extensively in recent work (Baldwin and Lyumkis 2020). It is also possible that an asymmetric object can have multiple preferential orientations distributed around Euler space. Strictly speaking the reconstruction will not be isotropic in such cases, but the interpretability of the maps should not be grossly affected.

4.6.4 Overcoming the effects of preferred orientation

To emphasize the point made in section 4.6.1, when a sample is characterized by preferred orientation there are three primary concerns: (i) denaturation of the protein; (ii) anisotropy and missing information in a direction-dependent manner; and (iii) attenuation of global resolution. Although a variety of approaches have been developed to address and/or overcome these distinct issues it often is difficult to find one that is effective, as is described in section 2.5, mentioned above. At the risk of some repetition, it nevertheless is worthwhile to elaborate on the options that are currently known to work in at least some cases.

The addition of surfactants can alter the surface tension properties and how the sample behaves with respect to the AWI. Surfactants can be phospholipids, detergents, amphipols, glycols, or other chemicals that form a surface barrier that would prevent the protein from interacting with the AWI. A diffusible surfactant at the AWI may also stabilize the thin aqueous film containing vitrified sample against dewetting, thus suppressing protein denaturation and aggregation (Glaeser *et al* 2016). An early observation showed that the addition of phospholipid vesicles to the sample of interest covers the AWI, acting as a cover slip for specimens embedded within a thin aqueous film (Frederik *et al* 1989). Surfactants can also displace proteins adhered to interfaces back into solution (MacRitchie 1998). The most common surfactant that experimentalists try is to add some detergent into the buffer solution. For example, the use of DDM completely altered the orientation distribution of HIV envelope trimers (Lyumkis *et al* 2013), and this behavior was titratable, i.e. directly proportional to the amount of DDM in the buffer (at concentrations below the critical micelle concentration). Other detergents can have similar effects

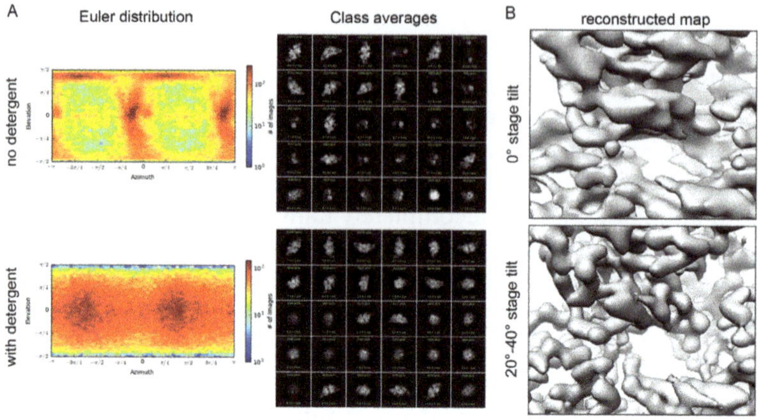

Figure 4.24. Altering an orientation distribution using surfactants and tilts. (A) (top-left) Orientation distribution and (top-right) 2D class averages of a grid containing bovine thyroglobulin sample prepared without the addition of detergent; (bottom-left) orientation distribution and (bottom right) 2D class averages of the same sample from a grid prepared after adding 0.1% CHAPS detergent. A more uniform distribution and new sets of views are apparent. (B) (top) Reconstructed density map from a dataset of HIV intasomes, collected without stage tilt; (bottom) reconstructed density map from the same dataset, but after applying a stage tilt ranging from 20° to 40° during data collection. The reconstructions were performed from a dataset adapted from Passos *et al* (2020).

(Chen *et al* 2019, Rollins *et al* 2019) (figure 4.24(A)), as might amphipols (Tribet *et al* 1996). Other approaches and the general utility of surfactants in cryo-EM has been recently discussed (Glaeser *et al* 2016). Adding a surfactant may appear to be simple to try, but it is clear that it is not guaranteed to be effective. In practice, the user needs to try different concentrations and/or types of surfactants, and there is no set rule as to what will work. Furthermore, even if the orientation distribution is altered, there is a chance that it remains pathological, but in a *new* way. The experimentalist then needs to try further approaches on a case-by-case basis. Finally, a surfactant also has the potential to negatively impact the integrity of the sample, which needs be taken into consideration.

The addition of a substrate support to cryo-EM grids may also alter the orientation distribution by providing a separate, sticky interface that detracts from the AWI. The support film will also provide a surface with which to concentrate the sample. Thin carbon films have traditionally been employed for this purpose (Stolinski and Gross 1969), while other supports such as graphene oxide (Pantelic *et al* 2010, Palovcak *et al* 2018) and graphene (D'Imprima *et al* 2019, Fan *et al* 2019, Pantelic *et al* 2012, Naydenova *et al* 2019) have gained popularity in recent years. When adhered to a support, the specimen also has an opportunity to avoid the AWI. The orientation of the specimen on the support film is often distinct from its orientation at the AWI. One can coat the support film with charged polymers such as poly-lysine, poly-arginine, poly-glutamate, or poly-aspartate, which can further alter the orientation bias on the surface (Chowdhury *et al* 2017) or other polymers such as polyethyleneimine (Patel *et al* 2018). Glow discharging using different ionic charges may induce distinct surface properties and also alter the orientation distribution. As above, none of these methods are generalizable solutions and usually require trial and error. The resulting orientation can be more problematic than helpful. For example, the same ribosome sample behaves very differently within open holes (presumably adhered to the AWI) compared to a carbon substrate support (Naydenova and Russo 2017). The former produces a high-resolution reconstruction, whereas the latter leads to a severely stretched and anisotropic low-resolution reconstruction. As with surfactants, the addition of supports may alter the surface binding properties and modify how much protein is denatured.

There has been increasing interest in monolayers and affinity tags on surface supports. For example, functionalization of graphene oxide coated grids with amino groups, with an optional polyethylene glycol spacer, helps to keep the particles away from both the graphene/water interface and the AWI (Wang *et al* 2020). This helps to protect the sample from denaturation and may also alter the orientation distribution. In variations of the above approach a highly specific, covalent affinity tag is appended to the polyethylene glycol chain, which additionally enables sample purification (Wang *et al* 2019). In another approach a monolayer of lipids is applied to a grid and is then coated with an additional layer of streptavidin; if surface-exposed lysines of the protein are chemically biotinylated (presumably with random labeling efficiency), the particle can adapt different orientations on the support film (Han *et al* 2012). Indeed, this was shown to increase the orientation distribution of thermosomes from *Sulfolobus solfataricus*, a complex that adopted highly preferential orientations over conventional grids (Han *et al* 2012). It also resulted in a more uniform orientation for the yeast RNA Pol II elongation complex, although the distribution was not completely random and was correlated to the extent of surface-exposed lysine residues (Lahiri *et al* 2019). These methods appear to pull particles away from the air–water interface and could provide an additional means by which to concentrate and/or purify rare proteins of interest. The drawbacks are the requirement for specialized grids and an increased thickness from the addition of extra layers of material. In the case of streptavidin-coated grids, an additional computational step to subtract the streptavidin diffraction pattern from the images is required. These new approaches are exciting, but have yet to be fully tested with a wide variety of specimens.

A simple and generalizable solution to address the effects of non-uniform sampling caused by preferred orientation is to tilt the grid during data collection. Due to the geometric configuration of the preferentially oriented sample with respect to the electron beam, for each cluster of orientations, tilting the grid generates a cone of illumination angles with respect to the cluster centroid, with half angle given by the tilt (Lyumkis 2019, Tan *et al* 2017, Baldwin and Lyumkis 2020). Tilting spreads out and improves the orientation distribution of the imaged object, which in turn improves Fourier space sampling (there are some exceptions to the case with high-angle tilts that will not be discussed here) (Baldwin and Lyumkis 2021). Numerous successful case studies have benefited from the approach (Passos *et al* 2020, Tan *et al* 2017, Haack *et al* 2019) (figure 4.24(B)), including samples as small as 40 kDa (Herzik *et al* 2019) and as large as the ~100 MDa nuclear pore complex (Zhang *et al*

2020, Huang *et al* 2020). An important consideration is that the sample concentration needs to be optimized for the tilt angle and thus should typically be slightly lower than the concentration used for data collection in the absence of stage tilt. Otherwise, tilting the grid is a simple and generalizable solution to the preferred orientation problem and does not require any additional work other than to set the stage tilt angle during data collection; all subsequent steps can be performed as normal (a complete protocol is available alongside the original publication (Tan *et al* 2017)).

There are several drawbacks to tilting the grid, however. There is generally more beam-induced movement in the images (Zheng *et al* 2017, Brilot *et al* 2012, Campbell *et al* 2012, Wright *et al* 2006). For this reason we always suggest using gold grids (Russo and Passmore 2014), or other grids that reduce beam-induced movement. Tilting could also lead to larger errors during image processing, for example, when estimating the contrast transfer function. Both of the above problems may be correctable computationally. An unavoidable drawback is the increase in ice thickness. The exact effect of these drawbacks on experimental reconstructions is currently not known. Although tilting may improve the isotropy and the global resolution of the reconstruction, it is important to emphasize that the sample adherence to the AWI remains unchanged. Therefore, if there is any denaturation or aggregation, tilting will not solve *this* problem and a different method must be employed.

An important recent development was the introduction of robotic inkjet dispensing devices for sample deposition and vitrification. These were pioneered with the Spotiton system (Razinkov *et al* 2016, Jain *et al* 2012) (now commercialized by SPT Labtech as 'Chameleon'). The technology was born out of the idea that only picoliter amounts of sample should be required for the actual imaging experiment, i.e. multiple orders of magnitude less than what we conventionally use. The developers also found that by adjusting the time between applying the sample (spotting) and plunging the spotted sample into liquid ethane, more of the protein is randomly distributed within the ice layer as opposed to being adsorbed to the AWI (Noble *et al* 2018b). However, even with a spot-to-plunge time of 100 ms, it was not possible to completely eliminate adsorption. Although theoretical calculations suggest that particles hit the AWI several orders of magnitude times faster than the time it takes to plunge the sample (Naydenova and Russo 2017), not all of these events will necessarily lead to absorption. As noted earlier, the adsorption rate depends on many factors and can vary from protein to protein by up to several orders of magnitude (Tripp *et al* 1995). Further experiments need to be performed to deduce the exact behavior of different samples on grids characterized by distinct ice thicknesses and with variable spot-to-plunge times. One important consideration is that spotting requires having a more concentrated sample at hand. This is because the AWI effectively concentrates the sample, as will be explained next. When complexes are rare or difficult to purify, spotting sample over holes may not be an option, although spotting over support films should not be affected. There are now many successful examples of how spotting proteins leads to either an improved distribution within the ice and/or less sample denaturation (Liu *et al* 2019, Lee *et al*

2019, Scapin *et al* 2018, Tan *et al* 2020). By keeping the protein away from the AWI Spotiton kills two birds with one stone—the sample integrity is maintained and the distributions are better randomized within the ice layer.

Under some circumstances the AWI can actually be beneficial. As was indicated earlier proteins may be concentrated on the AWI, sometimes by order(s) of magnitude (Kudryashova *et al* 2003). The practical effects of utilizing, for example, surfactants to cover the AWI or using inkjet dispensing devices is an apparent dilution of the on-grid concentration. But this 'diluted' concentration simply represents the more natural concentration of proteins in solution. Adsorption to the AWI leads to lower *apparent* concentrations required to obtain the desired number of protein particles within a field of view. For example, we had one case where we were not able to purify sufficient amounts of material to vitrify using an inkjet dispensing device. However, the sample was highly concentrated using conventional blotting techniques. In this scenario, simply tilting the grid addressed the orientation bias problem (Passos *et al* 2020). Because we were interested in the nature of small molecule binding within the center of the complex, any molecular disruptions at the AWI could be ignored. We could obtain reconstructions resolved to ~2.6 Å in the region where the small molecules reside for numerous case studies and are now using this general workflow for drug studies.

The above-mentioned techniques need not be mutually exclusive. If the protein is not degraded, it is possible to combine these approaches and/or combine data from different orientation distributions (Campbell *et al* 2020). However, care must be taken in the interpretation of such data to ensure that the sample was not affected by any individual method.

4.6.5 Areas of research

Preferred orientation continues to be a central and prevalent issue in the field, and its effects are incompletely understood. At the level of the specimen we need to better understand the prevalence and timescale of protein adherence to the AWI, the extent of protein denaturation, and the general behavior of specimens within the types of thin, aqueous film environments that are characteristic of cryo-EM grids. A comprehensive understanding of these (and related) issues will lead to a better appreciation of the implications of novel specimen vitrification devices, such as Spotiton (Razinkov *et al* 2016, Jain *et al* 2012). At the level of the reconstructed object we need to better understand the effects at a theoretical and experimental level of both anisotropic resolution and attenuation of global resolution.

We do not have a clear understanding of what happens to specimens within thin aqueous films and the extent to which adherence to the AWI damages the sample. What *is* clear is that the phenomenon is ubiquitous and so are the consequential detrimental effects. It would be useful to address the following questions: What are the time scales within which specimens adsorb to the AWI? How is adsorption, on a single-molecule level, affected by the (bio)chemical properties of the specimen? By buffer constituents? By surfactants? To what extent will it be possible to 'out-run' adsorption using short spot-to-plunge times on inkjet dispensing devices? If it is impossible to completely out-run adsorption, how close can we get, and what are the practical implications? Will novel surface supports (e.g. graphene) or surface treatments (e.g. functionalized graphene oxide) become widely adopted? Given that a hydrophobic graphene/water interface may remain (in the absence of functionalization), will it affect protein integrity? Preferred orientation? Anisotropy? What is the best strategy for minimizing adherence to the AWI? Will it involve surfactants, modified surfaces, novel vitrification devices, or combinations thereof?

For tilted data collection strategies it is necessary to experimentally define the true limitations imposed by factors associated with stage tilt, in practice. It may be possible to overcome some detrimental effects using computational methods, whereas increased ice thickness (and thus background noise) is unavoidable. An improvement in global resolution through improvements to sampling (as measured by the SCF) may be offset by the attenuation of information caused by decrements in image quality. An optimal tilt angle must therefore exist for every dataset characterized by a non-uniform orientation distribution. The correct approach to determining this requires taking into account experimental data. Further work needs to be done to address the experimental attenuation of resolution at tilt, and the interplay between the compensatory improvements by virtue of improving Fourier space sampling. An automated approach to identify an optimal tilt angle once a user begins to collect data, based on both theoretical and experimental considerations, would be beneficial.

We also do not have a complete understanding of how sampling affects resolution and the practical consequences on resolution attenuation. Numerous questions remain unanswered: How often are unsampled regions in the transform encountered in experimental cryo-EM reconstructions? How and to what extent do false positive orientation assignments elevate the SCF? If they occur within unsampled regions, what is the practical effect on resolution? To what extent is global resolution attenuated within experimental cryo-EM datasets? Addressing these questions will lead to a better understanding of the effects of non-uniform sampling caused by preferred orientation.

Most currently employed methods do not (completely) eliminate the preferred orientation problem. Rather than patching up the problem we need to get at the heart of the matter. Inkjet dispensing devices seem to offer the clearest path toward this goal. It will be interesting to see how further improvements using short spot-to-plunge times (Noble *et al* 2018b), perhaps in combination with novel substrate supports (D'Imprima *et al* 2019, Fan *et al* 2019, Pantelic *et al* 2012, Naydenova *et al* 2019), will improve sample distributions on grids. The gold standard must be to eliminate the preferred orientation problem altogether. There would be multiple benefits. We would dramatically decrease the size of cryo-EM datasets, since the absence of denaturation would lead to fewer discarded particles. Reconstructions would be characterized by isotropic resolution, since we will record full angular coverage from truly random orientation distributions within the ice layer. We would maximize the global resolution of reconstructed objects, since the geometrical SCF factor would be near unity. In turn, we can expect to see indirect benefits, such as being able to resolve more distinct structural states of heterogeneous mixtures of populations. While there are many ways to currently address the preferred orientation problem, its elimination needs to be the ultimate goal.

Acknowledgments

I would like to thank Mykhailo Kopylov and Bridget Carragher for providing figures for bovine thyroglobulin, Gennavieve Gray for help with figure preparation, and Sriram Aiyer, Philip Baldwin, and Bob Glaeser for critical reading and comments on the early drafts. Molecular graphics and analyses were performed with the USCF Chimera package (supported by NIH P41 GM103311). This work was supported by NIH grants R01 AI136680, R01 AI146017, U54 AI150472, NSF Award #1933864, and the Margaret T Morris Foundation.

4.7 B factors and map sharpening

Alexis Rohou
Genentech, 1 DNA Way, South San Francisco, CA 94080, USA

4.7.1 An ideal 3D reconstruction has a predictable radial amplitude spectrum

The ideal three-dimensional reconstruction of a macromolecule would be a map of its Coulomb potential at every point. The radially averaged amplitude spectrum, calculated from the Fourier transform (commonly referred to as the structure factor, *F*) of such an ideal map has predictable features. Its peak is expected at the origin and the low-resolution part of the spectrum (from 0 to ~0.1 Å^{-1}, corresponding to spacings greater than 10 Å) consists of an approximately exponential decay with modulations corresponding to the shape of the macromolecular envelope. At higher spatial frequencies (approximately 0.1 Å^{-1} and beyond), a uniformly random distribution of atom positions in space is a good approximation of most protein structures, leading to the expectation of an almost 'flat' (or 'white noise'-like) amplitude spectrum (figure 4.25).

4.7.2 Actual 3D reconstructions feature dampened amplitudes at high frequencies

At any given resolution a three-dimensional reconstruction obtained by single-particle averaging is often much blurrier (its features appear 'less sharp') than an ideal map of the macromolecule at that resolution, such as might be generated by a simulation from an atomic model.

In real space, this manifests as loss of resolvability of fine features. In some cases, say in maps at ~4.5 Å resolution, alpha helices may appear as almost-smooth tubular densities, lacking the helical grooves expected at that resolution and slight 'bumps' only where well-resolved bulky side chains would be expected at that resolution (see also section 6.4). In other cases, say maps at ~2.5 Å, the blurring effect will be less severe and it might lead to the aryl groups of aromatic side chains being resolved as flat continuous density rather than as holey rings, for example (figure 4.26).

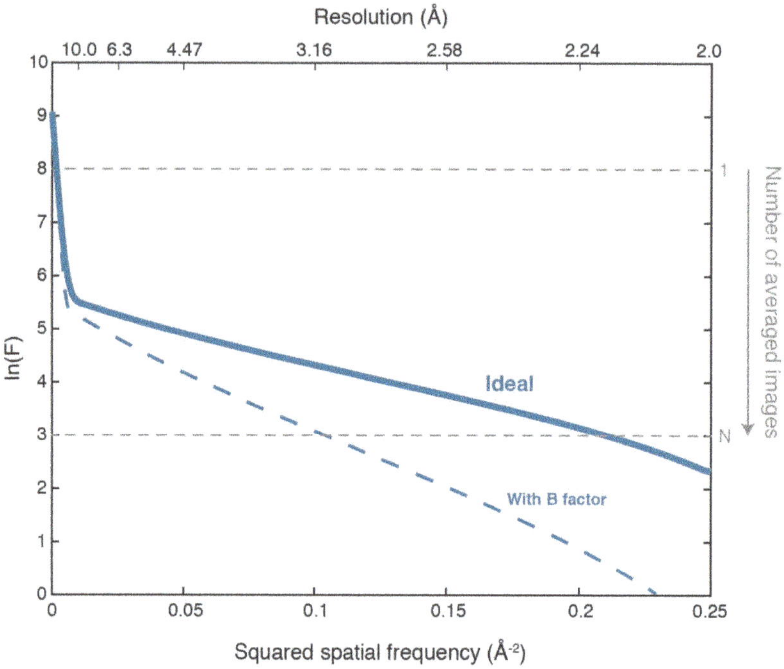

Figure 4.25. Schematic radial amplitude profile of an approximately 500 kDa protein. Solid line: ideal amplitude spectrum, based on the Guinier approximation at very low resolutions (>10 Å) and Wilson statistics (which assume random positions of atoms) at higher resolutions. This simplification ignores the effect of secondary-structure features, which, for some proteins, may significantly affect the spectrum up to ~4.5 Å. The overall decay in the Wilson regime is due to the shape of the atomic scattering factors used to simulate this curve. Dashed line: experimental imperfections result in attenuation of the amplitude spectrum, here simulated with a B factor of 50 $Å^2$, of approximately 3.3-fold at 3.16 Å, 12-fold at 2.24 Å, and 6000-fold at 1.2 Å (not shown). Gray dashed line: noise floor, drawn as a constant under the assumption that shot noise is the dominant source of noise. The effect of averaging during image processing is to lower the constant noise floor gradually, as a function of the number of images (right-hand side). As a first approximation, the final resolution of the reconstruction is the point at which the B factor-attenuated amplitude profile reaches the noise floor. In the case depicted above, a resolution of ~3.0 Å is reached, but the same number of particles in the absence of the experimental B factor would have led to a ~2.1 Å reconstruction. Experimentally tracking the attained resolution as a function of N can thus yield an estimate of the experimental B factor—see figure 4.26. Figure adapted from Rosenthal and Henderson (2003), copyright (2003), with permission from Elsevier.

We will explore the reasons for such 'blurriness' later in this section and in other parts of the book. Let us first consider how these effects can be measured and quantified.

To do so it is helpful to consider the 3D Fourier transform of the reconstructed map. Calculating and plotting the radial average of its complex amplitudes shows that the recovered molecular structure factors at, say, 4.5 Å are weaker than would be expected. In fact this loss of contrast becomes worse with higher spatial frequencies; rather than the expected slowly decaying spectrum (beyond ~0.1 $Å^{-1}$), one typically observes a steeper decay of amplitudes as a function of spatial frequency.

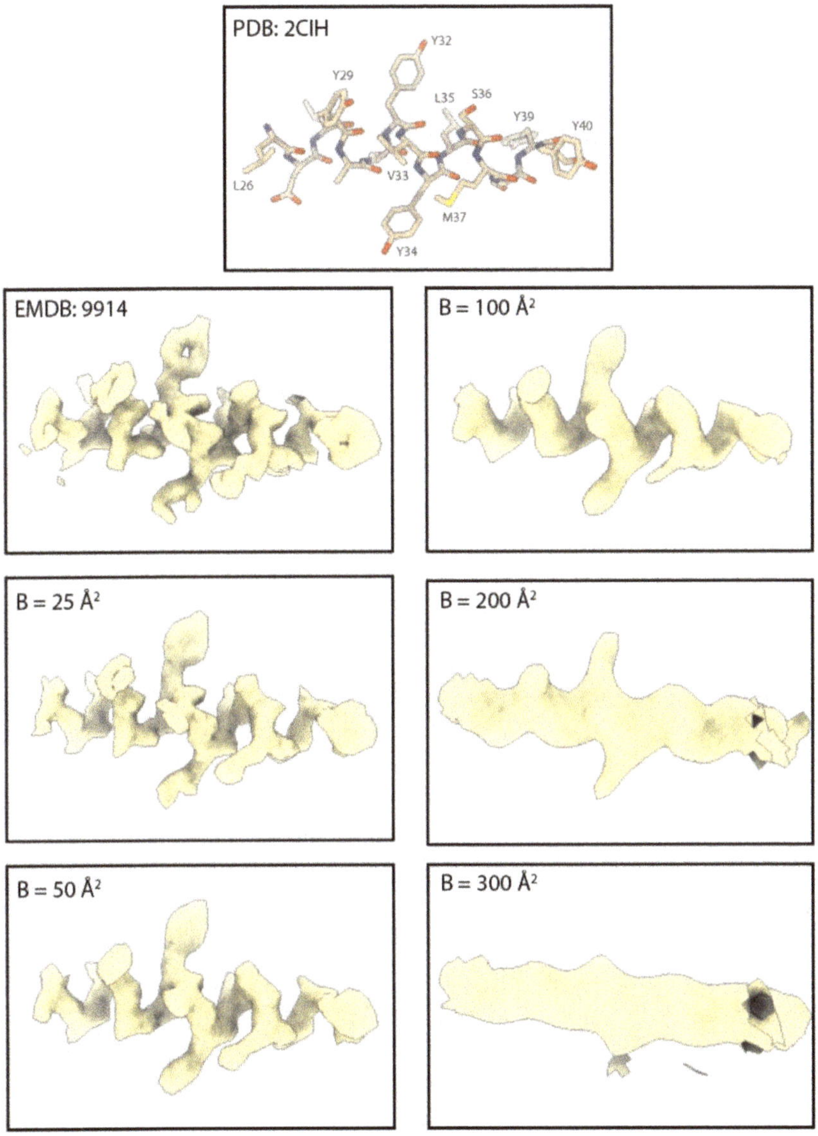

Figure 4.26. The effect of B factors on a 2.0 Å map. A tyrosine-rich alpha-helical stretch of 14 residues was selected (L26-Y40) from the structure of mouse heavy-chain apoferritin, and the corresponding region of a 2.0 Å cryo-EM map (Danev *et al* 2019) was segmented and is shown in isosurface rendering (top left). Increasing B factors were applied. At 50 Å^2 the holes in the tyrosine side chain arenes and the backbone carbonyls are no longer resolved; at 200 Å^2 the pitch of the alpha helix is barely resolved, and only the stubs of tyrosines 29, 32, and 34 are still visible—other slide chains are no longer visible. At 300 Å^2 all side chains, as well as the alpha-helical pitch, appear lost. Note, however, that the resolution of the map is unchanged and that restoration of all fine features may be achieved from any of the filtered maps in this simulation by applying the appropriate inverse Gaussian filter, a sharpening operation. Figure prepared with ChimeraX (Goddard *et al* 2018).

4.7.3 Several factors contribute to signal decay at high frequencies

Loss of high-resolution scattering amplitude can arise during imaging at the microscope, or later on during image processing.

During microscopy, any optical imperfection of the instrument (e.g. energy spread of the electron source, lens or specimen-stage instabilities) can dampen signal amplitudes in a resolution-dependent manner, affecting high frequencies more, as can specimen movement or charging, and radiation damage, among others.

Separately, one must expect some loss of contrast during the averaging step of the 3D reconstruction process. This is because the registration of noisy images with respect to each other is never error-free and thus the averaging is never perfectly coherent. Broadly speaking, the lower the signal-to-noise ratio of the experimental images, the greater the inaccuracies in their estimated positions and orientations, and the more incoherent the averaging into a final 3D reconstruction. For this reason any additional noise in the images, for example added solvent noise due to increased ice thickness or noise added by the electron detector, also leads to dampening the high-frequency amplitudes of the final, averaged 3D reconstruction.

Thus, the signal in the final 3D reconstruction is dampened at high frequencies not only because the original images themselves had dampened high-frequency amplitudes, but also because noise in the images leads to incoherent averaging and 'blurring out' of the features during 3D reconstruction.

The two types of effects are in fact entangled and often synergistic. Shot noise, the main source of noise in cryo-EM experiments, produces a flat (white spectrum) noise floor with constant power across all frequencies. Any signal loss in the high frequencies during image formation therefore also decreases the signal-to-noise ratio of the images, leading to a compounding reduction of high-frequency amplitudes in the final 3D reconstruction due to increased errors during image processing.

4.7.4 Gaussian falloff, parametrized by a B factor, is a useful model of signal loss

A useful way to quantify signal loss is to derive or measure attenuation envelopes, which describe, as a function of spatial frequency, the scaling factor applied to the signal. For example, an envelope value of 0.2 represents an 80% loss of signal at that frequency, meaning that the amplitude of the signal is multiplied by 0.2. Using this convention, the degradation of signal during imaging and averaging can be tracked, in principle, by successively multiplying the original signal's amplitude spectrum by a series of envelopes, each corresponding to one of the attenuations mentioned above. The aggregate envelope on the final 3D reconstruction is then the product of all the individual envelopes, encompassing everything from imperfections of the electron source to incoherent averaging during image processing.

What is the shape of these envelopes? Some envelope functions can be measured experimentally. For example, the envelope corresponding to optical imperfections in the imaging system can be measured using crystalline specimens as the ratio of amplitudes in the Fourier transform of an image to those in a diffraction pattern. Other envelopes, such as those associated with alignment errors during image processing, can be derived from first principles. As it turns out, a number of signal attenuation envelopes (for example, those due to optical image deterioration, or translational and rotational alignment errors) have the form of a Gaussian function of s, the spatial frequency: $E(s) = e^{-\frac{1}{4}Bs^2}$. This is analogous to the temperature factors used in x-ray crystallography to describe the effect of thermal displacements of atoms from their ideal crystalline position. The notation B, and the common usage of the phrase 'B factor' (units of $Å^2$) in cryo-EM, are thus used to describe these envelopes, even when they do not arise from thermal motions.

The use of Gaussian envelope functions and B factors has in fact been generalized in practice to measure total signal attenuation, even when some of the sources of attenuation cannot be described as being Gaussian functions since they can also vary as a function of higher powers of s. This approximation is largely justified on empirical grounds, because in many cases the radial amplitude profile of a 3D reconstruction decays approximately linearly when plotted on a log scale as a function of s^2 (figure 4.25). This suggests that the Gaussian approximation is often sufficient when considering the total attenuation envelope.

The simplification to a single B factor has numerous practical advantages, not least of which is the possibility of reporting, with a single parameter value, the magnitude of all imperfections and errors in processing that affect a cryo-EM experiment. In many cases, the B factor of a 3D reconstruction obtained from cryo-EM data gives a snapshot of the overall quality of the experimental data, its processing, and the final result.

Finally, this simple formulation allows us to make useful predictions, for example, about the number of particles which might be needed to reach a target resolution, given a set of experimental conditions. We will explore this in the next few sections.

4.7.5 Estimating B factors

The most common way to estimate the overall B factor of a cryo-EM experiment is to plot the log of the radial amplitude spectrum of the final 3D reconstruction as a function of s^2 and to then fit a linear function to this spectrum between ~0.1 $Å^{-1}$ and the frequency corresponding to the resolution of the map.

Another way to estimate the overall B factor is to compute many 3D reconstructions from the same experimental dataset, each time with a different number of particles N, and keep a record of the resolution reached by each reconstruction. Plotting the squared spatial frequency (s^2) corresponding to that resolution as a function of the logarithm of N, a graph called the 'ResLog plot' or 'Rosenthal–Henderson plot' (figure 11 in Rosenthal and Henderson 2003, Stagg *et al* 2014), yields a straight line whose slope is also referred to as the B factor (figure 4.27, left).

The two measurements are related. On a radial amplitude plot (figure 4.25), the averaging of aligned particle images can be depicted as a gradual lowering of the noise floor (dashed gray line). As more particles are averaged the point at which the B-factor-attenuated signal (dashed blue line) emerges over this noise floor moves to higher frequencies. This improvement in resolution as a function of the number of averaged images is depicted in the ResLog plot (figure 4.27, left). Attenuation by

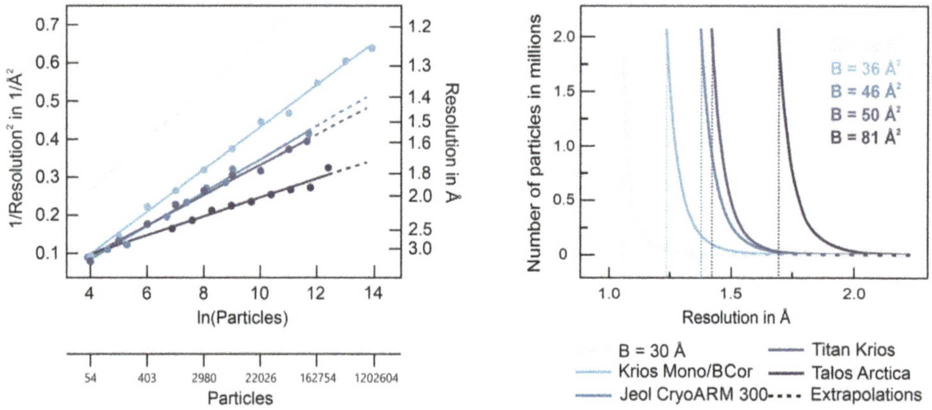

Figure 4.27. ResLog plots are useful predictors of the number of particle images required to achieve a desired map resolution. Left: ResLog plots obtained from four datasets of images of apoferritin collected on different microscopes. For each dataset, increasing numbers of particles are included in the 3D reconstruction, leading to maps of increasing resolution, each of which is represented by a dot on a plot of inverse squared resolution as a function of the logarithm of the number of particles contributing to the reconstruction. The B factor of each dataset can be computed from the slope of the linear fit to each dataset. Steeper fits, corresponding to smaller B factors, imply that fewer particles are required to reach a target resolution. Right: For practical purposes, a given B factor limits the attainable resolution of a reconstruction because of the logarithmic nature of the averaging process when Gaussian attenuation envelopes affect the averaged signal. For example, a dataset with a B factor of 81 Å^2 limits the attainable resolution to approximately 1.7 Å. Note that B factors originating from errors in image processing, protein dynamics, and other sources combine with those associated with instrumental imperfections to affect the final resolution. Figure reproduced with permission from Yip *et al* (2020). Copyright the authors, CC-BY-ND 4.0

larger B factors give amplitude spectra with steeper decays, so that the noise floor must be dropped 'lower' (i.e. more averaging is required) to attain an equivalent resolution. Correspondingly, higher B factors manifest as shallower ResLog plots (figure 4.27, left, dark blue curves).

Perhaps because of the computational cost associated with computing a ResLog plot, results from this method are seldom reported, even though they can be highly informative, for example, in making predictions of the number of particles that would be required to attain a desired resolution (figure 4.27, right). In contrast, the first method, which only requires a single reconstruction and its radial amplitude profile, is used routinely as the first step in a common protocol to correct for signal attenuation in the 3D reconstruction. This protocol, described in the following subsection, involves normalization, i.e. division, by a Gaussian function whose parameter is often reported as the 'sharpening' or 'visualization' *B* factor.

4.7.6 Sharpening a map

As mentioned at the beginning of this section, signal attenuation in the high frequencies manifests as loss of fine features in the real-space map. For example, side chains may not be visible or as 'well resolved' as expected at the map's resolution. Generally speaking, our capacity to interpret the map and extract biological meaning from it is hampered by this high-frequency attenuation.

To improve map interpretability and the accuracy of atomic models built into maps, one may perform filtering operations in Fourier space. For example, if the total attenuation envelope $E(s)$ is 0.2, multiplying the Fourier amplitudes of our map at frequency s by 5.0 will restore the power of our map to its ideal value, simulating a perfect experiment free of any attenuation. Because the output map does not look as 'blurry' anymore, this post-processing operation is commonly called 'map sharpening'.

Once the overall B factor has been estimated by inspection of the radial amplitude spectrum, the sharpening step consists in scaling each Fourier component of the map in reciprocal space by the inverse of the estimated attenuation Gaussian (mathematically, applying a negative B factor), leading to an approximately constant amplitude spectrum.

It may seem paradoxical, given the improved resolvability of real-space features after map sharpening, that the resolution of a map is actually unchanged by this filtering operation, but this is because map resolution is defined as a function of spectral signal-to-noise ratio (see section 5.2.2), which is left unchanged by Fourier filters.

At spatial frequencies that are dominated by noise (those beyond the resolution of the map), applying this type of filter would again leave the spectral signal-to-noise ratio unchanged, but increase the relative power of noise in the final real-space map. This deteriorates the interpretability of the map and it is often referred to as 'over-sharpening'. To prevent this, map sharpening programs typically apply a combination of the inverse Gaussian filter with a low-pass filter to ensure that frequency bands dominated by noise are attenuated and do not contribute to the final real-space map. These low-pass filters can either have a predefined functional form (e.g. a cosine-edged filter) or be derived from the map's FSC curve (itself an estimate of the map's spectral SNR, see section 5.2). In the latter case, this filter is called a figure of merit (FOM) filter (Rosenthal and Henderson 2003).

The final result of a sharpening operation should be a map whose amplitude spectrum is approximately flat from ~0.1 $Å^{-1}$ to the map's resolution, at which point it decays very rapidly to zero, and whose real-space interpretability is maximized.

4.7.7 A single inverse Gaussian filter using a global B factor does not always lead to the optimal map

Modeling the effects of instrumental and data processing imperfections as imposing a global Gaussian filter on the final 3D reconstruction remains to date the most common assumption used in map sharpening programs. Several alternative algorithms have been proposed and shown to lead to maps with improved features, however, suggesting that this approximation may not always be sufficiently accurate. This might be because some of the envelope functions (for example, those due to partial temporal coherence, or to defocus estimation errors) are functions of powers of s higher than s^2, so that the global signal decay envelope might not be quite Gaussian. In addition, the assumption that the amplitude spectrum is flat beyond ~0.1 Å^{-1} may not always hold. A major reason the Gaussian assumption might fail is that the 'blurriness' of maps varies locally, as a function of the flexibility of different domains, of secondary-structure elements, or of individual side chains (see section 5.2.4 on local resolution). A global Fourier filter of any kind (Gaussian or otherwise) cannot completely correct for such local variations.

Several algorithms have been proposed to overcome these limitations. For example, non-Gaussian dampening global envelopes can be corrected for by flattening the final spectrum without any functional fitting (Grant *et al* 2018) and a number of approaches can be used to sharpen local areas within maps to different extents so as to maximize local map interpretability. In one approach a number of alternative maps, each sharpened with a different B factor, are generated. Empirically, for a given region of the map one may use the strongest sharpening which still leaves map features connected with each other, since over-sharpening leads to the map turning to small, disconnected densities. One implementation formalizes this heuristic and automates the process to find the B factor which (globally or locally) maximizes details in the map while maintaining map connectivity (Terwilliger *et al* 2018). Another approach consists of estimating the local resolution (see section 5.2.4) and then filtering local areas of the sharpened map according to the estimated resolution (e.g. Cardone *et al* 2013, Ramírez-Aportela *et al* 2019) or using other denoising algorithms (e.g. Ramlaul *et al* 2019). Finally, the assumption that a flat spectrum should be expected can be dropped when an atomic model is available and methods are available to sharpen local areas of maps to match the local amplitude profile expected given the model (Jakobi *et al* 2017).

Related to all of the above in that it leads to improved map interpretability, but quite distinct in its theoretical underpinnings, density modification has been adopted from x-ray crystallography to the post-processing of cryo-EM maps (Terwilliger *et al* 2020). In this approach Fourier coefficients are modified to improve the likelihood of the map, as estimated by how closely the distributions of real-space map values inside and outside the macromolecule agree with expectations. In practice, this also leads to improved resolvability of features and the appearance that local filtering has been performed, although the Fourier coefficients need not have been modified in a manner similar to Fourier filters, Gaussian or otherwise, since their phases as well as their amplitudes are modified in the process.

In summary, many methods are now available, which are likely to yield better 3D maps than those obtained by global sharpening using a simple Gaussian filter. However, the use of global Gaussian envelope fitting via a single B factor remains very widespread because the B factor is thought of as a useful metric of the overall quality of a dataset and of the processes leading to the final 3D reconstruction.

Notably, the prediction that B factors estimated from amplitude spectra and those estimated from ResLog plots should agree is rarely confirmed experimentally. Where both values have been reported, the ResLog-estimated B factor has tended to be significantly greater than that estimated from the map's radial amplitude spectrum (table 4.2), suggesting caution might be warranted when using B factors estimated from amplitude decay plots for purposes other than map sharpening.

Table 4.2. Examples of reported *B* factors.

	B factor (Å^2)	Estimation method	Reference
Signal recovered with translational frame alignment for motion correction	160	Ratio of amplitude spectra after/before motion correction	Campbell *et al* (2012)
Signal loss due to radiation damage for every additional exposure of 1 electron /Å^2	4	Heuristic model	Bharat *et al* (2015)
E. coli β-galactosidase at 3.2 Å	85	Amplitude spectrum	Bartesaghi *et al* (2014)
E. coli β-galactosidase with PETG at 2.2 Å	75	Amplitude spectrum	Bartesaghi *et al* (2015)
Mouse apoferritin at 1.54 Å	46.9	ResLog	Kato *et al* (2019)
Human apoferritin at 1.65 Å	66	ResLog	Zivanov *et al* (2018)
Human apoferritin at 2.32 Å	45	Amplitude spectrum	Tegunov *et al* (2020) and pers. comm. from D Tegunov
In situ bacterial 70S ribosomal subunit at 3.7 Å	5 86	Amplitude spectrum ResLog	Tegunov *et al* (2020) and pers. comm. from D Tegunov
Mouse apoferritin at 1.22 Å	22.3 32.5	Amplitude spectrum ResLog	Nakane *et al* (2020)
Human apoferritin at 1.25 Å	36	ResLog	Yip *et al* (2020)
E. coli β-galactosidase at 1.8 Å	26	Amplitude spectrum	Merk *et al* (2020)
Urease at 1.98 Å	38 71.5	Amplitude spectrum ResLog	Righetto *et al* (2020) and pers. comm. from R Righetto
Apoferritin at 1.96 Å	50 114.3	Amplitude spectrum ResLog	Pers. comm. from R Righetto
Mycobacterial secretion system ESX-3 dimer at 3.8 Å	101 144	Amplitude spectrum ResLog	Famelis *et al* (2019) and pers. comm. from A Rivera Calzada

4.8 Optical aberrations and Ewald sphere curvature

Jasenko Zivanov[1,2]*, Takanori Nakane*[1] *and Sjors H W Scheres*[1]
[1]*MRC Laboratory of Molecular Biology, Cambridge, UK*
[2]*Current address: Laboratory of Biological Electron Microscopy, EPFL, Lausanne, Switzerland*

In this section, we will first examine the aberrations of the optical system more closely. We will begin with the relevant theory, followed by a description of the different types of aberrations, and a discussion of the effects of aberrations on cryo-EM structure determination. Then, we will also have a closer look at what happens when one of the earliest assumptions in section 1.2, that the object to be reconstructed is a thin one, no longer holds.

4.8.1 Further considerations on the aberration function $\gamma(\vec{s})$

In equation (1.4) in section 1.2 we encountered the aberration function $\gamma(\vec{s})$ that describes the phase delay of the electron as a function of 2D spatial frequency $\vec{s} = \{s_x, s_y\}$. There $\gamma(\vec{s})$ was introduced in a very specific form that corresponds to an idealized optical system. It consisted of only two terms: a quadratic term representing defocus and a fourth-order term representing the spherical aberration. The resulting $\gamma(\vec{s})$ was thus restricted to a small two-dimensional space of functions and it was, in particular, strictly *symmetrical*, i.e. $\gamma(\vec{s}) = \gamma(-\vec{s})$. The aberrations encountered in a real microscope, however, are not only potentially asymmetrical, but they also form a much more diverse space of functions. In the following, we will first elaborate on the effects of an arbitrary aberration function $\gamma(\vec{s})$ on the reconstructed image, and then we will describe the common types of aberrations and how they can be treated.

As the image wave propagates from the sample to the detector it is delayed by the aberration function $\gamma(\vec{s})$. Once it reaches the detector it can be approximated in reciprocal space as a function of the two structure factors $F(\vec{s})$ and $F_\mu(\vec{s})$ introduced in section 1.2:

$$W(\vec{s}) = \delta(\vec{s}) - \mathrm{i}F(\vec{s})\exp(\mathrm{i}\gamma(\vec{s})) - F_\mu(\vec{s})\exp(\mathrm{i}\gamma(\vec{s})). \tag{4.11}$$

This gives rise to an image I observed in real space that approximates the probability density of capturing an electron at a given detector pixel. This probability is equal to the squared magnitude of W, so its Fourier transform $\mathcal{F}\{I\}$ is a convolution (denoted by the symbol $\circledast$) of W with itself:

$$\begin{aligned}
\mathcal{F}\{I\} &= \mathcal{F}[\mathcal{F}^{-1}[W(\vec{s})]\mathcal{F}^{-1\dagger}[W(\vec{s})]] \\
&= \mathcal{F}[\mathcal{F}^{-1}[W(\vec{s})]]\circledast\mathcal{F}[\mathcal{F}^{-1\dagger}[W(\vec{s})]] \\
&= W(\vec{s})\circledast W^\dagger(-\vec{s}) \\
&= (\delta(\vec{s}) - \mathrm{i}F(\vec{s})\exp(\mathrm{i}\gamma(\vec{s})) - F_\mu(\vec{s})\exp(\mathrm{i}\gamma(\vec{s}))) \\
&\quad \circledast(\delta(-\vec{s}) + \mathrm{i}F^\dagger(-\vec{s})\exp(-\mathrm{i}\gamma(-\vec{s})) - F_\mu^\dagger(-\vec{s})\exp(-\mathrm{i}\gamma(-\vec{s}))),
\end{aligned}$$

where the superscript † denotes complex conjugation. Because we are assuming a weak phase-and-amplitude approximation, we can neglect higher order terms of the structure factors. The convolution above then simplifies to a product:

$$\begin{aligned}
\mathcal{F}\{I\} \approx \delta(\vec{s}) &+ \mathrm{i}F^\dagger(-\vec{s})\exp(-\mathrm{i}\gamma(-\vec{s})) - F_\mu^\dagger(-\vec{s})\exp(-\mathrm{i}\gamma(-\vec{s})) \\
&- \mathrm{i}F(\vec{s})\exp(\mathrm{i}\gamma(\vec{s})) - F_\mu(\vec{s})\exp(\mathrm{i}\gamma(\vec{s})).
\end{aligned}$$

Since the structure factors are both Fourier transforms of real-valued fields, they obey Friedel's law ($F(-\vec{s}) = F^\dagger(\vec{s})$):

$$\begin{aligned}
\mathcal{F}\{I\} \approx \delta(\vec{s}) &+ \mathrm{i}F(\vec{s})\exp(-\mathrm{i}\gamma(-\vec{s})) - F_\mu(\vec{s})\exp(-\mathrm{i}\gamma(-\vec{s})) \\
&- \mathrm{i}F(\vec{s})\exp(\mathrm{i}\gamma(\vec{s})) - F_\mu(\vec{s})\exp(\mathrm{i}\gamma(\vec{s})).
\end{aligned}$$

If we were to assume $\gamma(\vec{s})$ to be symmetrical, this would reproduce equation (1.5) from section 1.2. Instead, we perform an approximation that is common in cryo-EM, but not in material science: we assume that the structure factors are proportional to each other and that their ratio is independent of frequency. Making this approximation is not strictly necessary at this point, but it will simplify the final equations of this chapter. The true relation between the two structure factors depends on the material composition of the sample and lies beyond the scope of this book.

If we assume that there is a real-valued constant ξ such that $F_\mu(\vec{s}) = \xi F(\vec{s})$ for all frequencies s, then we can express the addition of F_μ to $\mathrm{i}F$ as a rotation and scaling of $\mathrm{i}F$. Specifically, $\mathrm{i}F$ can be rotated by $\gamma_\mu = -a\tan(\xi)$ and scaled by $\alpha = \sqrt{1+\xi^2}$ in order to equal $\mathrm{i}F + F_\mu$. Our image equation then becomes:

$$\mathcal{F}\{I\} \approx \delta(\vec{s}) + \mathrm{i}\alpha F(\vec{s})\exp(-\mathrm{i}\gamma(-\vec{s}) - \gamma_\mu) - \mathrm{i}\alpha F(\vec{s})\exp(\mathrm{i}\gamma(\vec{s}) + \gamma_\mu). \tag{4.12}$$

Finally, we observe that every function $f(\vec{s})$ can be separated into a symmetrical component $f_s(\vec{s})$ and an antisymmetrical one, $f_a(\vec{s})$, such that $f(\vec{s}) = f_s(\vec{s}) + f_a(\vec{s})$ for all s, while $f_s(-\vec{s}) = f_s(\vec{s})$ and $f_a(-\vec{s}) = -f_a(\vec{s})$. We therefore separate the aberration function γ analogously:

$$\begin{aligned}\mathcal{F}\{I\} &\approx \delta(\vec{s}) + \mathrm{i}\alpha F(\vec{s})\exp(-\mathrm{i}\gamma_s(\vec{s}) + \mathrm{i}\gamma_a(\vec{s}) - \gamma_\mu)\\ &\quad - \mathrm{i}\alpha F(\vec{s})\exp(\mathrm{i}\gamma_s(\vec{s}) + \mathrm{i}\gamma_a(\vec{s}) + \gamma_\mu)\\ &= \delta(\vec{s}) + \mathrm{i}\alpha F(\vec{s})(\exp(-\mathrm{i}\gamma_s(\vec{s}) - \mathrm{i}\gamma_\mu) - \exp(\mathrm{i}\gamma_s(\vec{s}) + \mathrm{i}\gamma_\mu))\exp(\mathrm{i}\gamma_a(\vec{s}))\\ &= \delta(\vec{s}) + 2\alpha F(\vec{s})\sin(\gamma_s(\vec{s}) - \gamma_\mu)\exp(\mathrm{i}\gamma_a(\vec{s})).\end{aligned}$$

In this form we can see that the symmetrical and the antisymmetrical components of the aberration function have different effects on the observed image: while the symmetrical component only scales the structure factors of the sample, the antisymmetrical component shifts their phases. The symmetry assumption made in chapter 1—which is still widely used in cryo-EM—corresponds to a real-valued CTF equal to $\sin(\gamma_s(\vec{s}) - \gamma_\mu)$. If we, however, allow for asymmetry in γ, then the CTF becomes a complex-valued function—its amplitude is controlled by γ_s and its phase angle by γ_a.

One place where we have already encountered an asymmetrical γ was in the case of beam tilt in section 3.4. Although a tilted beam affects both components of γ, its effect on the antisymmetrical component, axial coma, is much more pronounced than its effect on the symmetrical one. Because the phase shift due to axial coma is, however, invisible in the power spectrum of an image, it is the much weaker symmetrical effect that is typically used to measure beam tilt (e.g. using a Zemlin tableaux). As a consequence of the difficulty to achieve perfect alignment, some residual beam tilt is often observed in cryo-EM datasets that can impair the high frequencies of the reconstruction. Modern image processing algorithms allow this to be estimated and corrected for as part of the SPA refinement process.

4.8.2 Common types of aberrations

Apart from the distinction between symmetrical and antisymmetrical components of the phase shift, γ, we can also distinguish the *order* of aberrations. Because we are discussing arbitrary real-valued functions on a disc, it is helpful to use a mathematical basis set of orthonormal functions. One commonly used basis set are the Zernike polynomials $Z_n^m(\vec{s})$ (Zernike 1934) (figure 4.28). Those are ordered in bands of ascending complexity (order, n) and each band contains one more function with a different index m than the previous one, starting with one function, $Z_0^0(\vec{s})$, for the zeroth order. The bands alternate between containing only symmetrical and only antisymmetrical functions, so that we can express γ_s and γ_a as linear combinations of different sets of polynomials. In the following, we will use Zernike polynomials to discuss a number of specific aberrations, or aberration components. Please note that these are not mutually exclusive and that arbitrary superpositions of all of these aberrations can be encountered.

The zeroth order Zernike polynomial $Z_0^0(\vec{s})$ is a constant and thus symmetrical. It describes the constant phase shift of the image, in the case when an ideal phase plate has been used. The first order contains two linear functions, $Z_1^{-1}(\vec{s})$ and $Z_1^1(\vec{s})$, and both are antisymmetrical. They describe a shift of the entire image. Although all

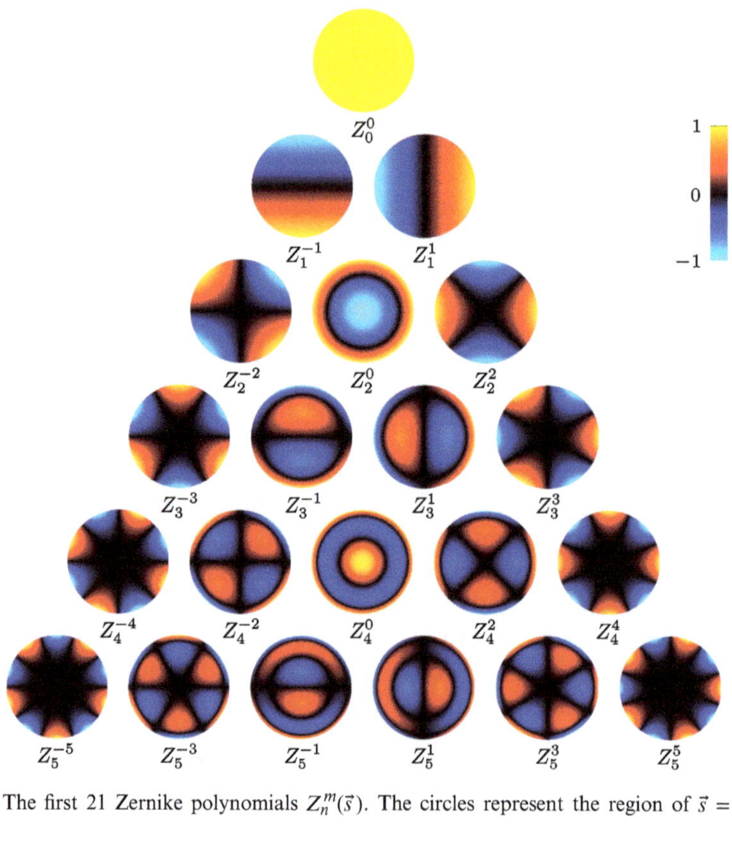

Figure 4.28. The first 21 Zernike polynomials $Z_n^m(\vec{s})$. The circles represent the region of $\vec{s} = \{s_x, s_y\}$ with $|\vec{s}| \leqslant 1$.

antisymmetrical aberrations shift the various Fourier components of the image, only linear combinations of these two functions (i.e. linear functions in 2D) shift all the waves in register, so that the image itself only appears to move as a whole, relative to a human observer.

The second order contains three symmetrical quadratic functions, $Z_2^{-2}(\vec{s})$, $Z_2^0(\vec{s})$, and $Z_2^2(\vec{s})$, that allow us to describe *astigmatism* and *defocus*. We have already encountered $Z_2^0(\vec{s})$ as the $|s|^2$ in the defocus term in equation (1.4) in section 1.2 (technically $|s|^2$ is only equal to $Z_2^0(\vec{s}) - \sqrt{3}$). Typically, we assume that the lenses of an electron microscope exhibit astigmatism, i.e. they do not always focus to the same extent in all image directions. This is usually modeled by extending equation (1.4) to include the other two quadratic polynomials. The space of aberrations spanned by all three polynomials allows for two arbitrary and independent defoci for two orthogonal directions at an arbitrary angle. In practice, such astigmatic defocus is usually not expressed in the Zernike basis, but instead by giving either the two extremal defocus values and one angle, or, equivalently, a real symmetrical 2×2 matrix A, such that the quadratic component of γ is equal to the quadratic form $\vec{s}^{A'} A\vec{s}$ (where the symbol A′ represents transposition).

The third order of Zernike polynomials contains the first non-trivial antisymmetrical aberrations: *axial coma*, corresponding to linear combinations of $Z_3^{-1}(\vec{s})$ and $Z_3^1(\vec{s})$, and *trefoil*, or *three-fold astigmatism*, corresponding to those of $Z_3^{-3}(\vec{s})$ and $Z_3^3(\vec{s})$. Axial coma produces a phase shift that grows with the third power along one specific direction, while leaving the phases on the line orthogonal to that direction unchanged. The result is a very strong shift in the high frequencies while the low ones remain mostly unaffected. Unnoticed beam tilt can therefore act as an impediment to reaching very high resolutions. Trefoil aberrations are also very common and they can also arise under beam tilt due to hexapolar lenses (Krivanek 1994). Under a trefoil aberration, the phase shift becomes positive three times and negative three times as one goes around the origin in Fourier space. This pattern is repeated in higher orders, as we encounter four-fold astigmatism in order 4 and five-fold astigmatism in order 5 and so on (figure 4.29).

The fourth order is again symmetrical and contains five polynomials. The center polynomial, $Z_4^0(\vec{s})$ is used to model *spherical aberration* and has also already appeared in equation (1.4) in section 1.2 as the s^4 term. Analogous to the case of Z_2^0, in order to obtain the term $|s|^4$ itself, we technically need to consider a linear combination of $Z_4^0(\vec{s})$, $Z_2^0(\vec{s})$, and $Z_0^0(\vec{s}) = 1$. The spherical aberration is almost always considered in cryo-EM, even if no other higher order aberrations are. This is necessary because considerable spherical aberrations are unavoidable when building microscopes. Although physical correctors do exist, they are expensive, while the extent of spherical aberration in a given microscope is very stable, so it can be corrected for during processing instead. The other type of fairly common fourth-order aberration is *four-fold astigmatism*, or *tetrafoil*, which can be expressed using $Z_4^{-4}(\vec{s})$ and $Z_4^4(\vec{s})$. Analogously to trefoil aberrations, here the phase shift becomes both positive and negative four times when going around the origin in a circle. Because four is an even number, however, this produces a symmetrical function. The

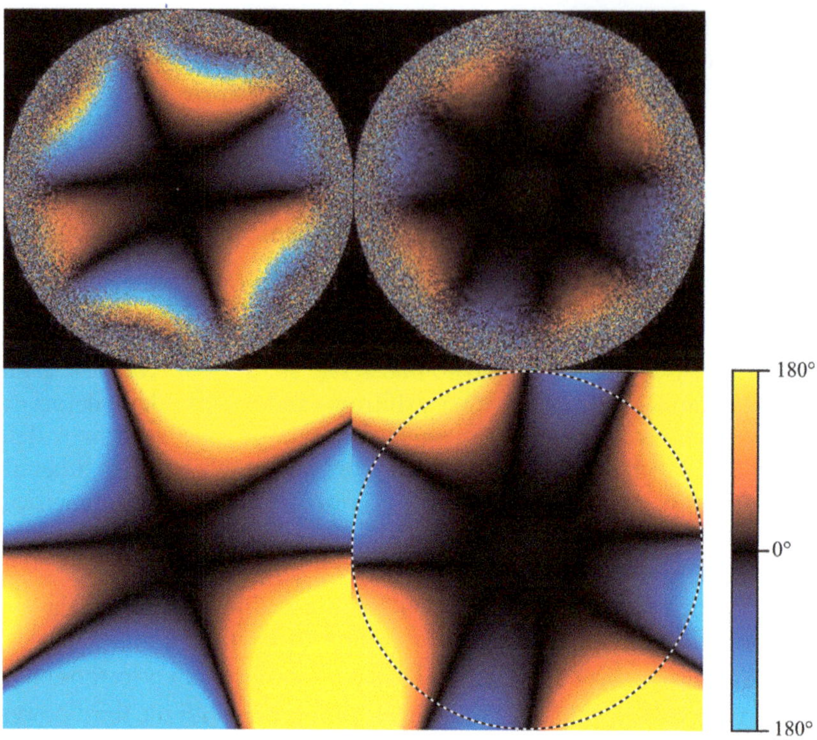

Figure 4.29. Examples of three-fold (left) and four-fold (right) astigmatism. The top images show the average phase shifts between projections of the reference reconstruction and the experimental images. Antisymmetrical aberrations are shown on the left; symmetrical aberrations on the right. The bottom images illustrate the corresponding models that consist of a linear combination of Zernike polynomials, as implemented in RELION. This image is reproduced from Zivanov *et al* (2020), CC BY 4.0. The dotted circle indicates the Nyquist frequency at 1.98 Å.

effect of a tetrafoil aberration is to slightly deform the rings of the CTF from an elliptical shape toward a square-like shape (figure 4.30). It is not uncommon, in fact, to see square-like Thon rings at high resolution, when the objective aperture is not perfectly clean, and it thus produces a charging effect.

Although the set of Zernike polynomials is infinite, the corresponding higher order aberrations are rarely considered in cryo-EM. Since, if they were present, they would become increasingly difficult to estimate, it is fortunate that their effects diminish rapidly as the order increases.

4.8.3 Practical considerations for aberration correction

Because symmetrical and antisymmetrical aberrations have very different effects on the resulting image, they also need to be estimated and corrected for in different ways. Specifically, the phase shift arising from antisymmetrical aberrations is invisible in the power spectrum of the image, so it cannot be measured without

Figure 4.30. Effects of symmetrical aberrations on the CTF. The image on the left shows a CTF expressed by the traditional model, while the one on the right shows the fit of a model that considers the higher order symmetrical aberrations that correspond to the bottom right image in figure 4.29. Note, in particular near the perimeter, that the four-fold astigmatism leads to a somewhat square-like shape of the Thon rings that cannot be expressed by the traditional model. This image is reproduced from Zivanov *et al* (2020), CC BY 4.0.

knowing the true phase angles of the structure factors, i.e. without prior knowledge about the sample.

As of RELION version 3.1 (Zivanov *et al* 2018) it has become possible to estimate the systematic phase shifts as part of a refinement procedure: once an initial estimate of the molecular structure has been obtained without correcting for aberrations, it can be used to predict the true phases of the structure factors. By comparing pairs of predicted and observed complex amplitudes over many particles, the systematic error associated with antisymmetrical aberrations can be inferred by averaging the phase discrepancy of each frequency.

Correction of such phase aberrations is possible in SPA, because the molecules are almost always observed under uniformly distributed in-plane rotation angles: while molecules often show a preferential orientation toward the water surface, their final rotation around the surface normal is always independent of the orientation of the pixel grid of the camera. Furthermore, every 3D structure factor of the reconstruction that we ultimately compute is a weighted average over certain 2D complex amplitudes of many particles. Because of the random in-plane rotation angles of the particles mentioned above, those source pixels are uniformly distributed along a circle in Fourier space. As a result, due to the antisymmetry of the aberration, the expectation value for the phase shift along that circle is always zero. An antisymmetrical aberration can therefore induce no systematic bias into the phase angles of the structure factors, so a reconstruction obtained without correcting for such aberrations can already be used as a ground truth for the estimation of phase shifts.

Once the phase shifts have been measured this way a parametric model, such as Zernike polynomials, is fitted to them. This allows us to extrapolate the estimated aberration into the higher frequencies of the image that could not be resolved in the initial reconstruction. Although antisymmetrical aberrations are more difficult to measure, they are trivial to correct. Once they are known their effect can be canceled out by applying the opposite phase shift to the observed image frequencies.

Symmetrical aberrations, on the other hand, can be measured more easily, but they are more challenging to correct for. Because they scale the complex magnitudes, they can be seen as Thon rings in the power spectra of micrographs. This typically only allows us to measure the first two orders of symmetrical aberrations: the defocus and astigmatism, as well as the extent of phase shift due to a uniform phase plate, if any. To estimate either higher order symmetrical aberrations or the defocus values of individual particles in a reliable way it is also necessary to perform a refinement procedure using a prior reference. Once the symmetrical aberrations are known they cannot be as easily cancelled out as can be done for the antisymmetrical ones. Instead, the correct magnitude of the CTF has to be considered throughout the refinement process.

When working with aberrations one important aspect to consider is the way in which the estimates influence each other. Wrong or missing estimates of higher order aberrations can compromise the estimates made of lower order aberrations—e.g. a wrong spherical-aberration assumption can produce a wrong defocus estimate—but only among either the symmetrical or the antisymmetrical aberrations, not between them. Incorrectly assumed antisymmetrical aberrations do not induce a bias in the estimates of the symmetrical ones, or vice versa.

One interesting fact to note is that, because image translations can also be expressed as order-1 antisymmetrical aberrations, antisymmetrical aberrations do interfere with determining the position of a particle. In particular, refining the particles without correcting for axial coma will systematically shift their positions to account for part of the coma-induced phase shift. Because the coma shifts the phases proportionally to the third power of the frequency, however, the high frequencies will be shifted insufficiently and the low ones excessively. When the coma is then itself estimated from those particles assuming the refined positions, the measured phase shift will contain a translational component running opposite to the coma (figure 4.31).

Although symmetrical aberrations do not interfere with determining the positions of molecules, they do depend on the scale of the image. Let the assumed pixel size be a times the true size. Then, the apparent spatial frequency $\vec{s}'$ is, in reality, equal to $\vec{s}/a$. If we substitute this into equation (1.4) in section 1.2,

$$\begin{aligned}\gamma(\vec{s}) &= 2\pi\left[\frac{C_s}{4}\lambda^3 s^4 - \frac{z}{2}\lambda s^2\right] \\ &= 2\pi\left[\frac{a^4 C_s}{4}\lambda^3 s'^4 - \frac{a^2 z}{2}\lambda s'^2\right] \\ &= 2\pi\left[\frac{C'_s}{4}\lambda^3 s'^4 - \frac{z'}{2}\lambda s'^2\right],\end{aligned} \tag{4.13}$$

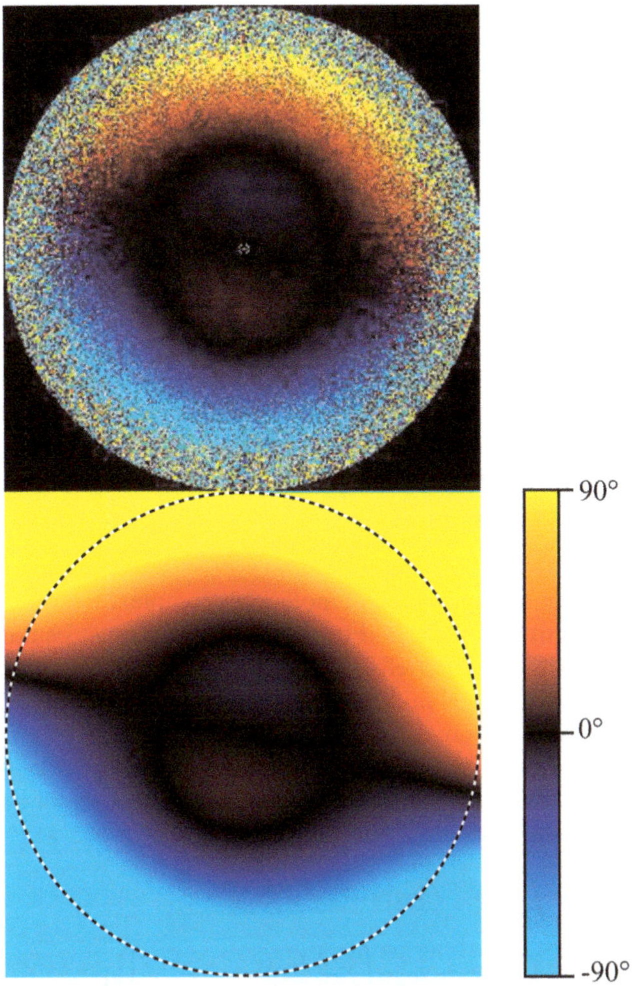

Figure 4.31. An example of the axial coma induced by strong beam tilt. The top image shows the average phase difference between the model and the experimental images; the bottom image shows a fit of the corresponding model that consists of a linear combination of Zernike polynomials, as implemented in RELION. Although the phase shift is small at low frequencies, due to its cubical growth, it exceeds 90° before reaching the Nyquist frequency (1.70 Å, indicated by the dotted circle). The two half-circles of opposite color at the center of the image indicate that the particles were refined without considering the axial coma: the particle positions exhibit a systematic error because the algorithm has attempted to explain away the coma by shifting the particles, which overcompensates the phase shift in the low frequencies.

we see that the apparent defocus z' and spherical aberration C'_s have increased by a factor of a^2 and a^4, respectively. If both z and C_s were estimated from the data, then this factor of a would be completely invisible. In practice, however, it is common to assume the value of C_s given by the microscope manufacturer and to only estimate the defoci. In that case, one would effectively assume an incorrect C'_s and that error would be absorbed by the estimated defoci in the low frequencies, which would produce an even more incorrect CTF in the higher frequencies. This source of error leads to a loss of resolution and it can often be seen as oscillations in the Fourier shell correlation (FSC) curve at higher resolutions. This issue can be corrected by refining the symmetrical aberrations and thus obtaining a value of C'_s consistent with the currently assumed pixel size—and then refining the defoci. Errors caused by an inaccurate nominal wavelength λ can be compensated for in an analogous manner. Note that in both cases, the true pixel size should still be calibrated by an external reference prior to building an atomic model.

4.8.4 Thick objects and the Ewald sphere

Until now, we have assumed that the object to be reconstructed is sufficiently thin for the top and bottom to be at the same defocus value, ΔZ. This has allowed to consider a single exit wave $\Psi_{\text{exitwave}}(x, y)$ for the entire object, which is represented by an object-dependent transmittance function $T(x, y)$, which has allowed integrating over the entire thickness of the object, $\int_z \mathrm{d}z$, as is done in equation (1.2) in section 1.2.

However, there are biological macromolecular complexes that are too thick for this 'projection' assumption to hold. For such objects, it must be admitted that the top of the object is at a different defocus from the bottom of the object. Intuitively[2], one could divide the object into multiple infinitely thin slices, each at its own defocus value. The image recorded on the detector would then result from adding up the contributions of all of the slices, each of which would be affected by its own defocus ΔZ_z. That way, the wave function at the image plane becomes an integral over the transmittance functions T_z of all slices:

$$\begin{aligned} W(s_x, s_y) &= \int_z T_z(s_x, s_y)\exp(\mathrm{i}\gamma(s_x, s_y, z))\mathrm{d}z \\ &= \int_z T_z(s_x, s_y)\exp\left(2\pi\mathrm{i}\left[\frac{C_s}{4}\lambda^3 s^4 - z\frac{\lambda}{2}s^2\right]\right)\mathrm{d}z \\ &= \exp\left(2\pi\mathrm{i}\frac{C_s}{4}\lambda^3 s^4\right)\int_z T_z(s_x, s_y)\exp\left(-2\pi\mathrm{i}z\frac{\lambda}{2}s^2\right)\mathrm{d}z. \end{aligned} \tag{4.14}$$

It then follows from the definition of the Fourier transform itself, $\mathcal{F}\{f_x\}(\nu) = \int_x f_x \exp(-2\pi\mathrm{i}x\nu)\mathrm{d}x$, that the wave function at the image plane acquires a third dimension, $s_z = \frac{\lambda}{2}s^2$, with $s^2 = s_x^2 + s_y^2$:

$$W(s_x, s_y) = \exp\left(2\pi\mathrm{i}\frac{C_s}{4}\lambda^3 s^4\right)F\left(s_x, s_y, \frac{\lambda}{2}s^2\right). \tag{4.15}$$

[2] But one might want to read Cowley and Moodie (1957).

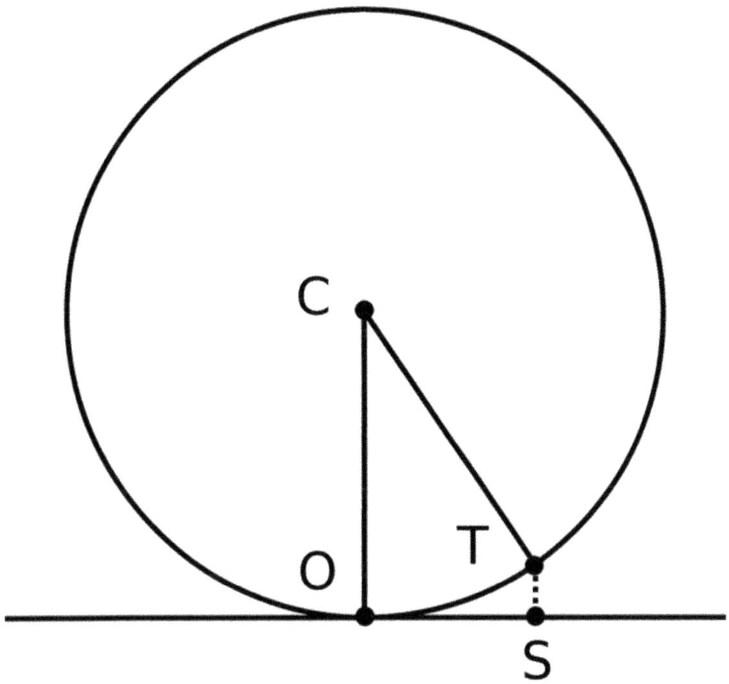

Figure 4.32. O is the origin of the reciprocal space and C is the center of the Ewald sphere, whose radius is λ^{-1}. When the Ewald sphere curvature can be ignored, S at $(s_x, s_y, 0)$ is sampled. Otherwise, T, which is above S and on the surface of the sphere, should be sampled. The z coordinate of T is $s_z = -\lambda^{-1} + (\lambda^{-2} - s^2)^{-1/2} = -\lambda^{-1} + \lambda^{-1}(1 - \lambda^2 s^2)^{-1/2} \approx -\lambda s^2/2 + O(s^4)$. Note that the curvature of the Ewald sphere is exaggerated in this figure. The resolution at which the distance between S and T becomes limiting for the reconstruction of particles of varying size is shown in figure 4.33.

This means that unlike the two-dimensional plane in Fourier space $F(s_x, s_y)$ in equation (1.5) in section 1.2, the wave function at the image plane now samples the surface of a sphere in the three-dimensional Fourier transform of the object (figure 4.32). This sphere is called the Ewald sphere (Ewald 1921). One consequence of the wave function at the image plane being sampled on the surface of a sphere is that it loses Friedel symmetry. At high resolution, two waves at $\vec{s}$ and $-\vec{s}$ are no longer each other's complex conjugate and they are inserted at positions that are no longer related by inversion symmetry. At low resolution the two positions overlap and the sum of the two components becomes the Fourier component of the original object multiplied by the familiar $2 \sin \gamma(\vec{s})$ term of the CTF as we have seen in equation (1.5) in section 1.2.

4.8.5 Ewald sphere correction

As is mentioned in section 4.3.2, defocusing causes delocalization of the signal in the real-space image. The contribution from any two Fourier components, at $\vec{s}$ and $-\vec{s}$, appear on opposite sides of the particle at a distance d from the center of the particle that depends on their resolution: $d = \Delta Z \lambda / |\vec{s}|$. Therefore, at sufficiently high resolution, and when using a high enough defocus, the two Fourier components are physically separated and they can thus be treated separately.

Under these conditions, Ewald sphere curvature can then be corrected by multiplying half of the Fourier components of an image by $\exp[i\gamma(\vec{s})]$ and the other half by $\exp[i\gamma(-\vec{s})]$, and then placing both components into their corresponding position on the surface of the Ewald sphere in the three-dimensional Fourier transform of the object. Russo and Henderson (2018) proposed to use real-space masks to further maximize the signal-to-noise ratio obtained from that operation. Their method was implemented in RELION and was used to improve the resolution of a reconstruction of the P22 virion capsid from 3.5 Å to 3.3 Å (Zivanov *et al* 2018).

Interestingly, already twelve years earlier, Wolf *et al* (2006) had suggested a procedure where no attempt to separate the two Fourier components was made, but the entire Fourier transform of the original image was multiplied by both $\exp[i\gamma(\vec{s})]$ and by $\exp[i\gamma(-\vec{s})]$, and the approximate Friedel pairs were inserted in both positions on the Ewald sphere. Half of these insertions are wrong and will contribute noise to the reconstruction. However, the other half is correct and their signal will add up constructively. Therefore, even though no attempt to separate the approximate Friedel pairs is made, this method can still lead to an improvement of the reconstruction. We note that if the improvement in the reconstruction is measured only by considering the FSC between reconstructions from half-sets, there will be no difference between the two methods since the incorrect half of the signal will be consistently incorrect in the two half-set reconstructions. Only by considering the improvement in the reconstruction against, for example, an atomic model or a map obtained independently, will the benefit of the real-space mask become apparent. The method without masking was implemented in FREALIGN (Grigorieff 2007) and it was used to improve the resolution of a reconstruction of the AAV2 capsid from 1.93 Å to 1.86 Å (Tan *et al* 2018).

Ewald sphere curvature correction has an additional benefit in that it can be used to determine the correct handedness of the reconstruction, something that is otherwise impossible from un-tilted images alone. One can perform Ewald sphere correction with either of two directions of curvature. If the reconstruction was done in the correct hand, then correction with opposite Ewald sphere curvature will lead to a worse result. However, we do note that, in practice, at resolutions where Ewald sphere curvature matters, features in the reconstructed density already often suffice to infer its correct handedness.

When ignoring the effects of the Ewald sphere in the reconstruction process, one introduces errors in the reconstruction. The errors depend on the diameter D of the object and the wavelength λ of the electrons. DeRosier proposed the following formula to calculate the resolution at which the average phase error becomes 66°: $d_{\text{lim}}^{-1} = \sqrt{(2 \times 0.7/D\lambda)}$ (DeRosier 2000). This estimate was later confirmed by independent work by Downing and Glaeser (2018). Figure 4.33 shows the corresponding resolutions for varying particle diameters at electron wavelengths corresponding to energies of 100, 200, and 300 keV, or wavelengths λ

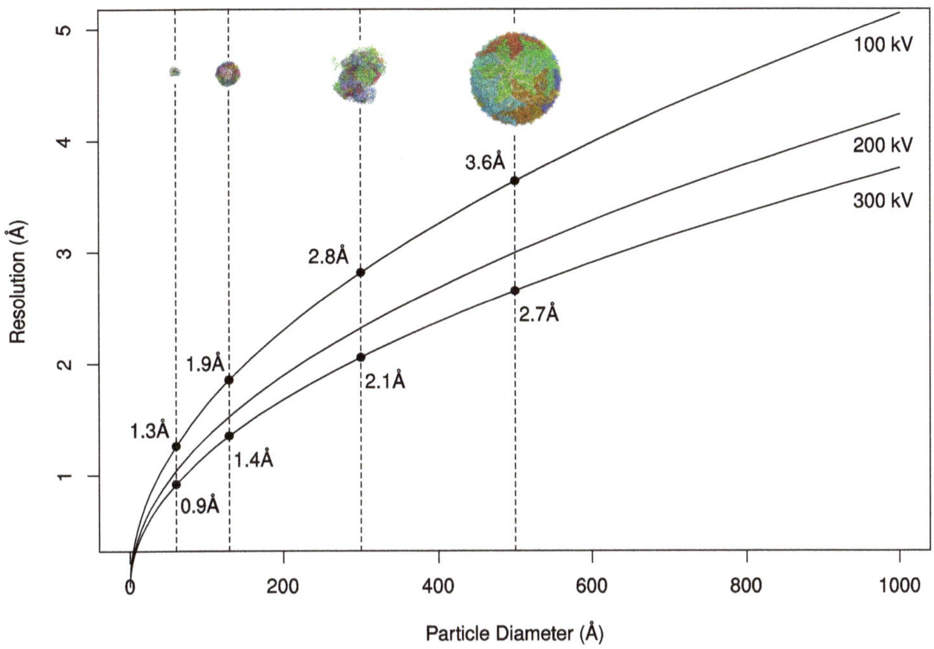

Figure 4.33. Resolution (d_{lim}) at which Ewald sphere curvature becomes relevant according to the equation proposed by DeRosier (2000). Four examples are shown: the streptavidin tetramer (PDB: 1STP) with diameter 60 Å, apoferritin (PDB: 2CIH) with 130 Å, 80S ribosome (PDB: 4UG0) with 300 Å, and Zika virus (PDB: 6CO8) with 500 Å.

of 0.037, 0.025, and 0.0196 Å, respectively. From this plot it follows that when using a 300 kV electron microscope, Ewald sphere correction can safely be ignored for macromolecular complexes of up to several hundreds of thousand daltons in molecular weight. However, for ribosomes, which have a diameter of approximately 300 Å, Ewald sphere correction becomes relevant at resolutions around 2 Å, while for virus particles with a diameter of around 700 Å the errors become limiting at 3 Å resolution. The P22 virion capsid mentioned above has a diameter of 700 Å; the AAV2 capsid has a diameter of 250 Å.

The effects of Ewald sphere curvature are more pronounced when using lower-energy electrons. It was recently postulated that using 100 keV electrons the amount of useful information in cryo-EM images of many biological objects is increased by 25% compared to using 300 keV electrons Peet *et al* (2019). For these low energies, Ewald sphere correction becomes relevant at approximately half the size of the particle relative to 300 keV. Therefore, if cheaper 100 kV electron microscopes are to be become popular for high-resolution reconstruction, Ewald sphere correction will become more important than it is currently. The same is true also for the aberrations described in section 4.8.2. For example, the amount of phase error introduced by beam tilt scales with the square of the wavelength of the electrons. Therefore, a given amount of beam tilt will result in 3.5 times larger phase errors when comparing 100–300 keV electrons.

References

Abeyrathne P D, Koh C S, Grant T, Grigorieff N and Korostelev A A 2016 Ensemble cryo-EM uncovers inchworm-like translocation of a viral IRES through the ribosome *eLife* **5** e14874

Afanasyev P, Seer-Linnemayr C, Ravelli R B G, Matadeen R, De Carlo S, Alewijnse B, Portugal R V, Pannu N S, Schatz M and van Heel M 2017 Single-particle cryo-EM using alignment by classification (ABC): the structure of *Lumbricus terrestris* haemoglobin *IUCrJ* **4** 678–94

Amunts A *et al* 2014 Structure of the yeast mitochondrial large ribosomal subunit *Science* **343** 1485–9

Bai X-C, McMullan G and Scheres S H W 2015a How cryo-EM is revolutionizing structural biology *Trends Biochem. Sci.* **40** 49–57

Bai X C, Rajendra E, Yang G, Shi Y and Scheres S H 2015b Sampling the conformational space of the catalytic subunit of human gamma-secretase *eLife* **4** e11182

Baldwin P R and Lyumkis D 2021 Non-uniformity of projection distributions attenuates resolution in cryo-EM *Prog. Biophys. Mol. Biol.* **150** 160–83

Baldwin P R and Lyumkis D 2020 Tools for visualizing and analyzing Fourier space sampling in cryo-EM *Prog. Biophys. Mol. Biol.* **160** 53–65

Barnes G and Gentle I 2011 *Interfacial Science: An Introduction* (Oxford: Oxford University Press)

Bartesaghi A, Matthies D, Banerjee S, Merk A and Subramaniam S 2014 Structure of β-galactosidase at 3.2-Å resolution obtained by cryo-electron microscopy *Proc. Natl Acad. Sci. USA* **111** 11709–14

Bartesaghi A, Merk A, Banerjee S, Matthies D, Wu X, Milne J L and Subramaniam S 2015 2.2 Å resolution cryo-EM structure of β-galactosidase in complex with a cell-permeant inhibitor *Science* **348** 1147–51

Bell J M, Chen M, Baldwin P R and Ludtke S J 2016 High resolution single particle refinement in EMAN2.1 *Methods* **100** 25–34

Bepler T *et al* 2019 Positive-unlabeled convolutional neural networks for particle picking in cryo-electron micrographs *Nat. Methods* **16** 1153–60

Bharat T A M, Christopher J R, Löwe J, Passmore L A and Scheres S H W 2015 Advances in single-particle electron cryomicroscopy structure determination applied to sub-tomogram averaging *Structure*

Booy F P and Pawley J B 1993 Cryo-crinkling: what happens to carbon films on copper grids at low temperature *Ultramicroscopy* **48** 273–80

Brilot A F, Chen J Z, Cheng A, Pan J, Harrison S C, Potter C S, Carragher B, Henderson R and Grigorieff N 2012 Beam-induced motion of vitrified specimen on holey carbon film *J. Struct. Biol.* **177** 630–7

Campbell M G *et al* 2012 Movies of ice-embedded particles enhance resolution in electron cryo-microscopy *Structure* **20** 1823–8

Campbell M G, Cormier A, Ito S, Seed R I, Bondesson A J, Lou J, Marks J D, Baron J L, Cheng Y and Nishimura S L 2020 Cryo-EM reveals integrin-mediated TGF-β activation without release from latent TGF-β *Cell* **180** 490–501.e16

Campbell M G, Veesler D, Cheng A, Potter C S and Carragher B 2015 2.8 Å resolution reconstruction of the *Thermoplasma acidophilum* 20S proteasome using cryo-electron microscopy *eLife* **4** e06380

Cardone G, Heymann J B and Steven A C 2013 One number does not fit all: mapping local variations in resolution in cryo-EM reconstructions *J. Struct. Biol.* **184** 226–36

Carragher B, Cheng Y, Frost A, Glaeser R M, Lander G C, Nogales E and Wang H-W 2019 Current outcomes when optimizing 'standard' sample preparation for single-particle cryo-EM *J. Microsc.* **276** 39–45

Chen J, Noble A J, Kang J Y and Darst S A 2019 Eliminating effects of particle adsorption to the air/water interface in single-particle cryo-electron microscopy: bacterial RNA polymerase and CHAPSO *J. Struct. Biol.:* X **1** 100005

Chowdhury S *et al* 2017 Structure reveals mechanisms of viral suppressors that intercept a CRISPR RNA-guided surveillance complex *Cell* **169** e11

Cowley J M and Moodie A F 1957 The scattering of electrons by atoms and crystals. I. A new theoretical approach *Acta Crystallogr.* **10** 609–19

Crowther R A, DeRosier D J and Klug A 1970 The reconstruction of a three-dimensional structure from projections and its application to electron microscopy *Proc. R. Soc. Lond.* A **317** 319–40

Danev R, Tegunov D and Baumeister W 2017 *eLife* **6** e23006

Danev R, Yanagisawa H and Kikkawa M 2019 Cryo-electron microscopy methodology: current aspects and future directions *Trends Biochem. Sci.* **44** 837–48

Dang S *et al* 2017 Cryo-EM structures of the TMEM16A calcium-activated chloride channel *Nature* **552** 426–9

De Jongh H H J, Kosters H A, Kudryashova E, Meinders M B J, Trofimova D and Wierenga P A 2004 Protein adsorption at air–water interfaces: a combination of details *Biopolymers* **74** 131–5

DeRosier D J 2000 Correction of high-resolution data for curvature of the Ewald sphere *Ultramicroscopy* **81** 83–98

De Rosier D J and Klug A 1968 Reconstruction of three dimensional structures from electron micrographs *Nature* **217** 130–4

Diaz R, Rice W J and Stokes D L 2010 Fourier–Bessel reconstruction of helical assemblies *Methods Enzymol.* **482** 131–65

Diebolder C A, Faas F G A, Koster A J and Koning R I 2015 Conical Fourier shell correlation applied to electron tomograms *J. Struct. Biol.* **190** 215–23

D'Imprima E, Floris D, Joppe M, Sánchez R, Grininger M and Kühlbrandt W 2019 Protein denaturation at the air–water interface and how to prevent it *eLife* **8** e42747

Downing K and Glaeser R 2008 Restoration of weak phase-contrast images recorded with a high degree of defocus: the 'twin image' problem associated with CTF correction *Ultramicroscopy* **108** 921–8

Downing K H and Glaeser R M 2018 Estimating the effect of finite depth of field in single-particle cryo-EM *Ultramicroscopy* **184** 94–9

Dubochet J, Adrian M, Chang J J, Homo J C, Lepault J, McDowall A W and Schultz P 1988 Cryo-electron microscopy of vitrified specimens *Q. Rev. Biophys.* **21** 129–228

Dubochet J and McDowall A W 1981 Vitrification of pure water for electron microscopy *J. Microsc.* **124** 3–4

Dudkina N V, Kudryashev M, Stahlberg H and Boekema E J 2011 Interaction of complexes I, III, and IV within the bovine respirasome by single particle cryoelectron tomography *Proc. Natl Acad. Sci. USA* **108** 15196–200

Déjardin P (ed) 2006 *Proteins at Solid–Liquid Interfaces* (Berlin: Springer)

Egelman E H 2007 The iterative helical real space reconstruction method: surmounting the problems posed by real polymers *J. Struct. Biol.* **157** 83–94

Egelman E H 2014 Ambiguities in helical reconstruction *eLife* **3** e04969

Elad N, Clare D K, Saibil H R and Orlova E V 2008 Detection and separation of heterogeneity in molecular complexes by statistical analysis of their two-dimensional projections *J. Struct. Biol.* **162** 108–20
Elmlund H, Elmlund D and Bengio S 2013 PRIME: probabilistic initial 3D model generation for single-particle cryo-electron microscopy *Structure* **21** 1299–306
Ewald P P 1921 Die Berechnung optischer und elektrostatischer Gitterpotentiale *Ann. Phys.* **369** 253–87
Famelis N *et al* 2019 Architecture of the mycobacterial type VII secretion system *Nature* **576** 321–5
Fan X, Wang J, Zhang X, Yang Z, Zhang J-C, Zhao L, Peng H-L, Lei J and Wang H-W 2019 Single particle cryo-EM reconstruction of 52-kDa streptavidin at 3.2 angstrom resolution *Nat. Commun.* **10** 2386–11
Fernandez I S *et al* 2013 Molecular architecture of a eukaryotic translational initiation complex *Science* **342** 1240585
Fernandez-Leiro R, Conrad J, Yang J C, Freund S M, Scheres S H and Lamers M H 2017 Self-correcting mismatches during high-fidelity DNA replication *Nat. Struct. Mol. Biol.* **24** 140–3
Fernandez-Leiro R and Scheres S H W 2016 Unravelling biological macromolecules with cryo-electron microscopy *Nature* **537** 339–46
Frank J 2008 Introduction: principles of electron tomography *Electron Tomography. Methods for Three-Dimensional Visualization of Structures in the Cell* 2nd edn ed J Frank (New York: Springer)
Frank J and Ourmazd A 2016 Continuous changes in structure mapped by manifold embedding of single-particle data in cryo-EM *Methods* **100** 61–7
Frank J, Radermacher M, Penczek P, Zhu J, Li Y, Ladjadj M and Leith A 1996 SPIDER and WEB: processing and visualization of images in 3D electron microscopy and related fields *J. Struct. Biol.* **116** 190–9
Frederik P M, Stuart M C A, Bomans P H H and Busing W M 1989 Phospholipid, nature's own slide and cover slip for cryo-electron microscopy *J. Microsc.* **153** 81–92
Ganzevles R A, Fokkink R, van Vliet T, Cohen Stuart M A and De Jongh H H J 2008 Structure of mixed β-lactoglobulin/pectin adsorbed layers at air/water interfaces; a spectroscopy study *J Colloid Interface Sci.* **317** 137–47
Gao H, Valle M, Ehrenberg M and Frank J 2004 Dynamics of EF-G interaction with the ribosome explored by classification of a heterogeneous cryo-EM dataset *J. Struct. Biol.* **147** 283–90
Gilbert P 1972 Iterative methods for the three-dimensional reconstruction of an object from projections *J. Theor. Biol.* **36** 105–17
Glaeser R M 1999 Review: electron crystallography: present excitement, a nod to the past, anticipating the future. *J. Struct. Biol.* **128** 3–14
Glaeser R M 2016 Specimen behavior in the electron beam *Methods Enzymol.* **579** 19–50
Glaeser R M 2019 How good can single-particle cryo-EM become? What remains before it approaches its physical limits? *Annu. Rev. Biophys.* **48** 45–61
Glaeser R M, Han B-G, Csencsits R, Killilea A, Pulk A and Cate J H D 2016 Factors that influence the formation and stability of thin, cryo-EM specimens *Biophys. J.* **110** 749–55
Glaeser R M, Typke D, Tiemeijer P C, Pulokas J and Cheng A 2011 Precise beam-tilt alignment and collimation are required to minimize the phase error associated with coma in high-resolution cryo-EM *J. Struct. Biol.* **174** 1–10

Goddard T D, Huang C C, Meng E C, Pettersen E F, Couch G S, Morris J H and Ferrin T E 2018 UCSF ChimeraX: meeting modern challenges in visualization and analysis *Protein Sci.* **27** 14–25

Grant T and Grigorieff N 2015 Automatic estimation and correction of anisotropic magnification distortion in electron microscopes *J. Struct. Biol.* **192** 204–8

Grant T, Rohou A and Grigorieff N 2018 cisTEM, user-friendly software for single-particle image processing *eLife* **7** e35383

Greber B J, Sutter M and Kerfeld C A 2019a The plasticity of molecular interactions governs bacterial microcompartment shell assembly *Structure* **27** 749–63.e4

Greber B J, Toso D B, Fang J and Nogales E 2019b The complete structure of the human TFIIH core complex *eLife* **8** e44771

Grigorieff N 2007 FREALIGN: high-resolution refinement of single particle structures *J. Struct. Biol.* **157** 117–25

Grigorieff N 2016 Frealign: an exploratory tool for single-particle cryo-EM *Methods Enzymol.* **579** 191–226

Gunning P A, Mackie A R, Gunning A P, Woodward N C, Wilde P J and Morris V J 2004 Effect of surfactant type on surfactant–protein interactions at the air–water interface *Biomacromolecules* **5** 984–91

Haack D B, Yan X, Zhang C, Hingey J, Lyumkis D, Baker T S and Toor N 2019 Cryo-EM structures of a group II intron reverse splicing into DNA *Cell* **178** 612–2

Han B-G, Walton R W, Song A, Hwu P, Stubbs M T, Yannone S M, Arbeláez P, Dong M and Glaeser R M 2012 Electron microscopy of biotinylated protein complexes bound to streptavidin monolayer crystals *J. Struct. Biol.* **180** 249–53

Han Y, Fan X, Wang H, Zhao F, Tully C G, Kong J, Yao N and Yan N 2019 High-yield monolayer graphene grids for near-atomic resolution cryoelectron microscopy *Proc. Natl Acad. Sci. USA* **117** 1009–14

Hanszen K J 1967 Neue Erkenntnisse über Auflösung und Kontrast im elektronenmikroskopischen Bild *Naturwissenschaften* **54** 125–33

Hanszen K J and Morgenstern B 1965 Die Phasenkontrast- und Amplitudenkontrast-Übertragung des elektronenmikroskopischen Objektivs *Z. Ange. Phys.* **19** 215–27

Harauz G and van Heel M 1986 Exact filters for general geometry 3-dimensional reconstruction *Optik* **73** 146–56

Henderson R 1995 The potential and limitations of neutrons, electrons and x-rays for atomic resolution microscopy of unstained biological molecules *Q. Rev. Biophys.* **28** 171–93

Henderson R 2013 Avoiding the pitfalls of single particle cryo-electron microscopy: Einstein from noise *Proc. Natl Acad. Sci. USA* **110** 18037–41

Henderson R, Chen S, Chen J Z, Grigorieff N, Passmore L A, Ciccarelli L, Rubinstein J L, Crowther R A, Stewart P L and Rosenthal P B 2011 Tilt-pair analysis of images from a range of different specimens in single-particle electron cryomicroscopy *J. Mol. Biol.* **413** 1028–46

Henderson R and McMullan G 2013 Problems in obtaining perfect images by single-particle electron cryomicroscopy of biological structures in amorphous ice *Microscopy* **62** 43–50

Herzik M A, Wu M and Lander G C 2019 High-resolution structure determination of sub-100 kDa complexes using conventional cryo-EM *Nat. Commun.* **10** 1032

Heymann J B, Conway J F and Steven A C 2004 Molecular dynamics of protein complexes from four-dimensional cryo-electron microscopy *J. Struct. Biol.* **147** 291–301

Huang G *et al* 2020 Structure of the cytoplasmic ring of the *Xenopus laevis* nuclear pore complex *Cell Res.* **30** 520–31

Ilca S L, Kotecha A, Sun X, Poranen M M, Stuart D I and Huiskonen J T 2015 Localized reconstruction of subunits from electron cryomicroscopy images of macromolecular complexes *Nat. Commun.* **6** 8843

Iudin A, Korir P K, Salavert-Torres J, Kleywegt G J and Patwardhan A 2016 EMPIAR: a public archive for raw electron microscopy image data *Nat. Methods* **13** 387–8

Jain T, Sheehan P, Crum J, Carragher B and Potter C S 2012 Spotiton: a prototype for an integrated inkjet dispense and vitrification system for cryo-TEM *J. Struct. Biol.* **179** 68–75

Jakobi A J, Wilmanns M and Sachse C 2017 Model-based local density sharpening of cryo-EM maps *eLife* **6** 1–26

Jensen G J 2001 Alignment error envelopes for single particle analysis *J. Struct. Biol.* **133** 143–55

Kato T, Makino F, Nakane T, Terahara N, Kaneko T, Shimizu Y, Motoki S, Ishikawa I, Yonekura K and Namba K 2019 CryoTEM with a cold field emission gun that moves structural biology into a new stage *Microsc. Microanal.* **25** 998–9

Kern D M, Sorum B, Hoel C M, Sridharan S, Remis J P, Toso D B and Brohawn S G 2020 Cryo-EM structure of the SARS-CoV-2 3a ion channel in lipid nanodiscs *bioRxiv* 2020.06.17. 156554

Khoshouei M, Radjainia M, Baumeister W and Danev R 2017 Cryo-EM structure of haemoglobin at 3.2 Å determined with the Volta phase plate *Nat. Commun.* **8** 16099

Kim J S, Saint-André C, Lim H S, Hwang C-S, Egly J-M and Cho Y 2015 Crystal structure of the Rad3/XPD regulatory domain of Ssl1/p44 *J. Biol. Chem.* **290** 8321–30

Kimanius D, Forsberg B, Scheres S and Lindahl E 2016 Accelerated cryo-EM structure determination with parallelisation using GPUs in RELION-2 *eLife* **5** e18722

Kriegeskorte N and Golan T 2019 Neural network models and deep learning—a primer for biologists *Curr. Biol.* **29** PR231–R236

Krivanek O L 1994 Three-fold astigmatism in high-resolution transmission electron microscopy *Ultramicroscopy* **55** 419–33

Krizhevsky A, Sutskever I and Hinton G E 2012 *Proc. of the 25th Int. Conf. on Neural Information Processing Systems* vol 1 pp 1097–105

Kudryashova E V, Meinders M B J, Visser A J W G, van Hoek A and de Jongh H H J 2003 Structure and dynamics of egg white ovalbumin adsorbed at the air/water interface *Eur. Biophys. J.* **32** 553–62

Lad M D, Birembaut F, Matthew J M, Frazier R A and Green R J 2006 The adsorbed conformation of globular proteins at the air/water interface *Phys. Chem. Phys.* **8** 2179–86

Lahiri I, Xu J, Han B-G, Oh J, Wang D, DiMaio F and Leschziner A E 2019 3.1 Å structure of yeast RNA polymerase II elongation complex stalled at a cyclobutane pyrimidine dimer lesion solved using streptavidin affinity grids *J. Struct. Biol.* **207** 270–8

LeBarron J, Grassucci R A, Shaikh T R, Baxter W T, Sengupta J and Frank J 2008 Exploration of parameters in cryo-EM leading to an improved density map of the *E. coli* ribosome *J. Struct. Biol.* **164** 24–32

LeCun Y *et al* 1989 Backpropagation applied to handwritten zip code recognition *Neural Comput.* **1** 541–51

LeCun Y, Bengio Y and Hinton G 2015 Deep learning *Nature* **521** 436–44

LeCun Y, Bottou L, Bengio Y and Haffner P 1998 Gradient-based learning applied to document recognition *Proc. IEEE* **86** 2278–324

Lee P D, Wei H, Tan D and Harrison S C 2019 Structure of the centromere binding factor 3 complex from *Kluyveromyces lactis* *J. Mol. Biol.* **431** 4444–54

Lehtinen J *et al* 2018 Noise2Noise: learning image restoration without clean data, *arXiv* 1803.04189

Leschziner A E and Nogales E 2006 The orthogonal tilt reconstruction method: an approach to generating single-class volumes with no missing cone for *ab initio* reconstruction of asymmetric particles *J. Struct. Biol.* **153** 284–99

Liao H Y and Frank J 2010 Definition and estimation of resolution in single-particle reconstructions *Structure* **18** 768–75

Liu Y, Zhou K, Zhang N, Wei H, Tan Y Z, Zhang Z, Carragher B, Potter C S, D'Arcy S and Luger K 2019 FACT caught in the act of manipulating the nucleosome *Nature* **577** 426–31

Ludtke S J, Baldwin P R and Chiu W 1999 EMAN: semiautomated software for high-resolution single-particle reconstructions *J. Struct. Biol.* **128** 82–97

Lyumkis D *et al* 2013 Cryo-EM structure of a fully glycosylated soluble cleaved HIV-1 envelope trimer *Science* **342** 1484–90

Lyumkis D 2019 Challenges and opportunities in cryo-EM single-particle analysis *J. Biol. Chem.* **294** 5181–97

Lyumkis D, Brilot A F, Theobald D L and Grigorieff N 2013 Likelihood-based classification of cryo-EM images using FREALIGN *J. Struct. Biol.* **183** 377–88

MacRitchie F 1998 Reversibility of protein adsorption *Stud. Interface Sci.* **7** 149–77

Marabini R, Herman G T and Carazo J M 1998 3D reconstruction in electron microscopy using ART with smooth spherically symmetric volume elements (blobs) *Ultramicroscopy* **72** 53–65

Marabini R, Rietzel E, Schroeder R, Herman G T and Carazo J M 1997 Three-dimensional reconstruction from reduced sets of very noisy images acquired following a single-axis tilt schema: application of a new three-dimensional reconstruction algorithm and objective comparison with weighted backprojection *J. Struct. Biol.* **120** 363–71

Martin A H, Cohen Stuart M A, Bos M A and van Vliet T 2005 Correlation between mechanical behavior of protein films at the air/water interface and intrinsic stability of protein molecules *Langmuir* **21** 4083–9

Martin A H, Grolle K, Bos M A, Cohen Stuart M A and van Vliet T 2002 Network forming properties of various proteins adsorbed at the air/water interface in relation to foam stability *J. Colloid Interface Sci.* **254** 175–83

Martin A H, Meinders M B J, Bos M A, Cohen Stuart M A and van Vliet T 2003 Conformational aspects of proteins at the air/water interface studied by infrared reflection–absorption spectroscopy *Langmuir* **19** 2922–8

McDowall A W, Chang J J, Freeman R, Lepault J, Walter C A and Dubochet J 1983 Electron microscopy of frozen hydrated sections of vitreous ice and vitrified biological samples *J. Microsc.* **131** 1–9

McMullan G, Faruqi A R, Clare D and Henderson R 2014 Comparison of optimal performance at 300 keV of three direct electron detectors for use in low dose electron microscopy *Ultramicroscopy* **147** 156–63

McMullan G, Vinothkumar K R and Henderson R 2015 Thon rings from amorphous ice and implications of beam-induced Brownian motion in single particle electron cryo-microscopy *Ultramicroscopy* **158** 26–32

Merk A, Fukumura T, Zhu X, Darling J E, Grisshammer R, Ognjenovic J and Subramaniam S 2020 1.8 Å resolution structure of β-galactosidase with a 200 kV CRYO ARM electron microscope *IUCrJ* **7** 1–5

Mindell J A and Grigorieff N 2003 Accurate determination of local defocus and specimen tilt in electron microscopy *J. Struct. Biol.* **142** 334–47

Morais M C *et al* 2003 Bacteriophage phi29 scaffolding protein gp7 before and after prohead assembly *Nat. Struct. Biol.* **10** 572–6

Moriya T, Saur M, Stabrin M, Merino F, Voicu H, Huang Z, Penczek P A, Raunser S and Gatsogiannis C 2017 High-resolution single particle analysis from electron cryo-microscopy images using SPHIRE *J. Vis. Exp.* **123** 55448

Murray S C, Flanagan J, Popova O B, Chiu W, Ludtke S J and Serysheva I I 2013 Validation of cryo-EM structure of IP3R1 channel *Structure* **21** 900–9

Nakane T *et al* 2020 Single-particle cryo-EM at atomic resolution *Nature* **587** 152–6

Nakane T, Kimanius D, Lindahl E and Scheres S H 2018 Characterisation of molecular motions in cryo-EM single-particle data by multi-body refinement in RELION *eLife* **7** e36861

Naydenova K, Peet M J and Russo C J 2019 Multifunctional graphene supports for electron cryomicroscopy *Proc. Natl Acad. Sci. USA* **116** 11718–24

Naydenova K and Russo C J 2017 Measuring the effects of particle orientation to improve the efficiency of electron cryomicroscopy *Nat. Commun.* **8** 629

Nguyen T H D *et al* 2016 Cryo-EM structure of the yeast U4/U6.U5 tri-snRNP at 3.7 Å resolution *Nature* **530** 298–302

Nicholson W and Glaeser R 2001 Review: automatic particle detection in electron microscopy *J. Struct. Biol.* **133** 90–101

Noble A J *et al* 2018a Routine single particle CryoEM sample and grid characterization by tomography *eLife* **7** 32

Noble A J, Wei H, Dandey V P, Zhang Z, Tan Y Z, Potter C S and Carragher B 2018b Reducing effects of particle adsorption to the air–water interface in cryo-EM *Nat. Methods* **15** 793–5

Nogales E 2016 The development of cryo-EM into a mainstream structural biology technique *Nat. Methods* **13** 24–7

Norde W 2011 Driving forces for protein adsorption at solid surfaces *Macromol. Symp.* **103** 5–18

Olah C *et al* 2018 The building blocks of interpretability *Distill* https://distill.pub/2018/building-blocks/

Orlova E V and Saibil H R 2004 Structure determination of macromolecular assemblies by single-particle analysis of cryo-electron micrographs *Curr. Opin. Struct. Biol.* **14** 584–90

Palovcak E, Wang F, Zheng S Q, Yu Z, Li S, Betegon M, Bulkley D, Agard D A and Cheng Y 2018 A simple and robust procedure for preparing graphene-oxide cryo-EM grids *J. Struct. Biol.* **204** 80–4

Pantelic R S, Meyer J C, Kaiser U, Baumeister W and Plitzko J M 2010 Graphene oxide: a substrate for optimizing preparations of frozen-hydrated samples *J. Struct. Biol.* **170** 152–6

Pantelic R S, Meyer J C, Kaiser U and Stahlberg H 2012 The application of graphene as a sample support in transmission electron microscopy *Lancet Microbe* **152** 1375–82

Passos D O *et al* 2020 Structural basis for strand-transfer inhibitor binding to HIV intasomes *Science* **367** 810–4

Patel A B, Louder R K, Greber B J, Grünberg S, Luo J, Fang J, Liu Y, Ranish J, Hahn S and Nogales E 2018 Structure of human TFIID and mechanism of TBP loading onto promoter DNA *Science* **262** eaau8872

Peet M J, Henderson R and Russo C J 2019 The energy dependence of contrast and damage in electron cryomicroscopy of biological molecules *Ultramicroscopy* **203** 125–31

Penczek P A 2010 Fundamentals of three-dimensional reconstruction from projections *Methods Enzymol.* **482** 1–33

Penczek P A and Asturias F J 2014 *Ab initio* cryo-EM structure determination as a validation problem *2014 IEEE Int. Conf. on Image Processing (ICIP) (Paris)*

Penczek P A, Fang J, Li X, Cheng Y, Loerke J and Spahn C M 2014 CTER-rapid estimation of CTF parameters with error assessment *Ultramicroscopy* **140** 9–19

Penczek P A, Grassucci R A and Frank J 1994 The ribosome at improved resolution: new techniques for merging and orientation refinement in 3D cryo-electron microscopy of biological particles *Ultramicroscopy* **53** 251–70

Penczek P A, Renka R and Schomberg H 2004 Gridding-based direct Fourier inversion of the three-dimensional ray transform *J. Opt. Soc. Am.* A **21** 499–509

Penczek P A, Zhu J and Frank J 1996 A common-lines based method for determining orientations for $N > 3$ particle projections simultaneously *Ultramicroscopy* **63** 205–18

Pettersen E F, Goddard T D, Huang C C, Couch G S, Greenblatt D M, Meng E C and Ferrin T E 2004 UCSF Chimera—a visualization system for exploratory research and analysis *J. Comput. Chem.* **25** 1605–12

Plaschka C *et al* 2015 Architecture of the RNA polymerase II-Mediator core initiation complex *Nature* **518** 376–80

Plaschka C, Lin P C and Nagai K 2017 Structure of a pre-catalytic spliceosome *Nature* **546** 617–21

Punjani A, Rubinstein J L, Fleet D J and Brubaker M A 2017 cryoSPARC: algorithms for rapid unsupervised cryo-EM structure determination *Nat. Methods* **14** 290–6

Radermacher M, Wagenknecht T, Verschoor A and Frank J 1987 Three-dimensional reconstruction from a single-exposure, random conical tilt series applied to the 50S ribosomal subunit of *Escherichia coli J. Microsc.* **146** 113–36

Ramírez-Aportela E *et al* 2019 Automatic local resolution-based sharpening of cryo-EM maps *Bioinformatics* **36** 765–72

Ramlaul K, Colin M P and Christopher H S A 2019 A local agreement filtering algorithm for transmission EM reconstructions *J. Struct. Biol.* **205** 30–40

Razinkov I, Dandey V, Wei H, Zhang Z, Melnekoff D, Rice W J, Wigge C, Potter C S and Carragher B 2016 A new method for vitrifying samples for cryoEM *J. Struct. Biol.* **195** 190–8

Razumovsky L and Damodaran S 1999 Surface activity–compressibility relationship of proteins at the air–water interface *Langmuir* **15** 1392–9

Righetto R D, Anton L, Adaixo R, Jakob R P, Zivanov J, Mahi M-A, Ringler P, Schwede T, Maier T and Stahlberg H 2020 High-resolution cryo-EM structure of urease from the pathogen *Yersinia enterocolitica Nat. Commun.* **11** 5101

Roh S H *et al* 2017 Subunit conformational variation within individual GroEL oligomers resolved by Cryo-EM *Proc. Natl Acad. Sci. USA* **114** 8259–64

Rohou A 2020 Fourier shell correlation criteria for local resolution estimation *bioRxiv* 2020.03.01.972067

Rohou A and Grigorieff N 2015 CTFFIND4: fast and accurate defocus estimation from electron micrographs *J. Struct. Biol.* **192** 216–21

Rollins M F, Chowdhury S, Carter J, Golden S M, Miettinen H M, Santiago-Frangos A, Faith D, Lawrence C M, Lander G C and Wiedenheft B 2019 Structure reveals a mechanism of CRISPR-RNA-guided nuclease recruitment and anti-CRISPR viral mimicry *Mol. Cell* **74** 132–42

Rosenthal P B and Henderson R 2003 Optimal determination of particle orientation, absolute hand, and contrast loss in single-particle electron cryomicroscopy *J. Mol. Biol.* **333** 721–45
Ruskin R S, Yu Z and Grigorieff N 2013 Quantitative characterization of electron detectors for transmission electron microscopy *J. Struct. Biol.* **184** 385–93
Russo C J and Henderson R 2018 Ewald sphere correction using a single side-band image processing algorithm *Ultramicroscopy* **187** 26–33
Russo C J and Passmore L A 2014 Electron microscopy: ultrastable gold substrates for electron cryomicroscopy *Science* **346** 1377–80
Russo C J and Passmore L A 2016 Ultrastable gold substrates: properties of a support for high-resolution electron cryomicroscopy of biological specimens *J. Struct. Biol.* **193** 33–44
Sander B, Golas M M and Stark H 2003 Automatic CTF correction for single particles based upon multivariate statistical analysis of individual power spectra *J. Struct. Biol.* **142** 392–401
Saxton W O 1978 *Computer Techniques for Image Processing in Electron Microscopy* (New York: Academic)
Scapin G, Dandey V P, Zhang Z, Prosise W, Hruza A, Kelly T, Mayhood T, Strickland C, Potter C S and Carragher B 2018 Structure of the insulin receptor-insulin complex by single-particle cryo-EM analysis *Nature* **556** 122–5
Scheres S H 2015 Semi-automated selection of cryo-EM particles in RELION-1.3 *J. Struct. Biol.* **189** 114–22
Scheres S H 2016 Processing of structurally heterogeneous cryo-EM data in RELION *Methods Enzymol.* **579** 125–57
Scheres S H *et al* 2007 Disentangling conformational states of macromolecules in 3D-EM through likelihood optimization *Nat. Methods* **4** 27–9
Scheres S H, Nunez-Ramirez R, Sorzano C O, Carazo J M and Marabini R 2008 Image processing for electron microscopy single-particle analysis using XMIPP *Nat. Protoc.* **3** 977–90
Scheres S H W 2012a A Bayesian view on cryo-EM structure determination *J. Mol. Biol.* **415** 406–18
Scheres S H W 2012b RELION: implementation of a Bayesian approach to cryo-EM structure determination *J. Struct. Biol.* **180** 519–30
Scheres S H W and Chen S 2012 Prevention of overfitting in cryo-EM structure determination *Nat. Methods* **9** 853–4
Scheres S H W, Valle M, Nuñez R, Sorzano C O S, Marabini R, Herman G T and Carazo J-M 2005 Maximum-likelihood multi-reference refinement for electron microscopy images *J. Mol. Biol.* **348** 139–49
Scherzer O 1949 The theoretical resolution limit of the electron microscope *J. Appl. Phys.* **20** 20–9
Schwartz O *et al* 2019 Laser phase plate for transmission electron microscopy *Nat. Methods* **16** 1016–20
Serna M 2019 Hands on methods for high resolution cryo-electron microscopy structures of heterogeneous macromolecular complexes *Front. Mol. Biosci.* **6** 33
Sigworth F J 1998 A maximum-likelihood approach to single-particle image refinement *J. Struct. Biol.* **122** 328–39
Sigworth F J, Doerschuk P C, Carazo J M and Scheres S H W 2010 An introduction to maximum-likelihood methods in cryo-EM *Methods Enzymol.* **482** 263–94
Sindelar C V and Grigorieff N 2012 Optimal noise reduction in 3D reconstructions of single particles using a volume-normalized filter *J. Struct. Biol.* **180** 26–38
Sorzano C O, Marabini R, Boisset N, Rietzel E, Schröder R, Herman G T and Carazo J M 2001 The effect of overabundant projection directions on 3D reconstruction algorithms *J. Mol. Biol.* **133** 108–18

Sorzano C O S, Marabini R, Herman G T, Censor Y and Carazo J M 2004a Transfer function restoration in 3D electron microscopy via iterative data refinement *Phys. Med. Biol.* **49** 509–22

Sorzano C O S, Marabini R, Velázquez-Muriel J, Bilbao-Castro J R, Scheres S H W, Carazo J M and Pascual-Montano A 2004b XMIPP: a new generation of an open-source image processing package for electron microscopy *J. Struct. Biol.* **148** 194–204

Spahn C M and Penczek P A 2009 Exploring conformational modes of macromolecular assemblies by multiparticle cryo-EM *Curr. Opin. Struct. Biol.* **19** 623–31

Stagg S M, Noble A J, Spilman M and Chapman M S 2014 ResLog plots as an empirical metric of the quality of cryo-EM reconstructions *J. Struct. Biol.* **185** 418–26

Stewart P L, Burnett R M, Cyrklaff M and Fuller S D 1991 Image reconstruction reveals the complex molecular organization of adenovirus *Cell* **67** 145–54

Stolinski C and Gross M 1969 A method for making thin, large surface area carbon supporting films for use in electron microscopy *Micron* **1** 340–52

Su M 2019 goCTF: geometrically optimized CTF determination for single-particle cryo-EM *J. Struct. Biol.* **205** 22–9

Subramaniam S 2013 Structure of trimeric HIV-1 envelope glycoproteins *Proc. Natl Acad. Sci. USA* **110** E4172–4

Tan Y Z *et al* 2020 Cryo-EM structures and regulation of arabinofuranosyltransferase AftD from mycobacteria *Mol. Cell* **78** 683–99.e11

Tan Y Z, Aiyer S, Mietzsch M, Hull J A, Mckenna R, Grieger J, Samulski R J, Baker T S, Agbandje-Mckenna M and Lyumkis D 2018 Sub-2 Å Ewald curvature corrected structure of an AAV2 capsid variant *Nat. Commun.* **9** 3628

Tan Y Z, Baldwin P R, Davis J H, Williamson J R, Potter C S, Carragher B and Lyumkis D 2017 Addressing preferred specimen orientation in single-particle cryo-EM through tilting *Nat. Methods* **14** 793–6

Taylor K A and Glaeser R M 1976 Electron microscopy of frozen hydrated biological specimens *J. Ultrastruct. Res.* **55** 448–56

Taylor K A and Glaeser R M 2008 Retrospective on the early development of cryoelectron microscopy of macromolecules and a prospective on opportunities for the future *J. Struct. Biol.* **163** 214–23

Tegunov D and Cramer P 2019 Real-time cryo-electron microscopy data preprocessing with Warp *Nat. Methods* **16** 1146–52

Tegunov D, Xue L, Dienemann C, Cramer P and Mahamid J 2020 Multi-particle cryo-EM refinement with *M* visualizes ribosome-antibiotic complex at 3.7 Å inside cells *Nat. Methods* **18** 186–93

Terwilliger T C, Oleg V S, Afonine P V and Adams P D 2018 Automated map sharpening by maximization of detail and connectivity *Acta Crystallogr.* D **74** 545–59

Terwilliger T C, Sobolev O V, Afonine P V, Adams P D and Read R J 2020 Density modification of cryo-EM maps *Acta Crystallogr.* D **76** 912–25

Thon F 1966 Zur Defokussierungsabhängigkeit des Phasenkontrastes bei der elektronenmikroskopischen Abbildung *Z. Naturforschung* **21a** 476–8

Thon F 1971 Phase contrast transfer function *Electron Microscopy in Material Science* ed U Valdrè (New York: Academic)

Tribet C, Audebert R and Popot J-L 1996 Amphipols: polymers that keep membrane proteins soluble in aqueous solutions *Proc. Natl Acad. Sci. USA* **93** 15047–50

Tripp B C, Magda J J and Andrade J D 1995 Adsorption of globular proteins at the air/water interface as measured via dynamic surface tension: concentration dependence, mass-transfer considerations, and adsorption kinetics *J. Colloid Interface Sci.* **173** 16–27
Trurnit H J 1960 A theory and method for the spreading of protein monolayers *J. Colloid Sci.* **15** 1–13
Valle M *et al* 2002 Cryo-EM reveals an active role for aminoacyl-tRNA in the accommodation process *EMBO J.* **21** 3557–67
van Heel M 1982 Detection of objects in quantum-noise-limited images *Ultramicroscopy* **7** 331–41
van Heel M 1987 Angular reconstitution: *a posteriori* assignment of projection directions for 3D reconstruction *Ultramicroscopy* **21** 111–23
van Heel M *et al* 2000 Single-particle electron cryo-microscopy: towards atomic resolution *Q. Rev. Biophys.* **33** 307–69
van Heel M 2013 Finding trimeric HIV-1 envelope glycoproteins in random noise *Proc. Natl Acad. Sci. USA* **110** E4175–7
van Heel M, Harauz G, Orlova E V, Schmidt R and Schatz M 1996 A new generation of the IMAGIC image processing system *J. Struct. Biol.* **116** 17–24
van Heel M and Schatz M 2005 Fourier shell correlation threshold criteria *J. Struct. Biol.* **151** 250–62
Vinothkumar K R and Henderson R 2016 Single particle electron cryomicroscopy: trends, issues and future perspective *Q. Rev. Biophys.* **49** 998
Vogel R H, Provencher S W, Von Bonsdorff C H, Adrian M and Dubochet J 1986 Envelope structure of Semliki Forest virus reconstructed from cryo-electron micrographs *Nature* **320** 533–35
Vonck J 2000 Parameters affecting specimen flatness of two-dimensional crystals for electron crystallography *Ultramicroscopy* **85** 123–9
Voss N, Yoshioka C, Radermacher M, Potter C and Carragher B 2009 DoG Picker and TiltPicker: software tools to facilitate particle selection in single particle electron microscopy *J. Struct. Biol.* **166** 205–13
Voss N R *et al* 2010 A toolbox for *ab initio* 3-D reconstructions in single-particle electron microscopy *J. Struct. Biol.* **169** 389–98
Wade R H 1992 A brief look at imaging and contrast transfer *Ultramicroscopy* **46** 145–56
Wagner T *et al* 2019 SPHIRE-crYOLO is a fast and accurate fully automated particle picker for cryo-EM *Commun. Biol.* **2** 218
Wang F, Liu Y, Yu Z, Li S, Cheng Y and Agard D A 2019 General and robust covalently linked graphene oxide affinity grids for high-resolution cryo-EM *Proc. Natl Acad. Sci. USA* **117** 24269–73
Wang F, Yu Z, Betegon M, Campbell M G, Aksel T, Zhao J, Li S, Douglas S M, Cheng Y and Agard D A 2020 Amino and PEG-amino graphene oxide grids enrich and protect samples for high-resolution single particle cryo-electron microscopy *J. Struct. Biol.* **209** 107437
Wolf M, DeRosier D J and Grigorieff N 2006 Ewald sphere correction for single-particle electron microscopy *Ultramicroscopy* **106** 376–82
Wong W *et al* 2014 Cryo-EM structure of the *Plasmodium falciparum* 80S ribosome bound to the anti-protozoan drug emetine *eLife* **3** e03080
Wright E R, Iancu C V, Tivol W F and Jensen G J 2006 Observations on the behavior of vitreous ice at approximately 82 and approximately 12 K *J. Struct. Biol.* **153** 241–52

Wu M, Lander G C and Herzik M A 2020 Sub-2 angstrom resolution structure determination using single-particle cryo-EM at 200 keV *J. Struct. Biol.:* X **4** 100020

Yip K M, Fischer N, Paknia E, Chari A and Stark H 2020 Atomic-resolution protein structure determination by cryo-EM *Nature* **587** 157–61

Yu G, Li K, Liu Y, Chen Z, Wang Z, Yan R, Klose T, Tang L and Jiang W 2016 An algorithm for estimation and correction of anisotropic magnification distortion of cryo-EM images without need of pre-calibration *J. Struct. Biol.* **195** 207–15

Zernike V F 1934 Beugungstheorie des Schneidenver-fahrens und seiner verbesserten Form, der Phasenkontrastmethode *Physica* **1** 689–704

Zhang C, Konermann S, Brideau N J, Lotfy P, Wu X, Novick S J, Strutzenberg T, Griffin P R, Hsu P D and Lyumkis D 2018 Structural basis for the RNA-guided ribonuclease activity of CRISPR-Cas13d *Cell* **175** 212–7

Zhang K 2016a Gctf: real-time CTF determination and correction *J. Struct. Biol.* **193** 1–12

Zhang K 2016b Gautomatch: fully automatic accurate, convenient and extremely fast particle picking for EM https://www2.mrc-lmb.cam.ac.uk/research/locally-developed-software/zhang-software/

Zhang K, Li S, Kappel K, Pintilie G, Su Z, Mou T-C, Schmid M F, Das R and Chiu W 2019 Cryo-EM structure of a 40 kDa SAM-IV riboswitch RNA at 3.7 Å resolution *Nat. Commun.* **10** 5511–6

Zhang W, Kimmel M, Spahn C M and Penczek P A 2008 Heterogeneity of large macromolecular complexes revealed by 3D cryo-EM variance analysis *Structure* **16** 1770–6

Zhang X, Jin L, Fang Q, Hui W H and Zhou Z H 2010 3.3 Å cryo-EM structure of a nonenveloped virus reveals a priming mechanism for cell entry *Cell* **141** 472–82

Zhang Y *et al* 2020 Molecular architecture of the luminal ring of the *Xenopus laevis* nuclear pore complex *Cell Res.* **30** 532–40

Zhang Y, Kostyuchenko V A and Rossmann M G 2007 Structural analysis of viral nucleocapsids by subtraction of partial projections *J. Struct. Biol.* **157** 356–64

Zhao J, Brubaker M A, Benlekbir S and Rubinstein J L 2015 Description and comparison of algorithms for correcting anisotropic magnification in cryo-EM images *J. Struct. Biol.* **192** 209–15

Zheng S Q, Palovcak E, Armache J-P, Verba K A, Cheng Y and Agard D A 2017 MotionCor2: anisotropic correction of beam-induced motion for improved cryo-electron microscopy *Nat. Methods* **14** 331–2

Zhou M *et al* 2015a Atomic structure of the apoptosome: mechanism of cytochrome c- and dATP-mediated activation of Apaf-1 *Genes Dev.* **29** 2349–61

Zhou Q, Huang X, Sun S, Li X, Wang H W and Sui S F 2015b Cryo-EM structure of SNAP-SNARE assembly in 20S particle *Cell Res.* **25** 551–60

Zhu J, Penczek P A, Schröder R and Frank J 1997 Three-dimensional reconstruction with contrast transfer function correction from energy-filtered cryoelectron micrographs: procedure and application to the 70S *Escherichia coli* ribosome *J. Struct. Biol.* **118** 197–219

Zivanov J, Nakane T, Forsberg B O, Kimanius D, Hagen W J, Lindahl E and Scheres S H 2018 New tools for automated high-resolution cryo-EM structure determination in RELION-3 *eLife* **7** e42166

Zivanov J, Nakane T and Scheres S H W 2020 Estimation of high-order aberrations and anisotropic magnification from cryo-EM data sets in RELION-3.1 *IUCrJ* **7** 253–67

Chapter 5

Map validation

5.1 Overview

Since cryo-EM images are very noisy it has long been a concern that the estimated resolution of a 3D reconstruction, and even the detailed features of a 3D density map, should not be trusted fully until there is independent evidence that the map is correct. In the early days, before high-resolution maps were obtained that could be compared to those obtained by x-ray crystallography, it occasionally happened that results published by different research groups were not consistent with each other, thus raising the question of which—if any—was correct. Even as the resolution improved, high-resolution maps continued to occasionally be published that were inconsistent with one another or with x-ray structures. As a result the software tools have been improved and new ones have been added in order to determine whether the maps are valid and up to what level of resolution. As long as the practices described in this chapter are used, the validity of the results obtained ought to no longer be a concern, although one must still remain cautious and self-critical at every stage of the work.

Section 5.2 describes the use of the Fourier shell correlation (FSC) function to estimate how consistent two independent reconstructions are with one another, which is a necessary requirement for the results to be valid. Section 5.3 then goes on to describe ways to exclude the possibility that bias or over-fitting might have affected the outcome. Finally, section 5.4 describes ways in which one can estimate the errors that remain after determining the translational alignment and the orientation angles for individual particles in the dataset.

The best criterion for validity of a map often proves to be whether it contains high-resolution features that are expected for the macromolecule of interest. Are there features that look like α-helices or β-sheets, for example, and are there densities for side-chain residues when they are expected to be there? While on the one hand this criterion can be very convincing, it also raises the question of whether the

doi:10.1088/978-0-7503-3039-8ch5

interpretation of a map is, itself valid. Further discussion of that question actually merits a chapter of its own, which follows this one.

5.2 Measures of resolution: FSC and local resolution

Steven Ludtke
Baylor College of Medicine, Houston, TX, USA

There are few more contentious topics in the cryo-EM community than that of resolution. This single number is often viewed as the most important parameter in describing the quality of a structure, or portion of a structure, and yet many have a poor understanding of what this number actually represents. The fact that resolution is often viewed by journals as a measure of the importance of a map can lead misguided practitioners to prioritize good resolution values over the quality of the actual reconstruction, so understanding the meaning and limitations of this concept are critical.

In optical microscopy the term resolution is generally defined as the minimum distance between two points at which they can still be individually distinguished. There are several associated mathematical definitions, the most widely used of which is the Rayleigh criterion (Rayleigh 1879). In a conventional optical microscope the resolution is ultimately limited by diffraction and thus the wavelength of the light. In electron microscopy, however, the electron wavelength is roughly 0.02 Å, a tiny fraction of the size of an atom, and diffraction does not play a limiting role. While other factors, such as lens aberrations, specimen motion, and spatial and temporal coherence of the electron beam, can impose their own limitations, in a high-end TEM their effect is normally small, permitting optical resolutions which can exceed 1 Å.

Despite the capabilities of the instrument optics, empirically we observe that high-resolution single-particle reconstructions typically achieve only 2–5 Å resolution. The reason for this is that radiation damage is ultimately the limiting factor in cryo-EM. As was explained previously in section 3.2, biomolecules can tolerate exposure to only a small number of $e_{-}/Å^2$ before suffering significant damage, thereby altering the structure of the object under observation. To prevent this damage from distorting structures the dose must be limited, which produces images which are extremely noisy. So, rather than being used as a measure of how blurry the map is, the word 'resolution' in cryo-EM refers to the highest spatial frequency to which reliable data extends within a structure.

Intuitively, resolution, used in this sense, is best represented by the spectral signal-to-noise ratio (SSNR), which tells us the ratio of signal intensity to noise intensity at a specific spatial frequency. A SSNR of 1.0 at 8 Å, for example, would mean that features with a periodicity of 8 Å consist of an equal level of accurate structural information and random noise. The SSNR thus provides a description of how trustworthy various features are in the map. However, for a number of reasons, rather than directly assessing the SSNR of a map, the FSC (Van Heel 1987) is the standard metric used to assess resolution.

As its name implies the FSC is a correlation, also known as a dot product, performed in Fourier space between two independently determined maps, which should ostensibly be identical except for added noise. While some familiarity with Fourier transforms is required to fully understand the FSC, the basic concept is fairly straightforward. Each spherical shell (in Fourier space) represents features with one specific spatial periodicity, in all possible directions. The radius of the shell defines the periodicity with larger shells representing smaller periodicities (finer details) and the position on the shell defines the direction (figure 5.1). When the average value of the correlation is computed between the values within the same shell in two different maps, a value ranging from 1 (perfect correlation) through 0 (no correlation) to −1 (perfect anticorrelation) is produced. This average value is computed for all shells and these are represented as a curve, with the spatial frequency (often represented as *s*) on the *X*-axis and the correlation coefficient on the *Y*-axis. A correlation coefficient of 1 indicates perfect agreement and thus no noise at all, and a coefficient of 0 indicates no correlation at all or pure noise at a particular spatial frequency. A typical FSC curve begins near 1.0 at low spatial frequency, falling to near 0.0 at higher resolutions, as shown in figure 5.2. Significant negative values (anticorrelation) should not be observed.

Counter-intuitively, resolution in cryo-EM is independent of the level of detail visible in the real-space map. If one were to take a map exhibiting clear 3 Å resolution features and apply a Gaussian low-pass filter (a blurring operation) to the map, the resulting map would have much less visible detail, but the resolution as assessed by FSC would be completely unchanged. That is to say, it is possible to

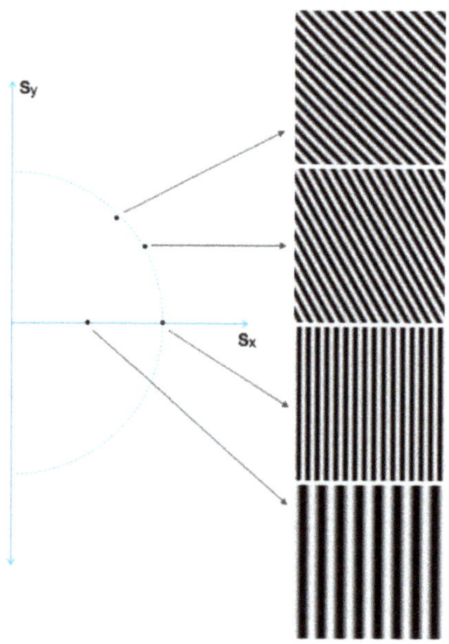

Figure 5.1. Relationship between points in Fourier space (left) and real-space features (right) in 2D. Points at the same radius represent the same periodicity in different directions. A smaller radius equates to longer periodicities. In 3D the waves are planar with Fourier points located on a spherical shell.

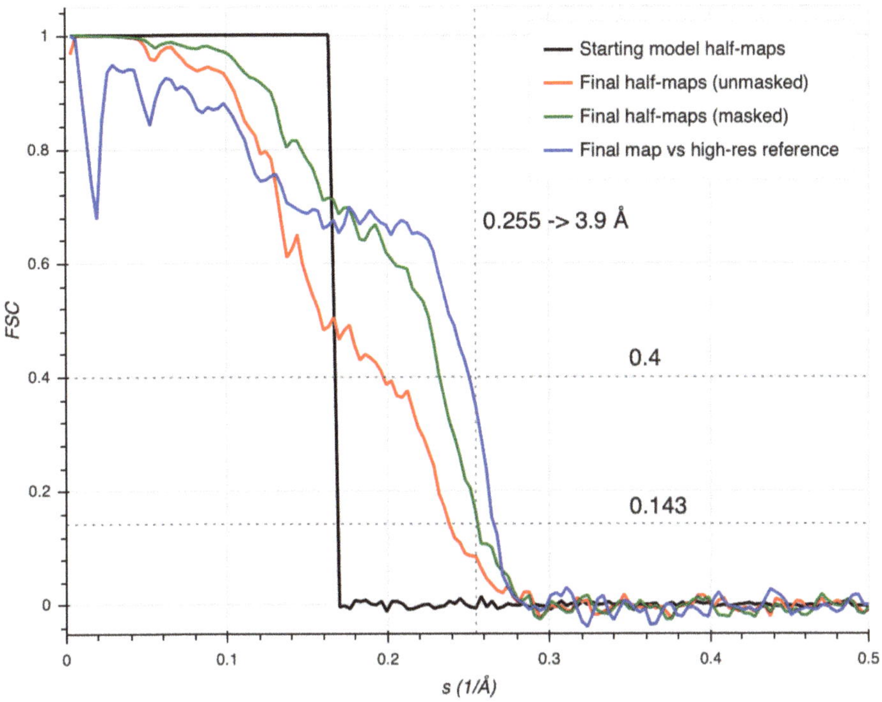

Figure 5.2. Examples of FSC curves from a real reconstruction. This reconstruction was not data-quality limited, meaning adding additional particles could result in further resolution improvements. The green and red curves are the FSC curves for the final refinement with and without a mask, respectively. The blue curve compares the final map against a high-resolution PDB model. The black curve is the FSC between the two initial models that were identical at low resolution, but then had the phases randomized at resolutions higher than 7.5 Å. Note that the 0.143 threshold applied to the masked 'gold-standard' FSC agrees well with the map versus model FSC at a threshold of 0.4.

have a map with features which are only visually resolved to 8 Å, but for the underlying map to still contain reliable data extending to 4 Å. This is, of course, not desirable, and considerable effort in the field is spent on producing maps visually representing all reliable features with as much detail as possible while suppressing all unreliable features. Strictly speaking this can only be achieved approximately, since noise exists at all spatial frequencies and the question of how much noise is permissible in a visual representation of a map is empirical.

It is also important to note that the spatial frequency, s, is expressed in units of $Å^{-1}$, not Å. When people refer to the resolution of the map it is conventional to take the reciprocal of the spatial frequency and thus talk about 4 Å resolution rather than 0.25 $Å^{-1}$ resolution. The meaning is the same. Furthermore, in cryo-EM the spatial frequency is the direct reciprocal of the real-space periodicity. Something at 1/5 $Å^{-1}$ spatial frequency corresponds to a real-space sine wave with a 5 Å periodicity. In related fields, such as SAXS, it is conventional for the X-axis to represent q or k, which are also a spatial frequency, but effectively differ from s by a factor of 2π.

The FSC has been considered to be the community standard for resolution assessment for several decades and for most of the last decade the method for computing the FSC has also become a generally agreed-upon standard. While the FSC does not provide a complete description of map quality it does provide one useful number for assessing the quality of a map or portion of a map. Other validation methods should also be employed whenever possible.

5.2.1 The 'gold-standard' FSC

For the FSC curve to truly measure the reliability or reproducibility of an averaged map, it is necessary for two different maps to be computed from the available data in a truly independent way. The standard strategy for computing the FSC for a reconstruction begins with splitting the raw particle data into even and odd numbered subpopulations, followed by independently refining the two maps reconstructed using each population. Prior to around 2010 many refinement strategies made use of a single common 3D starting model for the iterative refinement process, which introduced potential model bias into each refinement and led to somewhat exaggerated estimates of resolution in many, but not all, structures published prior to this time. Eventually the community largely agreed to adopt 'gold-standard' FSC procedures (Henderson *et al* 2012, Scheres 2012a), which simply require that the initial model, as well as the two ½ datasets used for FSC computation, be as truly independent from one another as possible. Having said that, there are other resolution estimation strategies which avoid this technique (Grant *et al* 2018, Herzik *et al* 2019).

The desired independence of two maps is generally achieved in one of two ways: in method one, first implemented by the Relion software package (Scheres 2012a), the same starting model is used for the even and odd datasets, but before use it is rigorously low-pass filtered to ~50 Å resolution, meaning that any features which emerge at resolutions higher than this emerge from the data rather than the initial model. The second method, first implemented in EMAN2 (Tang *et al* 2007, Ludtke and Serysheva 2013), takes the starting model and randomizes the Fourier phases beyond ~2× the targeted resolution. That is, if the resolution target is 5 Å, all of the phases beyond 10 Å will be randomized independently for the even and odd sets, producing starting models which are very different at high resolution (zero mean correlation). By starting with two maps which are very different at high resolution, rather than zero, the second approach presents a more stringent test for mutual convergence, but in most cases there is no practical difference. Note that if the refinement does not converge to a higher resolution than the cutoff/randomization resolution, then the result is not valid and the refinement must be re-run with a lower cutoff resolution.

5.2.2 Resolution thresholds

The value of 0.143 is the most commonly accepted threshold for reading the resolution from an FSC curve. This threshold value was derived from an attempt (Rosenthal and Henderson 2003) to produce resolution values which were similar to x-ray crystallography structures of similar quality. It is also important to note that 0.143 applies only when one is computing the FSC between even/odd maps, and the measured resolution is that of the final structure not that of each of the individual half-maps.

There are other situations where the FSC is used in cryo-EM and in these cases different threshold values are required. For example, another common resolution test is to compare the final averaged cryo-EM map with a higher resolution reference map of the same structure. Under a small set of assumptions, the FSC and SSNR can be interconverted using the simple formulas (Frank and Al-Ali 1975, Penczek 2002):

$$\mathrm{FSC} = \frac{\mathrm{SNR}}{1 + \mathrm{SNR}} \quad \text{or} \quad \mathrm{SNR} = \frac{\mathrm{FSC}}{1 - \mathrm{FSC}}.$$

From this we can see that FSC = 0.143 corresponds to a signal-to-noise ratio (SNR) of 0.167 in each ½ map. Since SSNR is additive when averaging is properly performed, the SNR is doubled in the averaged map and if we consider the situation where we are comparing a noisy map to a low-noise reference map the measured SNR will again be nearly doubled. Thus for comparisons between an averaged map and a low-noise reference model the threshold must be SNR = 0.667 → FSC = 0.4. If everything is in good agreement then the map versus reference FSC at 0.4 should be very close to the half-map FSC at a threshold of 0.143 (figure 5.1).

Another conceptually similar test is the 'map versus model' measurement often found when high-resolution maps are published. In this case the high-resolution cryo-EM map is used to produce an atomic model using any available technique, then that model is turned back into a now noise-free map and compared to the original. If the model building process were to ignore chemistry and simply place atoms such that they best match the original map, then it is possible to obtain very good agreement between the map and model. However, in this case model quality metrics, such as those produced by MolProbity (Chen *et al* 2010), are likely to be quite poor. Thus the map versus model FSC value must be considered together with the MolProbity results for the model. If both results are good it increases confidence that the map and model are reliable. If either yields poor results, then some aspect of the map or model is untrustworthy.

While we have now established the basics of resolution measurement some issues remain. The value of the threshold, or even whether the threshold should be a single specific value or vary with spatial frequency, remains a topic of debate in the community (van Heel and Schatz 2005, 2017). The question of how the independent maps are produced and how they can reasonably be treated prior to FSC calculation also remains somewhat controversial. Also, in some instances, the FSC curve may have unusual features (figure 5.3) which must be understood before blindly computing a resolution value.

5.2.3 FSC artifacts due to masking, filtration, and CTF

Even with the 'gold-standard' approach the FSC can still over-estimate the resolution if some additional factors are not considered. Counter-intuitively, some

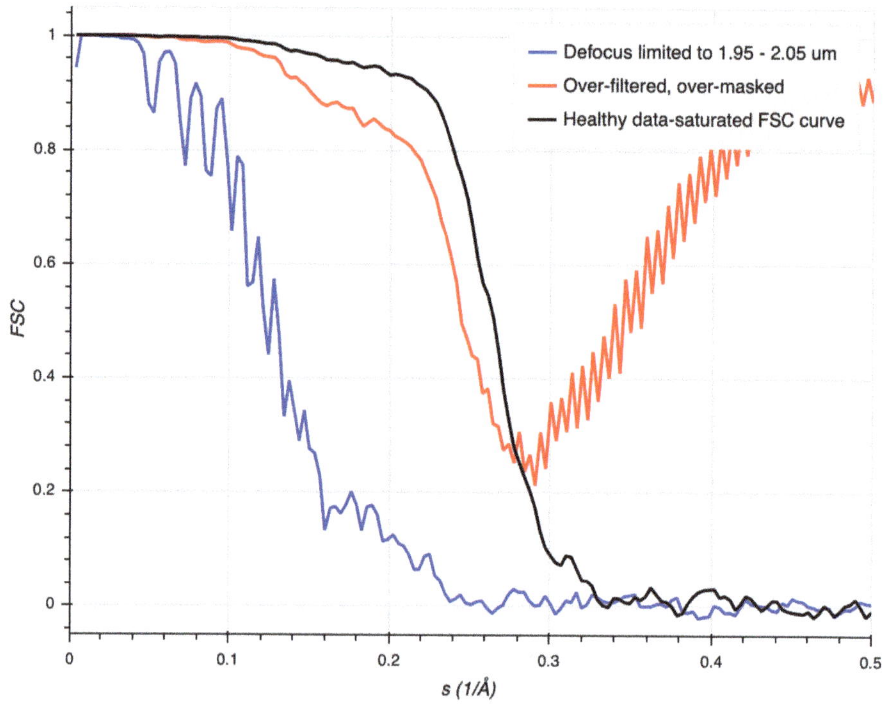

Figure 5.3. Examples of two potential problems. The black curve is a data-saturated 'healthy' FSC curve for reference. The red curve demonstrates an FSC with a low-resolution structure and a somewhat tight mask causing resolution exaggeration and the FSC rising toward 1.0 again. The high-resolution oscillations may be absent depending on the mask. Other refinement artifacts, such as over-fitting to noise, can cause a similar curve shape. In general, if the FSC initially falls then rises again something is wrong and the curve is not valid. The blue curve demonstrates a 'healthy' curve but with a narrow range of defocus values in the raw data. The FSC oscillations in this curve are real and will be correlated with modest real-space resolution dependent artifacts.

computational methods can degrade the quality of the map and yet produce higher resolution FSC curves so a proper understanding of these relatively common problems is critical.

If the FSC, or equivalently the SSNR, is computed between two completely unmasked volumes there is a strong sensitivity to particle box size, a point that is also addressed in section 5.4. The FSC is an average over the entire volume, so if the box size is increased then the average noise level increases and the FSC decreases. This effect encourages users to use a smaller box size, which unfortunately is associated with reconstruction artifacts, some of which can produce resolution exaggeration in addition to degrading the structure. A solution to this problem is to apply a mask prior to FSC calculation instead of reducing the box size. This is somewhat equivalent to the process of solvent-flattening in x-ray crystallography. In the latter, regions outside the reconstructed particle are set to a constant value equivalent to the mean solvent value. This eliminates the excess noise and focuses the FSC on the reconstructed particle, which is entirely appropriate.

Unfortunately, applying a mask prior to FSC calculation also introduces a new possibility, that of mask-correlation having an impact on the FSC. If the same mask is used on both of the volumes being compared and the mask is sharp (falling immediately from 1 to 0 at its edge), then the shape of the mask itself will produce some amount of high-resolution correlation leading to potential resolution exaggeration. To avoid this, the mask used must have a soft, typically Gaussian, edge with a width typically at least twice the targeted resolution. As long as the mask is sufficiently soft it will not generally cause exaggeration of the high-resolution FSC. Alas there are still some special cases where there could be a problem.

As an example of one problem which can still occur, consider an over-optimistic refinement targeting 5 Å resolution, which actually achieves only 15 Å resolution. If this reconstruction were directly or indirectly low-pass filtered, or blurred, to 15 Å, but the mask width was twice the target resolution (10 Å), the FSC may still be exaggerated, producing a resolution estimate which may even exceed 10 Å. While a soft-edged mask has very little power at high resolution it is generally not exactly zero. If the structure being masked has been filtered even more strongly than the mask then the mask may dominate the FSC and produce substantial resolution exaggeration (figure 5.3). This sort of problem is very easy to detect, as the structure itself will in no way look like a subnanometer resolution map and an unmasked FSC on the same structure will produce a dramatically lower resolution.

More is said in section 5.3 about how noise bias and over-fitting may generate resolution artifacts and in section 5.4 about estimating errors in the alignment of particles and assignment of orientation angles, which may, in turn, also generate deterministic errors in density maps which the FSC cannot detect.

The final issue we will consider is the influence of the contrast transfer function (CTF). Ideally, when performing a single-particle reconstruction, images will be collected over a range of different defocus values, typically spanning a ~1 μm range. Since the CTF has periodic zeros, if reconstructions used data with a single defocus, or a very narrow range of defocuses, the SNR will be dramatically reduced at these points. A similar problem can exist even with multiple defocus values if they are poorly selected and have coincident or near-coincident zeros (figure 5.3).

5.2.4 Local resolution

The vast majority of large molecules contain domains where conformational rearrangements are involved in molecular function, and molecules in solution are frequently exploring many such local conformations. In such situations, even when some particle classification is performed, dynamic regions of the molecule will still suffer from motion blurring, which will limit the high-resolution information in local regions within the structure.

For this reason, 'resolution maps' are frequently computed and these may then be used either to provide a visual representation of how the resolution varies in different parts of the map, to perform filtration differently on different parts of the map (Cardone *et al* 2013, Bell *et al* 2018), or to identify regions where focused classification should be performed, as described in section 4.5. One may see, for example, protein side chains in well-resolved regions of the map and at the same time see only smeared out secondary structure in other regions of the map. If local filtration is not applied, maps are typically filtered to represent the best resolved regions of the map. Doing this will cause the more dynamic regions to look like poorly structured noise because they have been under-filtered. When proper local filtration is applied (figure 5.4) these noisy regions will resolve into useful, if low-resolution, densities.

Several methods exist for computing resolution maps, based on different conceptual methods, and with different limitations. One easy-to-understand approach is the local FSC approach, where the map is subdivided into many smaller regions and the FSC is computed between matching regions (figure 5.4) (Cardone *et al* 2013). Unfortunately, similar to the argument against using small box sizes for reconstructions, if this is not done carefully over-estimates of the local resolution are possible. Even when this method is computed correctly the limited number of voxels in each local region means the local resolution numbers have much greater statistical uncertainty than the resolution computed for the overall map. Other methods, such as those implemented by ResMap (Kucukelbir *et al* 2014) and MonoRes (Vilas *et al* 2018), use alternative techniques to achieve the same aim. As a result, different local resolution estimates may measure different but related mathematical properties. While this means they will not perfectly agree, generally regions of high variability can be identified using any of these tools.

A general trend, often observed in local resolution maps, is for there to be a correlation with worsening resolution at higher radius. This does not occur in all cases and it may be software dependent. One cause for this is straightforward to understand: when particles are aligned translationally and rotationally in

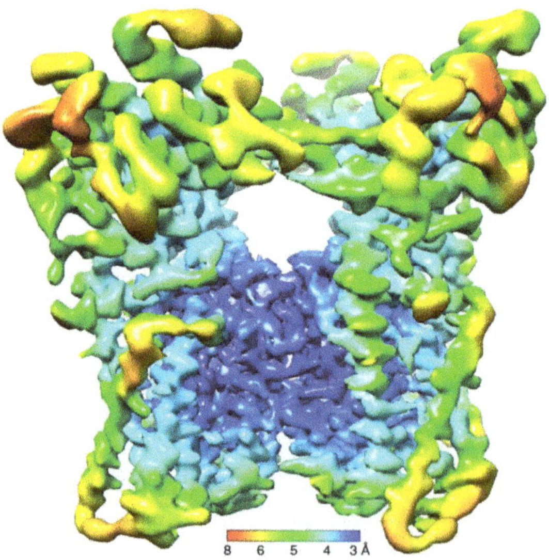

Figure 5.4. Representative local resolution computed for a map of TRPv1 with an overall resolution of 3.3 Å with applied local filtration. This was computed using the e2fsc tool in EMAN2.

single-particle reconstruction, there is some unavoidable uncertainty associated with the alignment. Translational uncertainty produces a uniform decrease in resolution throughout the map. However, rotational uncertainty causes increased blurring at higher radius, which may be reflected in local resolution estimates.

A method related to local resolution is the computation of variance maps (Penczek *et al* 2006, Zhang *et al* 2008). These maps do not provide local resolution, but rather they compute the local variance of the voxel values within the map in real-space at a single specified resolution. While this will not provide resolution values it can provide a measure of the reliability of features in local regions of the map and can also be used to identify portions of the structure which may require focused classification to resolve.

5.2.5 Resolution anisotropy

As described in section 4.6, preferred particle orientation can lead to a variety of undesirable effects. The most prominent of these is anisotropic resolution. Unfortunately, a structure with, for example, 5 Å resolution in the worst direction and 3 Å resolution in the best direction, makes presentation and analysis challenging. If the map is isotropically filtered to 5 Å resolution the observed features should be reliable and the structure will have a pleasing appearance. However, any useful information extending to 3 Å resolution, where side chains may begin being visible, will be eliminated. Conversely, if the map is filtered to 3 Å resolution then there will be artifacts along the lower resolution direction, appearing as directional smearing or sometimes a directional noise pattern. There are some proposed methods for assessing this preferred orientation problem (Baldwin and Lyumkis 2020, Naydenova *et al* 2019, Grigorieff 1998, Penczek 2002), but the only real solution is to collect data with more isotropic orientations, as also discussed in section 4.6.

5.3 Recognizing the effect of bias and over-fitting

Tim Grant
John W and Jeanne M Rowe Center for Research in Virology, Morgridge Institute for Research, Madison, WI, USA
Department of Biochemistry, University of Wisconsin-Madison, Madison, WI, USA

Peter B Rosenthal
Structural Biology of Cells and Viruses Laboratory, Francis Crick Institute, London, United Kingdom

5.3.1 Introduction and nature of the problem arising from iterative refinement

Structure determination of biological macromolecules by averaging projection images recorded in the electron microscope requires assignment of parameters describing the particle orientation and position in the image as well as microscope parameters (described in section 4.3) that affect the appearance of the particle in the image. Due to the radiation sensitivity of biological specimens, low exposure images must be recorded to avoid damaging high-resolution features (section 3.2).

The challenge in single-particle structure determination is therefore to assign these orientation parameters correctly in the resulting low SNR images. Parameter assignment is typically performed through an iterative refinement scheme (figure 5.5) in which projections of an initial map in different orientations are compared with the images to find the orientations that match each image. A new map is then calculated from all the images using the assigned orientations. The new map may be an improved structure compared to the starting map in the sense of possessing more detailed features and in the next refinement iteration projections of the map may have better agreement with the images and therefore assign more accurate orientation parameters to the particle images. The resolution limit of the data may be extended between iterations.

The accuracy of alignment will be determined by the SNR of the images, which depends on the amount of electron exposure and other factors. Errors in assigning orientations will blur map features and prevent map improvement (see section 5.4). In some cases the particle orientation assignment may be wrong, such as when the calculated map projection matches the wrong features in an image or matches noise rather than signal in the images.

For the case where the starting map does not match structure in the images, iterations of refinement may not bring the map into closer agreement with the images. The persistence during refinement of map features in the starting reference map that do not match signal in the data is called reference bias or model bias (Henderson 2013, Penczek *et al* 1992, Shatsky *et al* 2009, Sigworth 1998, van Heel 2013). A related problem occurs in the course of iterative refinement when noise in an image aligns to a reference map which biases the orientation parameters ('noise bias') (Chen *et al* 2013, Grigorieff 2000, Stewart and Grigorieff 2004). Depending on the refinement procedure, noise may be added as a feature of the map that is reinforced through refinement, particularly at high-resolution where noise may dominate over structural information. This build-up of extraneous map features due to aligned noise is called over-fitting or over-refinement.

A dramatic demonstration of the map bias problem, frequently discussed in the literature, occurs when a map is used to align images consisting of pure noise. When the image consists of pure noise the search for the optimal projection will always

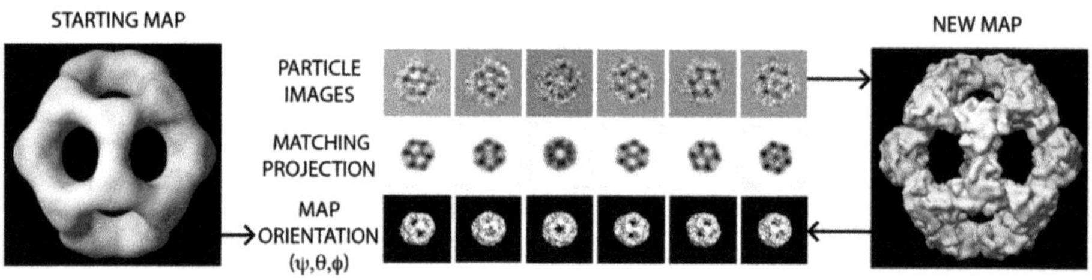

Figure 5.5. Iterative refinement of single-particle orientation parameters. Orientations are assigned to images by comparison to a low-resolution starting map. A new map is calculated from the new orientations and the procedure is repeated until no further improvement is observed. Based on the data described in Rosenthal and Henderson (2003). Reproduced with permission from Rosenthal and Rubinstein (2015). Copyright 2015 Elsevier.

identify a pattern in the noise that matches the map better than others. The average of the aligned noise images resembles the starting map, increasingly so as more noise images are added. In this case, there are no true structural features to guide the iterative refinement to any other outcome than the starting map. This is often called the 'Einstein from noise' problem from a famous example of template matching in two-dimensions, where images containing only noise are aligned against an image of Albert Einstein's face and the resulting average reveals the features of the template (Henderson 2013, Penczek *et al* 1992, Shatsky *et al* 2009, Sigworth 1998, van Heel 2013).

This map bias problem can potentially affect several important steps during image analysis. Its significance for any particular structure determination problem depends on the size of the particle, which is the major determinant of signal in the image, but is also influenced by heterogeneity and flexibility and also sample preparation and imaging conditions that reduce signal and alignment accuracy. Image comparison steps relying on cross-correlation are particularly susceptible and the advantages of using other target functions have been noted (Shatsky *et al* 2009, Stewart and Grigorieff 2004). More recent, widely adopted refinement protocols that use maximum likelihood target functions are more robust against over-fitting and, by increasing the radius of convergence for a wide range of starting maps, decrease sensitivity to initial map bias (Lyumkis *et al* 2013, Scheres 2012a, 2012b, Scheres and Carazo 2009, Sigworth 1998). Maximum likelihood 3D classification of heterogeneous specimens is discussed in section 4.5.

The importance of recognizing the effects of bias and over-fitting as well as the need for objective validation tools for structure determination in cryo-EM have been highlighted (Henderson *et al* 2012, Rosenthal and Rubinstein 2015), as they have been for other structural techniques. In x-ray crystallography a starting map with initially poor phases is improved by refinement of the coordinate model. Problems of model bias and over-fitting have been addressed by constraining degrees of freedom during refinement and assessing the progress of refinement by excluding reflection data for cross-validation (Brunger 1992). In single-particle cryo-EM, map determination and coordinate model refinement are separate steps. Mitigation of the bias problems are therefore not identical to those in crystallographic refinement, but related concepts of constrained refinement and cross-validation are useful.

5.3.2 Assessing the consistency of maps with projection data

Given the problem of map bias, demonstrating consistency of a map with projection data is important. While more powerful and automatic tools become available for the structure determination process, critical evaluation of the first steps in structure determination, including particle selection, 2D classification, and *de novo* map calculation is an opportunity to make sure that structure determination is on a firm ground and to avoid problems that could plague subsequent steps. The 3D map should agree with the image data used for reconstruction: map projections should agree with individual images (figure 5.5) and also with higher SNR 2D classes obtained by averaging.

5.3.2.1 Particle selection

The first step is selection of individual particles from the images (i.e. aligned movies). Large numbers of particle images are required to calculate a map at high resolution and millions of particles may be identified in images by a computer template matching procedure (section 4.2). In unfavorable cases, low dose images may be indistinguishable from noise in such procedures, particularly for smaller particles recorded very close to focus, and therefore may lead to an Einstein from noise problem. Verification that there is real particle signal in the image is vital to avoid bias leading to an artifactual structure. One recommended step is to record some images under conditions so that they may be visually identified as having the expected size and shape. This may be achieved by recording images at large defocus and electron exposure that enhance low-resolution features (Henderson 2013). The template used for selecting particles should be filtered to low resolution to avoid higher resolution detailed features matching noise at the outset. Failure to filter a high-resolution reference may have been responsible for at least one invalid structure in the literature (Henderson 2013, Subramaniam 2013, van Heel 2013). It is also essential that the image averages obtained in the subsequent 2D and 3D classification and averaging steps reveal information beyond the resolution limit used to filter the selection template (Scheres 2015). Absence of high-resolution features in image averages may suggest that the template is only finding matches in the noise up to the limit of the data included in the alignment and the outcome is simply the result of model bias.

5.3.2.2 2D classification and averaging

Classification and averaging of particles in 2D is the next step after particle selection with the goal of obtaining high signal-to-noise class averages of the particles. These averages ideally include particles in similar conformations and in similar orientations. Class averages that show detailed features suggest the particles from which they are calculated are structurally homogeneous and are the basis for selection of good particles for further analysis. These 2D image averages can also provide high signal-to-noise views for solving the problem of determining the relative orientation of different views in calculating a 3D structure. However, 2D averages obtained by iterative alignment may also be in part or in entirety a consequence of noise bias to the template. To address this problem 2D classification is started from a set of randomly generated references (e.g. an average of a random subset of particle images) that typically resemble featureless ‘blobs’. Iterative refinement producing featured class averages starting from a featureless reference suggests limited model bias. However, even with reference-free methods (Penczek *et al* 1992, Schatz and van Heel 1990, Sigworth 1998) the results of 2D classification and subsequent steps may be biased if the original particle picking was biased.

With improved SNR of images, the inclusion of large numbers of particles, and classification into different views and different conformers, the class averages should show high-resolution features of the molecule in projection including secondary structure (figure 5.6(A)). Although it is difficult to know whether the map has largely aligned to signal or noise, some examples of over-fitted noise are commonly

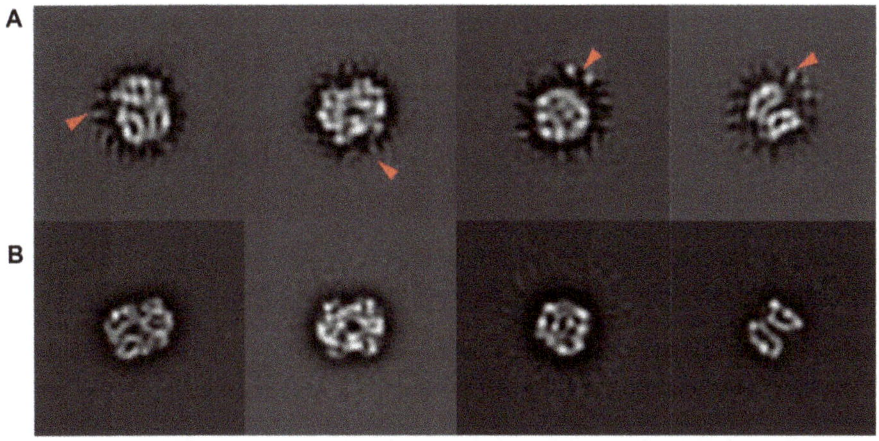

Figure 5.6. 2D image averages show structural features and noise. (A) Class averages for hemoglobin (64 kDa) following 20 iterations of classification show secondary structural features as well as spurious streaks (arrows) at the periphery within the radius of the circular mask. (B) These streaks are absent from the final average when an automasking procedure that suppresses noise in the class averages is applied between iterations. Class averages in (B) correspond to similar but not identical orientations to (A). Processing of EMPIAR 10250 (EMD-0409) in cisTEM. T Grant, unpublished.

recognized (Scheres 2016). The molecule in the averages may display correct low-resolution features, but may possess non-protein-like features in the form of density spikes including those that project from the particle periphery (figure 5.6(A)). These likely arise from over-fitted noise and can be suppressed by alignment routines that ensure that the strongest signal associated with the molecule is used in successive iterations (figure 5.6(B)).

There are always likely to be bad classes amongst the good ones that are largely the result of over-fitting. However, the absence of any detailed features in 2D images or the observation of over-fitting in some or all classes may be related to a lack of good quality images that support a strong classification of high-resolution features such as when images are excessively noisy due to the thickness of the ice film, poor imaging conditions that degrade high-resolution information, or excessive heterogeneity that prevents averaging of coherent and detailed features. Changing or improving 2D alignment procedures may affect this outcome. Possibly, a return to an experimental step of protein expression, purification, or improved imaging may be required to advance the structure determination.

5.3.2.3 Validating low-resolution 3D maps by tilt-pair analysis

There are several approaches to calculating *de novo* 3D structures from high SNR 2D class averages. Calculated projections of a 3D map should agree with class averages over a wide range of orientations. There should also be evidence for symmetry which affects the comparison of the 3D structure with projections and the reconstruction calculation. The case where an incomplete distribution of views is obtained such as with preferred orientations and associated problems for structure determination is discussed in section 4.6. A map can be wrong when 2D class averages are not assigned correct relative orientations or when the image classes themselves are wrong.

One of the few validation tests to demonstrate the agreement of a map with image data at low resolution is tilt-pair analysis (Henderson *et al* 2011, Rosenthal and Henderson 2003, Russo and Passmore 2014, Wasilewski and Rosenthal 2014). By recording a second image of the same particle field at a different orientation of the specimen tilt holder, the view of the particle should change in a predictable way, as expected from the 3D map (figure 5.7(A)). If the assigned orientations in the two images are related by the known tilt rotation of the specimen holder in the experiment, then the orientations are unlikely to be the result of chance alignment to noise. Tilt-pair analysis demonstrates the consistency of tilt-pair transformations for individual particles with the known tilt angle and tilt axis of the experiment. Tilt-pair analysis of a small number of particles (as few as 50–100) should show whether orientation parameters are consistent in the two images and indicate the accuracy of the assignment, which is better for large, homogeneous specimens (figure 5.7(B)). Failure of the test may indicate that the map is wrong, possibly only aligning to noise in the images and requiring further analysis. Map quality contributes to success in orientation determination and an improved map should perform better in the test. The tilt-pair test may be used to optimize parameters used in refinement, demonstrating the matching of signal rather than noise which is important to preventing over-fitting. For example, tilt-pair analysis shows that the highest resolution data in an image often makes little contribution to orientation determination (Henderson *et al* 2011) and can be excluded from refinement, while still contributing to the map reconstruction step. Analysis of tilted data is also required for experimental determination of absolute hand, though absolute hand may be determined from map features at high resolution.

5.3.3 Detecting over-fitting at high resolution in maps and effect on the FSC

If a map is correct at low resolution, successful iterative refinement and increasing numbers of particle images should improve the map with the appearance of structural features associated with quaternary structure, secondary structure, and ultimately amino acid side-chain features related to the protein sequence depending on the resolution obtained.

The appearance of the structural features uniquely specified by chemical features associated with a protein or nucleic acid sequence would seem to be sufficient to validate the map and the structure determination process, at least for the 3D class or conformer studied. This validation assumes that the high-resolution structural features were not introduced from external information such as by alignment of images using a map with high-resolution features where model bias is a concern. Much as in 2D classification, there is therefore an advantage in *de novo* reference-free determination of the initial reference maps. Where it is necessary to begin iterative refinement with a map possessing intermediate or high-resolution features, low-pass filtering of the starting map is essential to remove the features that could potentially bias the refinement, and to ensure any recovered high-resolution

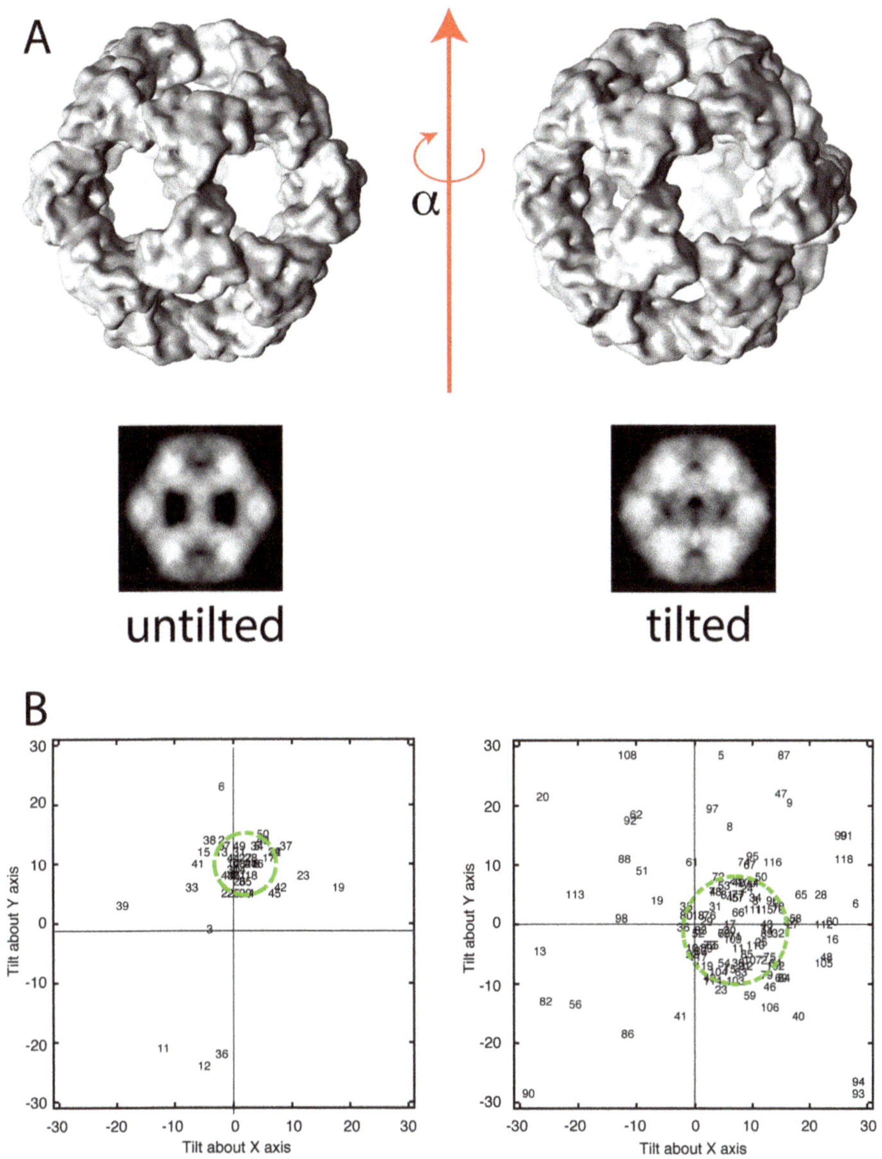

Figure 5.7. Tilt-pair analysis. (A) Schematic of tilt-pair experiment shows a 3D map (1.5 MDa *B. stearothermophilus* pyruvate dehydrogenase E2 core) in two orientations related by a tilt axis direction and tilt angle $\alpha = 15°$. Calculated projections are shown below the maps for 'untilted' and 'tilted' orientations. The direction of the electron beam is orthogonal to the plane of the figure. (B) Tilt-pair parameter plot for (left) 50 particles of the pyruvate dehydrogenase E2 core (1.5 MDa) and (right) 119 particles of β-galactosidase (450 kDa). The orientation of the experimental tilt axis is different for the two cases. The cluster of orientations at the tilt axis and tilt angle for β-galactosidase is wider than that for pyruvate dehydrogenase. Based on data in Henderson et al (2011). Reproduced with permission from figures 1, 4 and 5(c) of Wasilewski and Rosenthal (2014). Copyright 2014 Elsevier.

information is not the result of model bias. The appearance of high-resolution features then represents strong validation. However, over-fitted noise may still be a hazard in some cases because high-resolution noise features may resemble protein features such as side chains (Scheres and Chen 2012). Deciding what represents signal and what represents noise may be a challenge. Objective analysis of local features of the map with a fitted model should distinguish real protein features from spurious ones (section 6.4).

5.3.3.1 Detection of bias and over-fitting in the FSC

The progress of iterative map refinement is usually assessed by a global resolution improvement, the most widely reported measure being the FSC (Saxton and Baumeister 1982, van Heel and Harauz 1986, van Heel and Stoffler-Meilicke 1985). As discussed in detail in section 5.2, the FSC is a measure of the agreement or consistency between half-maps, each calculated from half the data, and is a measure of the SNR as a function of resolution. When images in both data half sets are aligned to the same map, there is potential for problems of bias and over-fitting to affect the calculated FSC (Grigorieff 2000, Scheres and Chen 2012). Noise in all of the particle images becomes correlated to the same reference and, when half-map reconstructions are calculated and compared with each other via the FSC, agreement between correlated noise can lead to an increase in the measured FSC values. Identical noise features build up on both half-maps and this can be re-enforced and amplified in subsequent rounds of refinement. The maps thus contain over-fitted noise features and the FSC is biased leading to exaggerated resolution claims.

The FSC calculated for a single map aligned against pure noise images will show strong correlations to the limit of the data resolution used in refinement and then a sharp decline with no signal at higher resolution. This type of FSC curve is sometimes observed in real single-particle cases, suggesting the map is primarily matching noise in the images, and is one example of many problems that can be revealed by detailed inspection of the FSC curve (Penczek 2010).

Resolution-limited refinement places an upper resolution limit on the data used for refinement and is an important tool for validating the FSC. While data used for refinement below this limit is potentially subject to over-fitting, the high-resolution data beyond the limit is not used for particle alignment and is a useful measure of whether the agreement between half-datasets persists to higher resolution. These 'free shells' are analogous to reflection data excluded in x-ray crystallographic refinement (Grant *et al* 2018, Rosenthal and Henderson 2003, Scheres and Chen 2012, Shaikh *et al* 2003). When used for iterative refinement in image processing programs, the resolution limit may be extended in successive iterations to include higher resolution data when there is demonstrated signal present.

The most commonly applied approach is now to use two independent maps and particle sets (gold-standard refinement, section 5.2). While each half-map may align to both structural signal and noise, only common signal between the maps during averaging will be reinforced. Thus, while noise bias will exist in each of the half sets, this bias will be different between the half sets and so will not result in increased FSC values.

During refinement the resolution used in refining particle parameters is typically limited to strong signal by filtering or down-weighting resolution shells by their signal content. Different algorithms may do this slightly differently (Grant *et al* 2018, Punjani *et al* 2017, Scheres 2012a), but in all cases this weighting of the signal is related to the FSC between the two half-maps. Ideally this means that only resolution shells with strong SNR will be used in alignments and alignment of high-resolution noise features or over-fitting will be reduced. Resolution-limited refinement and gold-standard refinement approaches to reducing over-fitting and validating the FSC exploit the observation that the lower resolution data, or similarly, the signal in a map calculated from half the data, are sufficient for accurate alignments (Grant *et al* 2018, Henderson *et al* 2011, Scheres 2012a, 2012b, Scheres and Chen 2012).

Chen *et al* (2013) have proposed the noise-substitution test to demonstrate and measure the amount of over-fitting in the FSC in any refinement procedure. The test is performed alongside the usual refinement of particle orientations in images, but in which data shells in the images higher than an arbitrarily chosen resolution are replaced with structure factors of noise. This is achieved by randomizing image structure factor phases in these higher resolution shells. As there is no signal beyond the chosen resolution in the phase randomized images, the FSC measured between half-maps reconstructed from them is expected to be zero for the randomized shells (as in the aforementioned case for the precipitous drop of FSC when images consist of pure noise only), and any signal detected in the FSC beyond the chosen resolution must be due to over-fitting. By the arguments above, signal in the FSC should be detected just to this limit. Figure 5.8(A) shows an example where the test of the refinement procedure indicated over-fitting of noise and an exaggerated resolution estimate from the FSC, for all choices of threshold. The noise-substitution test in figure 5.8(C) shows that gold-standard refinement avoids bias in the FSC. Note that phase randomization is also used to assess FSC artifacts caused by masking as discussed in section 5.2.

5.3.4 Local over-fitting

Progress in determining cryo-EM structures at high-resolution has also shown that map resolution may not be uniform, particularly due to flexible or disordered regions or due to other sources of heterogeneity. Measures of local resolution are discussed in section 5.2. Where local resolution measures are high, the strong map signal may be validated by coordinate models that match the local map features.

Because the FSC is a global resolution indicator, its use as a global filter during refinement treats all parts of a map in the same way. It may prevent over-fitting in the strongest, best resolved parts of the map which define the resolution limit of map features used in refinement. However, filtering to the global resolution limit may leave excessive noise in the poorly resolved regions of a map. Iterative refinement may result in the local build-up of spurious features in these weak regions that can still lead to incorrect biological interpretation and also contributes to poorer alignment parameters in each iteration.

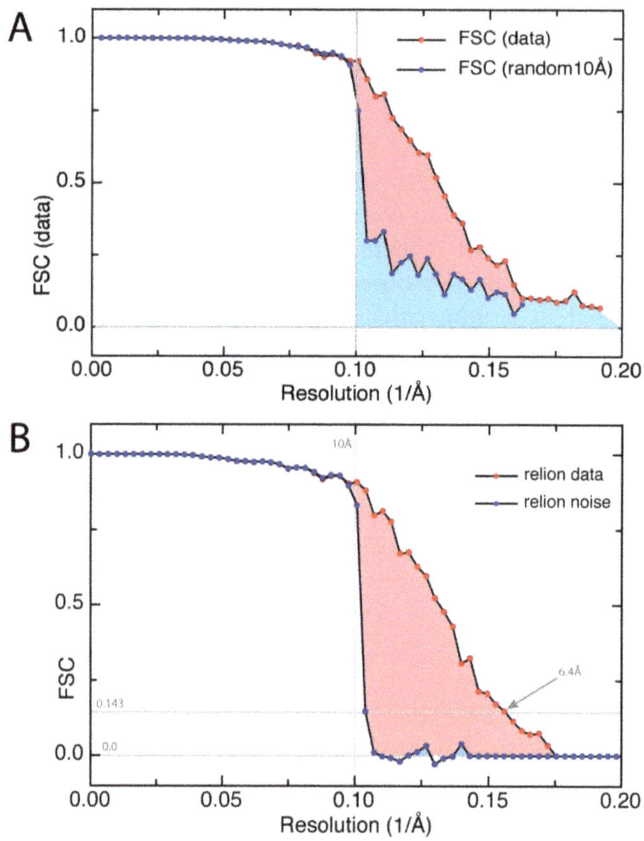

Figure 5.8. Assessment of bias in the FSC by the high-resolution noise-substitution test. (A) A case where suboptimal particle orientation refinement leads to over-fitting. The FSC is shown between half-maps for original image data (red circles) and image data with randomized phases beyond 10 Å (blue circles). The portion of FSC signal due to over-fitting is shown by the shaded blue region and the portion due to true structural signal is given by the shaded pink region. (B) The FSC between half-maps refined using independent half-maps and FSC-weighting (program *Relion*) for original image data (red circles) or image data with randomized phases beyond 10 Å (blue circles). Reproduced from panels (b) and (c) of figure 3 of Chen *et al* (2013). CC BY 3.0 unported, https://creativecommons.org/licenses/by/3.0/.

A striking example is the case of a map of a membrane protein containing a micellar transmembrane region, such as a detergent-solubilized membrane protein. Due to difficulty in crystallizing membrane proteins, detergent-solubilized membrane proteins are an important application for single-particle analysis. Because the micelle contains many disordered detergent molecules, it should appear smooth and devoid of strong high-resolution features. However, in many cases noise builds up in the micellar region creating highly textured rather than smooth features indicating over-fitting (figure 5.9(A)). Because this is a spatially separated region with distinct structural properties, a mask may be employed to remove its effect in refinement or to filter out high-resolution information if the low-resolution signal of the micelle is

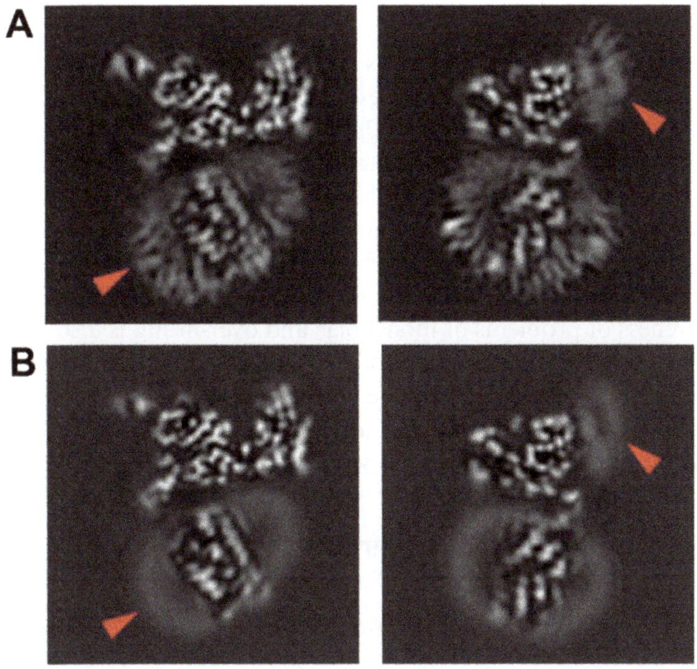

Figure 5.9. Appearance of over-fitted regions in a 3D map of a detergent-solubilized membrane protein. (A) Slice of a 3D map of a cannabinoid receptor in two different orientations refined using a standard refinement procedure showing a streaky appearance of the micellar region (arrow, left) and disordered extra-membrane region (arrow, right). (B) Slices of the 3D map refined using local weighting procedure shows more uniform features in the micelle region and for disordered region. Processing of EMPIAR 10288 (EMD-0339) in cisTEM. T Grant, unpublished.

beneficial to particle alignment (Grant *et al* 2018, Oldham *et al* 2016). Similarly, disordered polypeptide regions may develop high-resolution artifacts.

Recently, new objective procedures have been developed to address the variable signal within the map and to reduce over-fitting in weaker regions, such as non-uniform refinement (Punjani *et al* 2020) and LAFTER/SIDESPLITTER (Ramlaul *et al* 2019, Ramlaul *et al* 2020). While using different methodologies, both procedures down-weight spatial frequencies in map regions that are inconsistent between subsets of the data (e.g. half-maps). This is similar to the global filtering in standard refinements, but allows regions within a map to be weighted differently. As show in figure 5.9(B), local signal-to-noise weighting prevents over-fitting in micellar or disordered extracellular regions of a membrane protein map.

As discussed earlier, an improved map that better represents the structural features in the image data should also determine more accurate orientations leading to better structures from refinement. Indeed, improved orientation determination has been reported when local over-fitting is reduced (Punjani *et al* 2020). Thus, for high-resolution study of many specimens it may be essential to address the problem of local over-fitting. Ideally these procedures should not abandon advantages of successful global procedures.

5.3.5 Conclusion

Structure determination by cryo-EM has had success over many years, but in recent years it has dramatically improved in resolution, expanding the range of issues that must be considered in map validation. Progress has also increased understanding of the problems of map bias and over-fitting during particle selection and iterative map refinement, but nevertheless these problems are always present to some degree. Students of the subject should make sure that all steps in structure determination are robust against potential problems of bias and over-fitting to avoid the worst mistakes, but also to obtain the optimal results when structure determination is largely correct. Awareness of problems of model bias and over-fitting is also important when interpreting the maps by molecular models. Similar validation approaches include excluding data from atomic coordinate refinement, or fitting models against one half-map and evaluating the agreement against the second map. In addition, global scores should be supplemented by local scores of model and map agreement.

5.4 Estimates of alignment accuracy

J Bernard Heymann
Laboratory for Structural Biology Research, NIAMS, NIH, Bethesda, MD 20892, USA

The history of cryo-EM is marked by specific advances in technology that increased the achievable resolution. These include the development of the field emission gun (FEG) (Crewe *et al* 1968) and direct electron detectors (Faruqi and Henderson 2007). The first improved the coherence of the electron source and the second raised the detector SNR as well as allowed for beam-induced movement compensation (Li *et al* 2013). At each stage one of the key reasons for the limit in resolution is the accuracy of particle image alignment. Yet we still have a relatively poor understanding of how to estimate and address alignment accuracy.

Single-particle analysis (SPA) for cryo-EM suffers from a chicken-and-egg causality dilemma. To achieve a proper 3D reconstruction from 2D particle images, we need the correct orientations of those images. To obtain the correct orientations we need a reference map with sufficiently accurate detail. The most appropriate reference map is the one reconstructed from the particle images. The implication is that, in an absolute sense, we can never know the correct orientations of the particle images, nor whether we reconstructed a correct map.

However, the assumption that the particles of interest adopt one or few discrete structures means that there must be some coherence between the images. Both 2D and 3D alignment is based on classifying the images according to consistent features. Significant correlation is typically taken as an indication of correctness, but this can be misleading. An incorrect reference map may produce similar correlation results compared to when using a correct map, indicating a local minimum on the correlation landscape. Identification of such a situation requires proper validation approaches. This means that our understanding of alignment accuracy is indirect and we have to be careful to avoid bad choices.

5.4.1 Correlation and the signal-to-noise ratio (SNR)

Alignment algorithms employ a search strategy of some sort (grid, gradient descent, or stochastic) and an objective function to compare images. Whatever strategy is used, the accuracy of the alignment is crucially dependent on the objective function and how it is interpreted.

All objective functions are related to correlation and suffer from the same pragmatic issues. The assessment of resolution is also based on correlation in the form of FSC (Harauz and van Heel 1986). Furthermore, the SNR of the particle images relates to alignment and the SNR of the reconstruction relates to the FSC. We therefore talk about very similar statistical measures, only differing in the particular contexts in which they are used. In image processing we manipulate the statistical distributions to reveal consistencies from which we can derive reasonable results such as 2D averages and 3D reconstructions. How we perform these manipulations determines the outcomes.

If we assume that every image is composed of a desirable signal, S, and the rest is undesirable noise, N, the most common model is a simple addition of the two (Heymann 2018b, Penczek 2002):

$$F = S + N. \tag{5.1}$$

This is typically calculated in frequency space (also referred to as Fourier or reciprocal space) to allow filtering or analysis at different spatial frequencies (resolutions). The signal is taken as the part that is consistent between particle images (Heymann 2018b) and ideally should reflect the actual structure of the particle.

If we compare two images with an underlying 'true' signal S, the cross-correlation function is

$$\begin{aligned} CC &= \frac{\sum_{\Omega}(S + N_1)(S^* + N_2^*)}{\sqrt{\sum_{\Omega}|S + N_1|^2 \sum_{\Omega}|S + N_2|^2}} \\ &= \frac{\sum_{\Omega}|S|^2 + \sum_{\Omega}SN_2^* + \sum_{\Omega}S^*N_1 + \sum_{\Omega}N_1N_2^*}{\sqrt{\sum_{\Omega}|S + N_1|^2 \sum_{\Omega}|S + N_2|^2}}. \end{aligned} \tag{5.2}$$

The summation is over any user-defined, or algorithm-defined, domain Ω (single voxels, shells, or whole images in frequency space) and * denotes the complex conjugate. Because the signal decays with spatial frequency, most algorithms use an effective low-pass filter to eliminate the dominant noise at high frequencies (decreasing the contribution made by the last three terms in the numerator of equation (5.2)). Strong features in the micrograph, such as crystalline ice, give high correlation coefficients with typical templates used for particle picking and alignment, showing that correlation is often an ambiguous tool. Alignment accuracy is therefore a delicate function of the balance between the signal and noise contributions.

When data are merged from enough particle images, 2D averaging and 3D reconstruction average out the noise to a point that it can be neglected. We can then eliminate one of the noise terms in equation (5.2) and obtain increased correlation coefficients. This works well if the signal contribution is high enough and the average images represent the actual structure of the particles. However, if the noise is dominant the averaging may produce any arbitrary results that eventually lead to incorrect reconstructions. The skill in image processing for cryo-EM lies in understanding how to deal with this and validate the result, more than it is the particular algorithms or software used (Heymann 2018a, Heymann *et al* 2018).

5.4.2 Analysis of alignment accuracy with synthetic data

We can only obtain accurate estimates of alignment errors using synthetic data. This is with the understanding that synthetic data are simplified compared to real data because of assumptions that must be made about the nature of particle images and background noise. We typically neglect the presence of neighboring molecules and other undesirable objects in the images. We also often use the exact map from which the images were obtained as a reference. This corresponds to equation (5.2) where the signal is exactly known and only one noise term remains. This simplification means that the results are informative only as to the best possible outcome from alignment.

Because the SNR is ill-defined (box 5.1 and figure 5.10) we cannot have an absolute scale for it. Nevertheless, the relative values can be of use to obtain an idea of how the SNR influences alignment. A simple experiment is to add specific amounts of Gaussian noise to noise-less projections of a map and align them to the map as reference. This eliminates many other uncertainties, leaving only the relationship between the SNR and alignment. Figure 5.11 shows the errors obtained for two cases, one asymmetric and the other icosahedral (an extended analysis is presented in Heymann 2015). The translational error decreases gradually with SNR, while the rotational error shows a strong dependence on the SNR within a ten-fold change. The difference in positions of the curves for the two cases results from the uncertainty in the definition of the SNR.

Such experiments are typically only done when evaluating alignment algorithms. Regardless of the algorithm, the common trend is a decrease in alignment accuracy that is strongly dependent on the amount of noise within a specific transition region (Cong *et al* 2003, Heymann 2015, Joyeux and Penczek 2002, Sander *et al* 2003, Scheres *et al* 2005). The transition, which happens to occur in the range of SNR values estimated for cryo-electron micrographs, has a strong influence on alignment. The main difference between the various alignment methods is a shift on the SNR scale, while the shape remains similar. It is difficult to assess whether the performances of the algorithms are inherent in their fundamental approaches, or whether they need to be refined to properly address the universal statistical properties of the images.

Efforts to model the alignment errors with respect to the SNR resulted in only a limited understanding of the quantitative relationship (Baldwin and Penczek 2005,

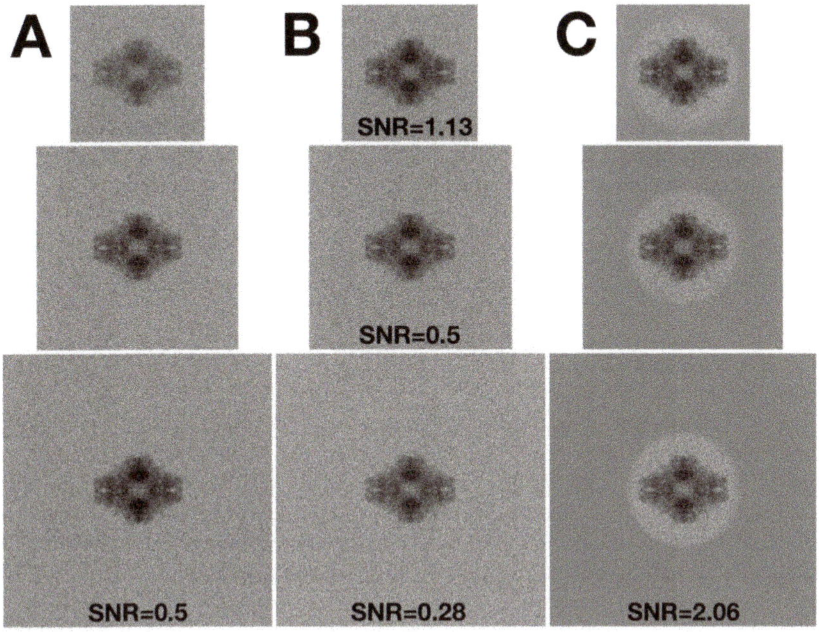

Figure 5.10. Illustration that the numerical value of the SNR depends upon the box size, and upon masking (if any) because the particle is limited in size (has compact support) while the noise extends to the image edge. (A) Projections rendered with the same SNR but different box sizes, giving different contrast between the particle and the background. (B) Projections rendered to adjust the SNR to the box size, resulting in the same local SNR for the particle and contrast. (C) The images in (B) masked with a soft mask to produce the same SNR, regardless of box size. The box edge sizes are 264, 396, and 528. Note that the SNR values chosen here are for illustrative purposes, while micrograph SNR values are usually much lower.

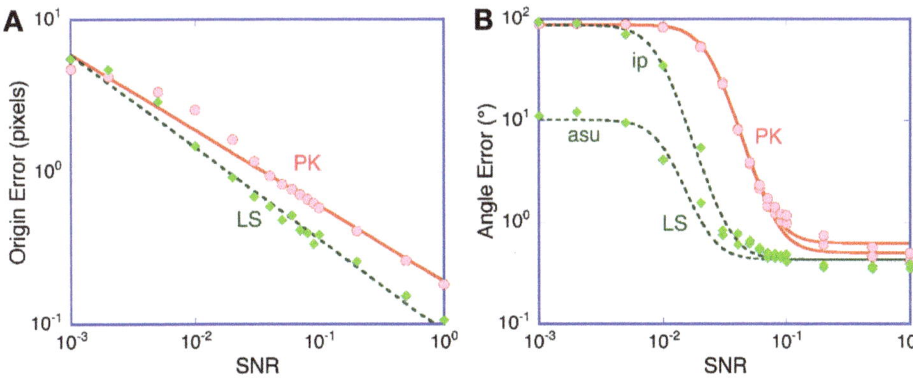

Figure 5.11. Alignment accuracy for two synthetic cases: asymmetric proteinase K (3DYB)—dots (PK); icosahedral lumazine synthase (1NQX)—diamonds (LS). (A) Translational error decreases gradually with increased SNR. (B) Rotational error decreases abruptly with increased SNR. For each case two curves are shown: angular error of the projection view within an asymmetric unit (asu) and in-plane rotation error (ip) (see Heymann (2015) for a description of how it was done).

Box 5.1. The definition of the SNR

Although the SNR is such an important, widely referenced parameter, it is actually poorly defined. The reason is that the signal (the particle) has a finite size (or compact support in mathematical terms), while the area in which there is noise is infinite (or at least extends to the image edge). Figure 5.10 shows an *in silico* experiment to illustrate what the SNR means. We calculate a synthetic image where we add an image of the particle to an image of Gaussian noise, such that the SNR is defined as the ratio of the signal and noise variances:

$$\alpha = \frac{\sum_i (s_i - \langle s \rangle)^2}{\sum_i (n_i - \langle n \rangle)^2}, \tag{5.3}$$

where s_i and n_i are the signal and noise values at pixel i, and $\langle s \rangle$ and $\langle n \rangle$ denote the average signal and the noise averages, with the latter approximately zero. Figure 5.10(A) shows what happens when we use a fixed SNR but change the box size. It is clear that the particle stands out better in the larger box size. However, in an electron micrograph we can pick any box size, but the particle contrast relative to the background will remain the same. Figure 5.10(B) shows a simulation where the particle to background intensity is constant. However, now the calculated SNR depends on the box size, because we include more noise in the bigger box. In SPA it is common to mask the particle, thus eliminating a lot of background noise and effectively increasing the SNR. Figure 5.10(C) shows the effect of masking the images in figure 5.10(B), obtaining the same SNR and contrast independent of box size. The user is responsible for choosing appropriate box and mask sizes, decisions that may impact the final quality of reconstruction achieved.

An important contribution to the SNR is the thickness of the specimen, which will therefore also affect alignment accuracy. Obtaining a specimen with the ice as thin as possible (close to the size of the particle) is a prerequisite to achieving the best possible resolution, as has been discussed in section 2.4.

Jensen 2001). The main problem is that the alignment parameters (translational and rotational) are coupled, resulting in a complex situation. What is clear is that these errors lead to a suppression of information at high resolution and can be empirically modeled as a B factor (see the next section) (Henderson *et al* 2011, Heymann 2019).

5.4.3 The relationship between alignment accuracy and resolution

The first attempt to derive an estimate of the resolution of a reconstruction from the amount and distribution of data was by Crowther *et al* (1970). Their formula estimates the resolution from a tilt series reconstruction (such as in tomography) with an even spacing of the tilt angles of Δa (in radians), giving the resolution as a function of the particle diameter, D:

$$d \sim \Delta a D. \tag{5.4}$$

While widely quoted (see, for example, section 1.2.7), this formula does not take into account the effect of other influences such as noise, spectral signal decay, contrast transfer, and most importantly, alignment accuracy. Nevertheless, it reflects the required angular step size to cover a shell at a particular spatial frequency. We can therefore not expect the resolution to be better than this.

We can simulate the effect of alignment errors on resolution in a synthetic case. When we impose errors on exact particle orientations and do a reconstruction, the suppression of the signal at high frequency becomes evident (figures 5.12(A) and (B)) (Heymann 2019). The signal power relative to the error-free reconstruction can then be fitted to a Gaussian decay curve, parameterized either by a B factor or a characteristic width, $\sigma = 1/B$. While the relationship of this curve to the FSC curve is not straightforward, where it approaches zero is the maximum resolution we can expect for a reconstruction. Figures 5.12(C) and (D) shows how it was applied in a real case (the 450 kDa β-galactosidase) (Heymann 2019), where the angular error inferred from the synthetic case is about 2° for a reconstruction that gave a resolution of 3.4 Å ($FSC_{0.143}$). Using equation (5.4) and taking the particle diameter as ~120 Å, we estimate the angular error corresponding to 3.4 Å as 1.6°.

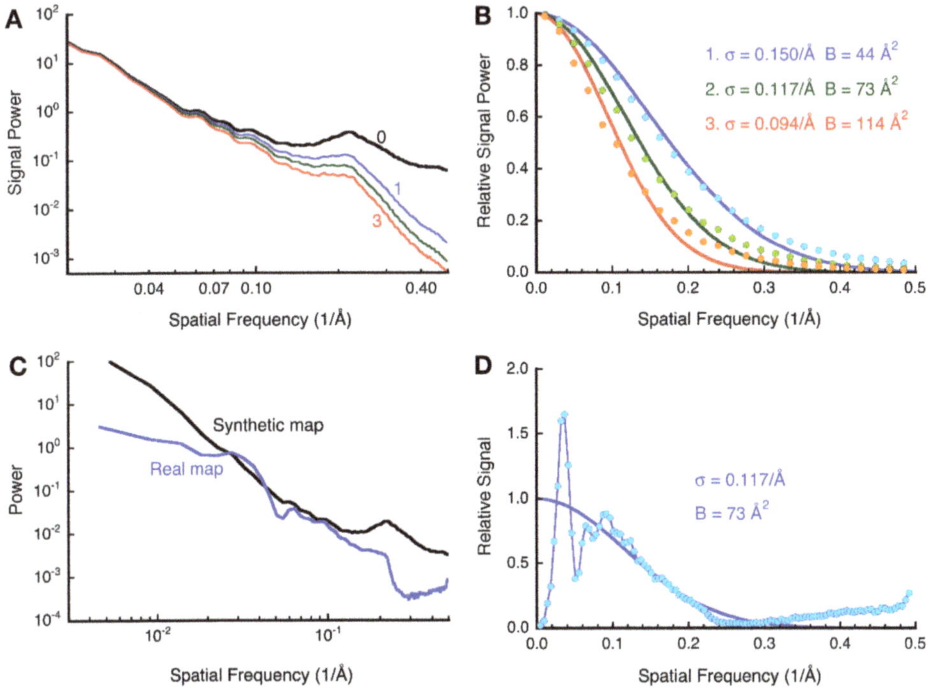

Figure 5.12. Estimating the suppression of the reconstruction signal as a result of alignment errors and deriving representative B factors. (A) Radial power spectra of reconstructions from synthetic particles with increasing imposed errors (in degrees). (B) The relative signal calculated as the ratio of the error-imposed to error-free radial power spectra in (A). (C) Radial power spectrum of a real map (blue) compared to a that of a synthetic map (black). (D) The relative signal calculated as the ratio of the real and synthetic map radial power spectra in (C). (See Heymann (2019) for a detailed description.)

5.4.4 Estimating alignment accuracy from tilt pairs

As indicated before, to calculate exact errors we need an exact reference ('a gold-standard'), which we do not have for real data. If we acquire two or more images of the same particle, the agreement in their alignment to an appropriate reference could be used to determine angular accuracy. Rosenthal and Henderson proposed the use of tilt pairs of micrographs to assess alignment errors (Rosenthal and Henderson 2003). Figure 5.13 is based on the analysis of angular error between tilt pairs of several cases by Henderson *et al* (2011). The angular error shows a clear trend relative to the size of the particle, i.e. the amount of coherent signal.

Note that this study (Henderson *et al* 2011) was done before the widespread use of direct detectors and the results should not be taken as representative of the alignment accuracy that can be achieved today. For instance, the smallest molecule in their study, β-galactosidase, was imaged on a CCD camera and gave an alignment error of ~10°. Later work with direct detectors produced much higher quality data with corresponding smaller alignment errors. The case shown in figures 5.12(C) and (D) is again β-galactosidase, with an inferred alignment error of 2°. The angular errors in figure 5.13 should therefore be much lower for micrographs acquired with current direct detectors.

5.4.5 Estimating alignment accuracy from the reconstructed map

Alignment of particle images with respect to a reference map relies on the detail present in the map. One way to test this is to assess the effect of errors in alignment on the calculation of a matching function. Scheres implemented a scheme in Relion to evaluation the magnitude of imposed error required to decrease the similarity between error-free and error-imposed projections from a map (Scheres 2012a). Scheres also showed that there is a good correspondence between these angular accuracy estimates and those of Henderson *et al* (2011) shown in figure 5.13.

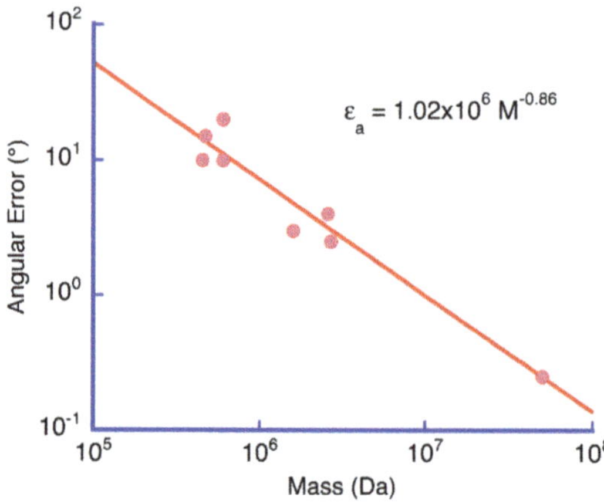

Figure 5.13. Angular error estimates from micrograph tilt-pair analysis shows a clear relationship to the signal power as represented by the specimen molecular mass. (Data from Henderson *et al* 2011.)

Nevertheless, it is an indirect assessment of overall alignment error and strongly related to the detail in the map.

5.4.6 Estimating alignment accuracy from projection-matching results

Alignment involves calculating scores for matching the particle images to a set of reference projections covering all the orientations. In principle, the accuracy should be related to how much better the best score is compared to other scores. Vargas *et al* (2016) proposed a measure calculating the angular spread of the highest seven scores for each particle (they call it 'particle precision'). In addition, they calculate the same kind of spread, but compared to a score from the previous iteration of alignment (they call it 'particle accuracy'). The first measures the correspondence to the current reference map, while the second measures whether the orientational assignment is stable from iteration to iteration. As expected, there is a correspondence between these measures for good particles. Together, these two measures can be used to better select particles for inclusion in reconstructions. An important feature of this approach is that it distinguishes alignment errors that are close to the correct orientations from those that are due to spurious correlations with noise. Unfortunately, it does not produce actual estimates of the errors.

5.5 Discussion

The success of a cryo-EM experiment primarily lies in obtaining a sufficient quantity of data with a good SNR and random particle orientations. The subsequent image processing is then subject to the limitations imposed by imaging conditions. Assuming that we have a uniform rigid structure, or that we can classify a dataset into a set of rigid structures, the SNR of the data determines alignment accuracy.

Many software packages have been developed for SPA (https://en.wikibooks.org/wiki/Software_Tools_For_Molecular_Microscopy). Given a good reference map, all packages are able to produce about the same quality reconstructions assuming proper use (Heymann *et al* 2018). The main reason is that the underlying statistics of the data remain the same and the algorithms work as long as they follow good statistical practices. Unfortunately, only a few reports of algorithm development for cryo-EM include an analysis of alignment errors as a function of SNR (for an example, see Joyeux and Penczek 2002). The most common assessment of an algorithm is in producing a reconstruction with a good FSC curve. While this is a valid target, it misses the opportunity to understand how the algorithms treat alignment errors.

Extracting the most information out of a cryo-EM dataset depends on how the user handles the SNR in the particle images (Heymann 2018a, Heymann *et al* 2018). As shown in figure 5.10, the user can manipulate the effective SNR by changing the box size and imposing a mask. Care must be taken not to use a mask that, itself, introduces high frequency components that may influence alignment and cause over-fitting (Heymann 2018a).

The user also provides a reference map that may bias the alignment and result in converging on a local minimum in alignment parametric space. The reference map is often masked as well and the same issues arise as with masking the particle images.

There is a risk that an arbitrary map may be produced from noise when too many particles are used for alignment and reconstruction without first having a well-established reference map. To detect this situation, the user can run a coherence test, comparing the map quality obtained from real particles to that obtained from noise (Heymann 2015). The hope is that, as our understanding of SPA improves, we will eventually eliminate the user effect.

Care must be taken not to overinterpret the estimates for alignment accuracy. These estimates are valuable to assess how well the alignment progresses, but the specific values depend on an understanding what is actually calculated. The tilt-pair analysis (Rosenthal and Henderson 2003) comes the closest to a true alignment accuracy estimate but it relies on a reference map with sufficiently detailed information to support the alignment. The Relion method (Scheres 2012a) is to a large extent a reflection of the reconstruction quality and is therefore related to the FSC assessment. The method of Vargas *et al* (2017) is a pragmatic approach to understand the relative accuracy and to improve particle selection, but it does not offer estimates of translational and rotational errors.

Ultimately, we understand that we will never have a true measure of alignment accuracy for real data. The assessments developed so far are valuable in providing guidance to the user. The hope is that this effort will continue and we will have better alignment accuracy estimates in the future.

Acknowledgements

This work was supported by the Intramural Research Program of the National Institute for Arthritis, Musculoskeletal and Skin Diseases, NIH.

References

Baldwin P R and Lyumkis D 2020 Non-uniformity of projection distributions attenuates resolution in cryo-EM *Prog. Biophys. Mol. Biol.* **150** 160–83

Baldwin P R and Penczek P A 2005 Estimating alignment errors in sets of 2-D images *J. Struct. Biol.* **150** 211–25

Bell J M, Chen M, Durmaz T, Fluty A C and Ludtke S J 2018 New software tools in EMAN2 inspired by EMDatabank map challenge *J. Struct. Biol.* **204** 283–90

Brunger A T 1992 Free *R* value: a novel statistical quantity for assessing the accuracy of crystal structures *Nature* **355** 472–5

Cardone G, Heymann J B and Steven A C 2013 One number does not fit all: mapping local variations in resolution in cryo-EM reconstructions *J. Struct. Biol.* **184** 226–36

Chen S, Mcmullan G, Faruqi A R, Murshudov G N, Short J M, Scheres S H and Henderson R 2013 High-resolution noise substitution to measure overfitting and validate resolution in 3D structure determination by single particle electron cryomicroscopy *Ultramicroscopy* **135** 24–35

Chen V B *et al* 2010 MolProbity: all-atom structure validation for macromolecular crystallography *Acta Crystallogr.* D **66** 12–21

Cong Y, Kovacs J A and Wriggers W 2003 2D fast rotational matching for image processing of biophysical data *J. Struct. Biol.* **144** 51–60

Crewe A V, Eggenberger D N, Wall J and Welter L M 1968 Electron gun using a field emission source *Rev. Sci. Instrum.* **39** 576–83

Crowther R A, De Rosier D J and Klug A 1970 The reconstruction of a three-dimensional structure from projections and its application to electron microscopy *Proc. R. Soc. Lond.* A **317** 319–40

Faruqi A R and Henderson R 2007 Electronic detectors for electron microscopy *Curr. Opin. Struct. Biol.* **17** 549–55

Frank J and Al-Ali L 1975 Signal-to-noise ratio of electron micrographs obtained by cross correlation *Nature* **256** 376–9

Grant T, Rohou A and Grigorieff N 2018 cisTEM, user-friendly software for single-particle image processing *eLife* **7** e35383

Grigorieff N 1998 Three-dimensional structure of bovine NADH:ubiquinone oxidoreductase (complex I) at 22 A in ice *J. Mol. Biol.* **277** 1033–46

Grigorieff N 2000 Resolution measurement in structures derived from single particles *Acta Crystallogr.* D **56** 1270–7

Harauz G and van Heel M 1986 Exact filters for general geometry three dimensional reconstruction *Optik* **73** 146–56

Henderson R *et al* 2012 Outcome of the first electron microscopy validation task force meeting *Structure* **20** 205–14

Henderson R 2013 Avoiding the pitfalls of single particle cryo-electron microscopy: Einstein from noise *Proc. Natl Acad. Sci. USA* **110** 18037–41

Henderson R, Chen S, Chen J Z, Grigorieff N, Passmore L A, Ciccarelli L, Rubinstein J L, Crowther R A, Stewart P L and Rosenthal P B 2011 Tilt-pair analysis of images from a range of different specimens in single-particle electron cryomicroscopy *J. Mol. Biol.* **413** 1028–46

Herzik M A, Fraser J S and Lander G C 2019 A multi-model approach to assessing local and global cryo-EM map quality *Structure* **27** 344–58.e3

Heymann J B 2015 Validation of 3D EM reconstructions: the phantom in the noise *AIMS Biophys.* **2** 21–35

Heymann J B 2018a Map challenge assessment: fair comparison of single particle cryoEM reconstructions *J. Struct. Biol.* **204** 360–7

Heymann J B 2018b Using reconstruction statistics to predict the number of images required for single particle analysis *Microsc. Microanal.* **24** 1216–7

Heymann J B 2019 Single-particle reconstruction statistics: a diagnostic tool in solving biomolecular structures by cryo-EM *Acta Crystallogr.* F **75** 33–44

Heymann J B *et al* 2018 The first single particle analysis map challenge: a summary of the assessments *J. Struct. Biol.* **204** 291–300

Jensen G J 2001 Alignment error envelopes for single particle analysis *J. Struct. Biol.* **133** 143–55

Joyeux L and Penczek P A 2002 Efficiency of 2D alignment methods *Ultramicroscopy* **92** 33–46

Kucukelbir A, Sigworth F J and Tagare H D 2014 Quantifying the local resolution of cryo-EM density maps *Nat. Methods* **11** 63–5

Li X, Mooney P, Zheng S, Booth C R, Braunfeld M B, Gubbens S, Agard D A and Cheng Y 2013 Electron counting and beam-induced motion correction enable near-atomic-resolution single-particle cryo-EM *Nat. Methods* **10** 584–90

Ludtke S J and Serysheva I I 2013 Single-particle cryo-EM of calcium release channels: structural validation *Curr. Opin. Struct. Biol.* **23** 755–62

Lyumkis D, Brilot A F, Theobald D L and Grigorieff N 2013 Likelihood-based classification of cryo-EM images using FREALIGN *J. Struct. Biol.* **183** 377–88

Naydenova K, Peet M J and Russo C J 2019 Multifunctional graphene supports for electron cryomicroscopy *Proc. Natl Acad. Sci. USA* **116** 11718–24

Oldham M L, Grigorieff N and Chen J 2016 Structure of the transporter associated with antigen processing trapped by herpes simplex virus *eLife* **5** e21829

Penczek P, Radermacher M and Frank J 1992 Three-dimensional reconstruction of single particles embedded in ice *Ultramicroscopy* **40** 33–53

Penczek P A 2002 Three-dimensional spectral signal-to-noise ratio for a class of reconstruction algorithms *J. Struct. Biol.* **138** 34–46

Penczek P A *et al* 2006 Estimation of variance in single-particle reconstruction using the bootstrap technique *J. Struct. Biol.* **154** 168–83

Penczek P A 2010 Resolution measures in molecular electron microscopy *Methods Enzymol.* **482** 73–100

Punjani A, Rubinstein J L, Fleet D J and Brubaker M A 2017 cryoSPARC: algorithms for rapid unsupervised cryo-EM structure determination *Nat. Methods* **14** 290–6

Punjani A, Zhang H and Fleet D J 2020 Non-uniform refinement: adaptive regularization improves single particle cryo-EM reconstruction *Nat. Methods* **17** 1214–21

Ramlaul K, Palmer C M and Aylett C H S 2019 A local agreement filtering algorithm for transmission EM reconstructions *J. Struct. Biol.* **205** 30–40

Ramlaul K, Palmer C M, Nakane T and Aylett C H S 2020 Mitigating local over-fitting during single particle reconstruction with SIDESPLITTER *J. Struct. Biol.* **211** 107545

Rayleigh F R S 1879 XXXI. Investigations in optics, with special reference to the spectroscope *Lond. Edinb. Dubl. Philos. Mag. J. Sci.* **8** 261–74

Rosenthal P B and Henderson R 2003 Optimal determination of particle orientation, absolute hand, and contrast loss in single-particle electron cryomicroscopy *J. Mol. Biol.* **333** 721–45

Rosenthal P B and Rubinstein J L 2015 Validating maps from single particle electron cryomicroscopy *Curr. Opin. Struct. Biol.* **34** 135–44

Russo C J and Passmore L A 2014 Robust evaluation of 3D electron cryomicroscopy data using tilt-pairs *J. Struct. Biol.* **187** 112–8

Sander B, Golas M M and Stark H 2003 Corrim-based alignment for improved speed in single-particle image processing *J. Struct. Biol.* **143** 219–28

Saxton W O and Baumeister W 1982 The correlation averaging of a regularly arranged bacterial cell envelope protein *J. Microsc.* **127** 127–38

Schatz M and van Heel M 1990 Invariant classification of molecular views in electron micrographs *Ultramicroscopy* **32** 255–64

Scheres S H 2012a RELION: implementation of a Bayesian approach to cryo-EM structure determination *J. Struct. Biol.* **180** 519–30

Scheres S H 2012b A Bayesian view on cryo-EM structure determination *J. Mol. Biol.* **415** 406–18

Scheres S H 2015 Semi-automated selection of cryo-EM particles in RELION-1.3 *J. Struct. Biol.* **189** 114–22

Scheres S H 2016 Processing of structurally heterogeneous cryo-EM data in RELION *Methods Enzymol.* **579** 125–57

Scheres S H and Carazo J M 2009 Introducing robustness to maximum-likelihood refinement of electron-microscopy data *Acta Crystallogr.* D **65** 672–8

Scheres S H and Chen S 2012 Prevention of overfitting in cryo-EM structure determination *Nat. Methods* **9** 853–4

Scheres S H, Valle M, Nunez R, Sorzano C O, Marabini R, Herman G T and Carazo J M 2005 Maximum-likelihood multi-reference refinement for electron microscopy images *J. Mol. Biol.* **348** 139–49

Shaikh T R, Hegerl R and Frank J 2003 An approach to examining model dependence in EM reconstructions using cross-validation *J. Struct. Biol.* **142** 301–10

Shatsky M, Hall R J, Brenner S E and Glaeser R M 2009 A method for the alignment of heterogeneous macromolecules from electron microscopy *J. Struct. Biol.* **166** 67–78

Sigworth F J 1998 A maximum-likelihood approach to single-particle image refinement *J. Struct. Biol.* **122** 328–39

Stewart A and Grigorieff N 2004 Noise bias in the refinement of structures derived from single particles *Ultramicroscopy* **102** 67–84

Subramaniam S 2013 Structure of trimeric HIV-1 envelope glycoproteins *Proc. Natl Acad. Sci. USA* **110** E4172–4

Tang G *et al* 2007 EMAN2: an extensible image processing suite for electron microscopy *J. Struct. Biol.* **157** 38–46

van Heel M 1987 Similarity measures between images *Ultramicroscopy* **21** 95–100

van Heel M 2013 Finding trimeric HIV-1 envelope glycoproteins in random noise *Proc. Natl Acad. Sci. USA* **110** E4175–7

van Heel M and Harauz G 1986 Resolution criteria for three dimensional reconstruction *Optik* **73** 119–22

van Heel M and Schatz M 2005 Fourier shell correlation threshold criteria *J. Struct. Biol.* **151** 250–62

van Heel M and Schatz M 2017 Reassessing the revolutions resolutions *bioRxiv* 224402

van Heel M and Stoffler-Meilicke M 1985 Characteristic views of *E. coli* and *B. stearothermophilus* 30S ribosomal subunits in the electron microscope *EMBO J.* **4** 2389–95

Vargas J, Melero R, Gomez-Blanco J, Carazo J M and Sorzano C O S 2017 Quantitative analysis of 3D alignment quality: its impact on soft-validation, particle pruning and homogeneity analysis *Sci. Rep.* **7** 6307

Vargas J, Oton J, Marabini R, Carazo J M and Sorzano C O 2016 Particle alignment reliability in single particle electron cryomicroscopy: a general approach *Sci. Rep.* **6** 21626

Vilas J L *et al* 2018 MonoRes: automatic and accurate estimation of local resolution for electron microscopy maps *Structure* **26** 337–44.e4

Wasilewski S and Rosenthal P B 2014 Web server for tilt-pair validation of single particle maps from electron cryomicroscopy *J. Struct. Biol.* **186** 122–31

Zhang W *et al* 2008 Heterogeneity of large macromolecular complexes revealed by 3D cryo-EM variance analysis *Structure* **16** 1770–6

Chapter 6

Model building and validation

6.1 Overview

This chapter describes (1) the processes used to build atomic-resolution models, which represent an interpretation of the density maps produced by cryo-EM, and (2) the computational tools available for evaluating and validating these models. Section 6.2 begins with the situation that high-resolution models are known for one or more subunits, or perhaps for homologous macromolecules, which then need to be fitted into a cryo-EM density map. Next, section 6.3 goes on to describe the tools that are used to build an atomic-resolution model of the structure, without there being a previously determined structure to which to refer. Both topics are again taken up in section 6.5. Many of these software tools have been adapted from ones that were initially developed for use in x-ray crystallography, or were at least inspired by them.

Conceptually, fitting a known, atomic-resolution model into a new density map, or even building an *ab initio* model that reflects the known chemical structure of a macromolecule, should be a very reliable and forgiving process. In reality, ambiguities often arise as to how to interpret the density map and thus mistakes in the model are always possible. This is particularly true when the resolution of the map is not high. Section 6.4 thus discusses ways in which not only the quality of the map but also the quality of the model can be evaluated, and these topics are discussed again in section 6.5.

This chapter then concludes with section 6.6, which describes how both cryo-EM structures and data can be archived for open access to the entire community. Availability of atomic models is invaluable to the biological scientific community to move forward in their understanding of the system being described, as well as to biotech and pharma that often rely on structure-based drug design and improvement to produce effective therapeutics. Years ago, with the boom of crystal structures, the funding agencies, publishers, and scientific community agreed that such deposition must be an obligation upon publication. Beyond atomic coordinates, density maps (or structure factors for x-ray crystallography) are also invaluable tools to judge the

doi:10.1088/978-0-7503-3039-8ch6

accuracy and/or reliability of the models. In the case of cryo-EM, repositories exist also for the raw data that have proven very useful for algorithm developers to test new software, or for the community to improve previous density maps and models as these data analysis tools improve. The shear volume of data being produced by the cryo-EM field poses an important logistic and even economic challenge that image repositories must face moving forward.

6.2 Using known components or homologs: model building

Sony Malhotra[1,2]*, Agnel Praveen Joseph*[2] *and Maya Topf*[3]
[1]*Institute of Structural and Molecular Biology, Department of Biological Sciences, Birkbeck College, University of London, Malet Street, London WC1E 7HX, UK*
[2]*Scientific Computing Department, Science and Technology Facilities Council, Research Complex at Harwell, Didcot OX11 0FA, UK*
[3]*Centre for Structural Systems Biology, Heinrich-Pette-Institut, Leibniz-Institut fur Experimentelle Virologie, Hamburg, Germany, and Universitat klinikum Hamburg Eppendorf, Hamburg, Germany*

For a macromolecular assembly represented by a cryo-EM map at less than atomic resolution, if atomic models of the assembly or its components are available it is usually helpful to use them to gain more information on the structure. The initial model of the assembly is generated by fitting (or docking) the models rigidly into the cryo-EM map. This step is often followed by a flexible fitting (or refinement). The choice of technique(s) to be used in this process is governed by the resolution of the map (figure 6.1).

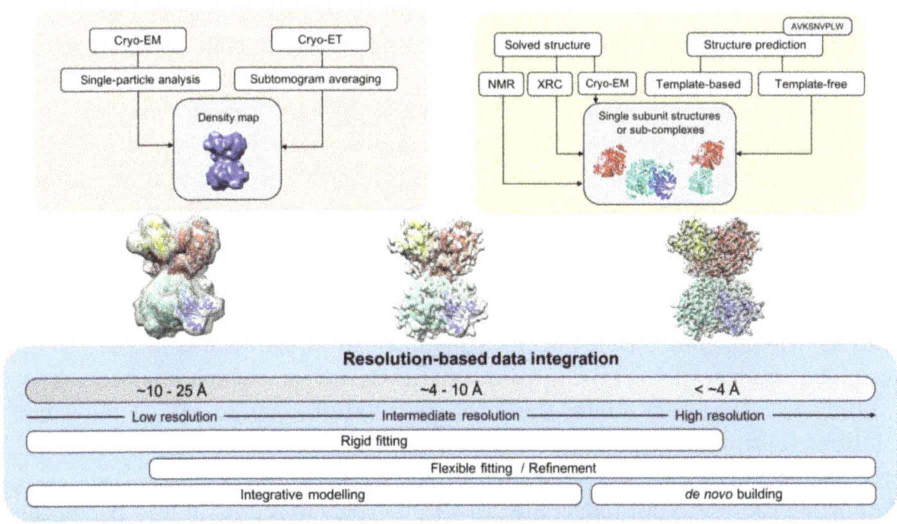

Figure 6.1. Approaches to determine structures of macromolecular complexes using cryo-EM at different resolution ranges. The structure of a macromolecular assembly can be determined using cryo electron microscopy or tomography (top-left box). The atomic structure of the individual components or subcomplexes (top right box) may have been previously solved experimentally, or they may need to be predicted (either with or without a template). The structures of the components/subcomplexes are then integrated using an approach that is generally dictated by the resolution of the density map (bottom box). EM = electron microscopy; ET = electron tomography; XRC = x-ray crystallography; NMR = nuclear magnetic resonance.

6.2.1 Identifying known/modeled structures of individual subunits

Structures of components can be derived by techniques such as x-ray crystallography, NMR spectroscopy, and increasingly high-resolution cryo-EM (Bai *et al* 2015). When no experimental structure is available, models of the components can sometimes be predicted using template-based or template-free structure-prediction approaches (figure 6.1). Template-based modeling (or homology modeling) refers to the calculation of a protein structure using the knowledge of a structurally related protein ('template'). Template-free (*ab initio*) modeling aims to predict the structure without having such knowledge (Kuhlman and Bradley 2019). The latter approach is becoming increasingly more accurate and is gaining popularity due to recent advances in contact prediction (Abriata *et al* 2018) and machine learning (Senior *et al* 2020) methods. Methods that integrate these different concepts, such as RaptorX (Peng and Xu 2011) and I-TASSER (Roy *et al* 2010), have been very successful in achieving high accuracy in the modeling of difficult protein targets (Kc 2017, Kryshtafovych *et al* 2019).

Prior to fitting the known/modeled structural component(s) in the cryo-EM map, it is useful to subject them to quality assessment. Methods that assess the geometry of the structure and identify regions with clashes and outliers in the Ramachandran plot include MolProbity (Chen *et al* 2010) and CaBLAM (Richardson *et al* 2018a). Models can also be assessed by statistical or knowledge-based potentials derived from an analysis of known protein structures in the Protein Data Bank (Berman *et al* 2000), such as DOPE(MODELLER) (Shen and Sali 2006, Sali and Blundell 1993), QMEAN (Benkert *et al* 2008) and ProQ3D (Uziela *et al* 2017), as well as by hybrid approaches, such as ModFold (McGuffin 2008). The step of model assessment can often help to avoid errors that could become aggravated during the refinement step, for example by excluding unreliable regions of the model (Joseph *et al* 2016, Topf *et al* 2006). Next, the goodness-of fit between the input model and the maps is evaluated during optimization of the position and orientation of the model in the density map (described in section 6.2.2) and the most commonly used score is the cross-correlation. This is also useful for comparing a given model to other possible models with different conformations (which could either be generated during fitting or originating from different experiments and/or structure-prediction methods) and can be expressed in a statistical form (Topf *et al* 2006, Farabella *et al* 2015, Velázquez-Muriel *et al* 2005). The TEMPy software (Farabella *et al* 2015) provides other global scores, such as the local mutual information, which has been shown to be more discriminatory than the cross-correlation coefficient at intermediate resolutions (and its combination with a map/model overlap information gives best precision and accuracy across a range of resolutions (Joseph *et al* 2017a)). Since global fitting scores do not reflect any localized errors in model fit it is also necessary to calculate the quality of the fit 'locally', i.e. in defined regions of the map and model, such as a secondary-structure element (TEMPy-SCCC; Farabella *et al* 2015) or a residue (e.g. TEMPy-SMOC (Joseph *et al* 2016), Q-scores (atoms, residues, or entire models) (Pintilie *et al* 2020) and PHENIX CC per residue (Afonine *et al* 2018a)). Efforts to model different targets using current software packages and tools, and a systematic analysis of metrics used for assessing the quality of fits, is provided in the cryo-EM model challenges (www.emdataresource.org).

6.2.2 Rigid-body fitting

Finding an optimal solution to fitting the components in the map is not a trivial task. Success is dependent on several factors including the map resolution (figure 6.1), the size of the component relative to the map, the accuracy of the structural model of the component, and the scoring function used to measure the goodness-of-fit. One option is to rigidly fit each individual component by applying a translational/rotational search, involving six degrees of freedom within the cryo-EM map. This can be done locally (e.g. UCSF Chimera, Goddard *et al* 2007; DOCKEM, Roseman 2000) or globally, using FFT-accelerated docking (e.g. Situs, Wriggers 2010; HADDOCK-EM, van Zundert *et al* 2015), spherical harmonics acceleration (e.g. ADP_EM, Garzón *et al* 2007), Gaussian mixture models (GMMs) (e.g. gmfit, Kawabata 2008) or heuristic approaches (e.g. MODELLER/Mod-EM, Topf *et al* 2005).

One way to avoid misplacing the model in the cryo-EM map is to segment the map (Volkmann 2002, Pintilie *et al* 2010) and focus on the region associated with the specific component, for example when prior knowledge exists from labeling experiments (Yi *et al* 2015). Another approach is to use multiple fitting scores and apply clustering and/or consensus analysis to identify the most probable position of the components in the map (Farabella *et al* 2015). In assembly fitting, multiple components are docked simultaneously in the map (Lasker *et al* 2009, Pandurangan *et al* 2015). Although this approach is computationally expensive, it can be integrated with additional restraints, such as cross-linking mass-spectrometry (Kim *et al* 2018). Software like IMP (Saltzberg *et al* 2019), HADDOCK (van Zundert *et al* 2015), and Power (Degiacomi and Dal Peraro 2013) allow such integration of data from various sources.

6.2.3 Flexible fitting

The conformation of the structural component/s that are rigidly fitted into the cryo-EM map is often different from the conformation represented by the map itself. Additionally, if the initial model was generated using a protein structure prediction method it may suffer from local loop distortions, movement of secondary-structure elements, and other errors that are introduced during the modeling process. Flexible-fitting/refinement approaches can be very useful in these scenarios, in particular at intermediate (~4.5–10 Å) to near-atomic resolution (figure 6.1), whereby the fit to the map is optimized without distorting model geometry.

At intermediate resolution, the positions of secondary-structure elements are likely to be recognizable, with helices seen at lower resolution than beta sheets and beta sheets at lower resolution that beta strands. Therefore, methods have been developed to automatically detect such secondary-structure elements (Baker *et al* 2007). On the other hand, fitting of individual atoms and residues is often not possible without the risk of overfitting, as the map may not contain sufficient information.

Different approaches for flexible fitting have been developed over the years. MDFF (Trabuco *et al* 2008) uses a standard force field with a potential based on the density, while programs such as Flex-EM (Joseph *et al* 2016, Topf *et al* 2008) and CDMD (Igaev *et al* 2019) try to improve the correlation between the modeled structure and the cryo-EM map. In Flex-EM/RIBFIND, rigid-body dynamics is combined with simulated annealing to avoid overfitting, where the rigid bodies can be applied to any group of atoms in a hierarchical fashion (depending on the resolution) (Pandurangan and Topf 2012). MDFF uses molecular dynamics, and restraining of secondary structures, domains, or symmetry can be applied. By investigating dynamics and allowing flexibility of the full protein rather than rigid elements, the high-resolution model can be optimized in the cryo-EM map (Orzechowski and Tama 2008). At resolution of ~4.5 Å, upon integration of Monte Carlo sampling, local density-based optimization in Rosetta refinement can achieve models of atomic level accuracy (DiMaio *et al* 2015). Other approaches use elastic-network models combined with various optimization protocols (e.g. Direx (Wang and Schröder 2012), iMODFIT (Lopéz-Blanco and Chacón 2013), and normal modes (NMFF) (Tama *et al* 2004a)).

At a resolution of ~4–4.5 Å the backbone can be traced and other features such as large side chains start to appear (Terwilliger *et al* 2018b). However, it is still very useful to start with pre-calculated model/s (if available) and refine them. When large conformational changes are observed between the model and the map, flexible-fitting approaches (described above) can be applied first (Joseph *et al* 2016) and this can be followed by further local refinement with software such as ISOLDE (Croll 2018) and common crystallographic software like Refmac (Murshudov *et al* 2011) (combined with COOT, Emsley *et al* 2010) and PHENIX (*phenix.real_space_refine*) (Afonine *et al* 2018b).

Fitting and refinement of atomic models at intermediate-to-low resolution is usually challenging, where the above-mentioned approaches can be adopted to optimize the model. If the initial model is of poor quality, errors can be propagated during further stages of refinement. In particular at low resolution, integrative modeling (Joseph *et al* 2017b) can prove very useful and provide insights into modeling of multi-component assemblies in the cryo-EM maps. Consensus derived using multiple modeling approaches can be used to compare the fitted models at low resolution (Ahmed and Tama 2013, Pandurangan *et al* 2014). To tackle the uncertainty, in particular in regions of poor local resolution (e.g. corresponding to mobile loops) despite a higher global resolution (Kucukelbir *et al* 2014), multiple conformers (ensemble) are sometimes refined and provided in the cryo-EM map (Farabella *et al* 2015, Sachse *et al* 2007, Herzik *et al* 2019).

6.3 Building atomistic models in cryo-EM density maps

Corey F Hryc and Matthew L Baker
Department of Biochemistry and Molecular Biology, Structural Biology Imaging Center, McGovern Medical School at The University of Texas Health Science Center at Houston, 6431 Fannin Street, Houston, TX 77030, USA

6.3.1 Introduction

Over the course of the last two decades, single-particle electron cryo-microscopy (cryo-EM) has emerged from 'blob-ology' and can now routinely produce structures at resolutions that readily allow for atomistic interpretations, directly from a density map without the aid of structural templates (Frank 2009, 2017). This rapid improvement in obtainable resolutions is in part due to the technological advancements in electron microscopes, imaging hardware, and tools for data processing (Kühlbrandt 2014, Weis and Hagen 2020, Faruqi and Henderson 2007, Milazzo *et al* 2011, Bammes *et al* 2011, Cheng *et al* 2016, Mendez *et al* 2019, Zhang *et al* 2010b, Tan *et al* 2016, Lander *et al* 2009, Stagg *et al* 2006). Additionally, new model-building tools have been developed, enabling fast, robust model generation in complex density maps at resolutions better than 5 Å (Kryshtafovych *et al* 2018, Lawson *et al* 2021).

Prior to obtaining near-atomic resolutions (better than 4.5 Å resolution), model building in cryo-EM was essentially restricted to feature-recognition methods (secondary-structure identification) (Zhou *et al* 2000, Kong and Ma 2003, Kong *et al* 2004, Jiang *et al* 2001, Baker *et al* 2007), template-based methods (rigid-body and flexible fitting) (Chiu *et al* 2005, Malhotra *et al* 2019, Kim and Sanbonmatsu 2017, Lindert *et al* 2009), or density-constrained modeling (Topf *et al* 2005, 2006, Topf and Sali 2005, DiMaio *et al* 2009). Established almost a decade prior to obtaining near-atomic-resolution cryo-EM density maps, computational tools could recognize the characteristic cylindrical α-helices and planar β-sheets (Zhou *et al* 2000, Kong and Ma 2003, Kong *et al* 2004, Jiang *et al* 2001, Baker *et al* 2007); the spatial arrangement of secondary structures could be used as a mechanism to search for known structures with homologous folds, providing some basic insight into organization and potential function of the protein (Bowman *et al* 2003, Baker *et al* 2005, Zhou *et al* 2001). Along a similar timeline, a number of exhaustive rigid-body fitting techniques were developed, in which a known structure could be optimally localized to a density map through an exhaustive search process (reviewed in Baker *et al* 2010a). As resolution began to improve, flexible-fitting tools were developed that expanded on rigid-body fitting, allowing a structure of interest to be fitted to a density map and then be deformed to produce a model that better fit the map.

The first forays into computational modeling using cryo-EM sought to integrate the density map itself as a constraint in both *ab initio* and comparative modeling (Topf *et al* 2005, 2006, Topf and Sali 2005, Baker *et al* 2006). Essentially, a density map served as an envelope in fold space that could be used by standard computational modeling tools to constrain model building or enhance the ability to identify the 'correct' structure from an ensemble of potential structures/decoys. Accordingly, the ability of constrained modeling to produce 'correct' structures was related to the resolution of the map. At resolutions less than 10 Å, constrained modeling could identify protein topologies, albeit with relatively poor sequence-to-structure assignment, while higher-resolution maps (5–10 Å) offered improved structural details, although falling short of being able to provide accurate atomistic information. Given higher-resolution density maps (better than 5 Å), along with a few methodological caveats, constrained density modeling could produce high-quality, resolution-appropriate models.

In 2008 the first near-atomic-resolution structures were determined using single-particle cryo-EM; three were viruses (rotavirus (3.8 Å resolution) (Zhang *et al* 2008), cytoplasmic polyhedrosis virus (3.88 Å resolution) (Yu *et al* 2008) and epsilon15 (4.5 Å resolution) (Jiang *et al* 2008)), while the fourth structure was GroEL (4 Å resolution), a chaperonin (Ludtke *et al* 2008) (figure 6.2(A)–(D)). These structures offered unprecedented map detail, clearly resolving the complete backbone path and large, bulky side chains (Zhou 2008). While GroEL and rotavirus had known structures determined by x-ray crystallography, epsilon15 and CPV represented unique challenges for modeling, as they had no known structures or only low sequence-similarity structural homologs. In these cases, new model-building methods had to be developed in order to derive a structural model for the individual components (Yu *et al* 2008, Jiang *et al* 2008, Ludtke *et al* 2008). In both of these works, the key step to building a structural model was the development of a sequence-to-structure alignment based on secondary-structure elements in the map and those predicted in the sequence of the individual proteins. As previously established (Bowman *et al* 2003, Baker *et al* 2005, Zhou *et al* 2001), secondary structures could be used as 'anchors' when determining protein folds. Adapting this

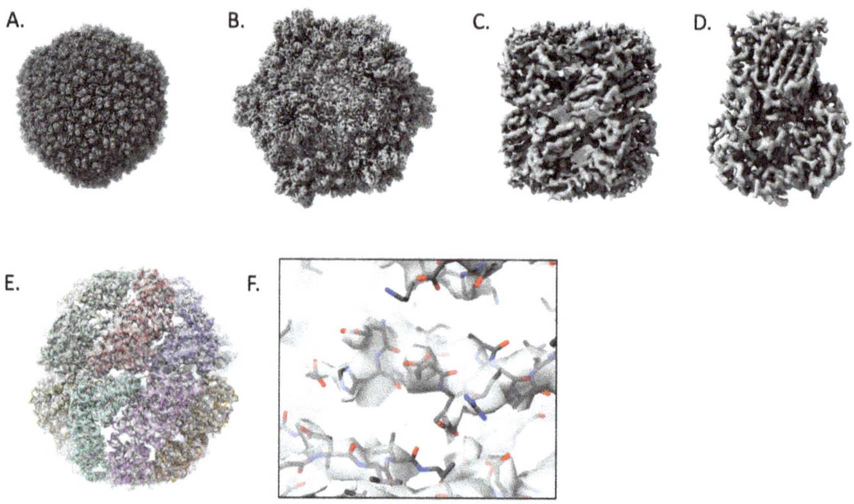

Figure 6.2. Near-atomic-resolution density maps. Four of the earliest near-atomic-resolution density maps from single-particle cryo-EM are shown. The maps are (A) bacteriophage epsilon15 (4.5 Å resolution), (B) cytoplasmic polyhedrosis virus (3.88 Å resolution), (C) GroEL (4 Å resolution), and (D) rotavirus VP6 (3.8 Å resolution). In (E) MM-CPN (4.3 Å resolution) is shown with the model superimposed on the density map. A zoomed-in view of the MM-CPN map and model are shown in (F).

idea for *de novo* modeling, the secondary-structure elements became anchors for mapping the sequence to the observed secondary structure in a density map (Abeysinghe *et al* 2008a, 2008b). The intervening loops could then be connected by tracing the connectivity in the density map assuming an idealized Cα–Cα distance; the bulky side-chain densities could serve as local registration points. It should be noted that, while some large side chains could be seen, the vast majority of side-chain density could not be discerned. As a result, only Cα backbone traces were generated and optimized to fit the density map (Baker *et al* 2010b).

While the methods used to generate models were initially *ad hoc*, formalization of the approach was shown with the GroEL reconstruction and later appeared as part of the first interactive cryo-EM model-building toolkit, Gorgon (Baker *et al* 2011). Rather than using an exhaustive sequence-to-structure matching algorithm, which becomes computationally intractable if more than a few dozen secondary-structure elements are involved, Gorgon implemented an inexact graph-matching approach that could quickly and efficiently generate a sequence-to-structure alignment (Abeysinghe *et al* 2008a, 2008b). Gorgon also featured a number of computational tools to connect loops, optimize model bond distances and generate Cα backbone models.

The first four near-atomic-resolution reconstructions represented a milestone in single-particle cryo-EM, paving the way for more near-atomic-resolution density maps and a new generation of modeling software. Two years later another significant milestone was achieved. Using a map–model refinement procedure, the structure of another chaperonin, MM-CPN, was determined to a resolution such that almost 50% of the non-glycine/alanine/proline side chains could be seen in the density (Zhang *et al* 2010a) (figure 6.2(E) and (F)). Using Gorgon an initial Cα-backbone model was generated. Adapting x-ray crystallographic utilities, such as MolProbity and COOT (Davis *et al* 2007, Emsley *et al* 2010), the model was optimized to fit the density map while maintaining proper stereochemistry. The resulting structural model for MM-CPN was the first all-atom model to be constructed directly from a single-particle cryo-EM density map without the aid of a structural template. Similar to that of earlier modeling tools, this approach would prove to be foundational in the development of new model-building and refinement utilities.

With map resolution improving and modeling tools becoming more prevalent (figure 6.3), data standards, archiving, and deposition became increasingly important (Henderson *et al* 2012, Lawson *et al* 2011, 2016). Arising from a round-table discussion at the January 2010 'Modeling of Cryo-EM Maps' workshop, developers of modeling tools requested a set of 'standards', i.e. maps that would be suitable for them to use in methods development and testing, while users requested more comprehensive knowledge of how to use these tools. In response, a special session, the Cryo-EM Modeling Challenge, was organized at the Pacific Symposium for Biocomputing (PSB), for which method developers were provided with a set of test maps and encouraged to present their methodology and results at the meeting (Ludtke *et al* 2011). Unlike structure-prediction competitions (e.g. Critical Assessment of Structure Prediction (CASP) Moult 2005), the 'ground truth' structures were not necessarily known. Thus, the challenge was a showcase, highlighting six

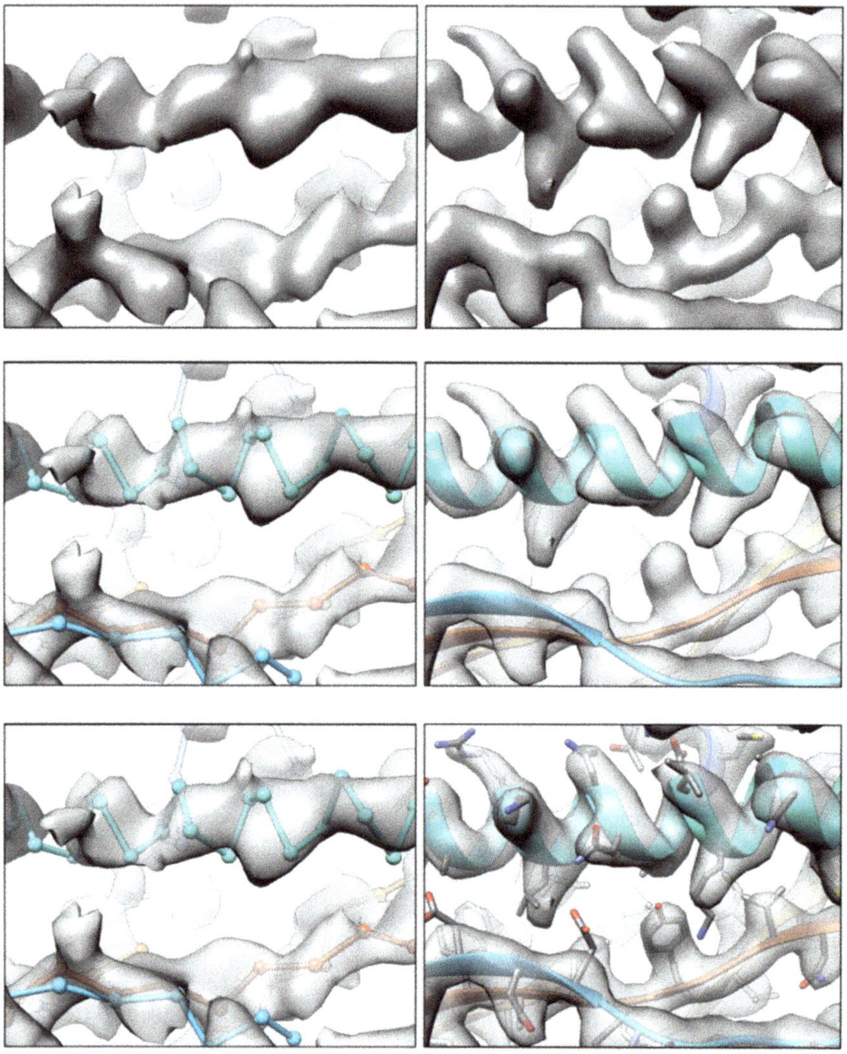

Figure 6.3. Maps and models at near-atomic resolutions. The panels illustrate an example of the interpretations that can be made at subnanometer resolution. The left side panels are the P22 map published at 4 Å resolution (Chen *et al* 2011) (note: using current resolution criteria, the density map is ~5 Å). The right side panels are the latest 3.3 Å resolution P22 map (Hryc *et al* 2017). The top row shows a portion of the density map (α-helix and β-sheet below). The middle row reveals the tracing of the backbone. The bottom row displays the resolution features that can be interpreted in the 3.3 Å map but not in the lower resolution data.

topics: map segmentation, secondary-structure element annotation, protein backbone tracing, rigid-body fitting, flexible fitting, and *de novo* modeling. For this challenge, 135 entries for the 13 datasets (at ~24–2.5 Å resolution) were submitted; all maps and models were subsequently made available through the challenge server.

While there were no 'winners' in the competition, it is recognized that this provided an important entry point to expand the development of computational modeling in near-atomic-resolution cryo-EM density maps.

As a follow-up to the 2010 Cryo-EM Modeling Challenge, a second cryo-EM map/model challenge was sponsored by the EMDataBank (Kryshtafovych *et al* 2018, Heymann 2018, Lawson and Chiu 2018, Heymann *et al* 2018, Richardson *et al* 2018b). Like the first challenge, the 2015/2016 Model Challenge provided developers and users with a set of cryo-EM density maps to benchmark their methods. Taking the original challenge a step further, a comprehensive, independent assessment was also provided and made available through the challenge server. A number of new utilities were demonstrated in the challenge, showcasing everything from refinement of known structures in a map to fully automated model building (Chen and Baker 2018, Yu *et al* 2018, Wang *et al* 2018, Terwilliger *et al* 2018a, Terashi and Kihara 2018). Whereas the first challenge was an introduction to cryo-EM modeling, this challenge truly provided a picture of the emerging utilities and a clear path toward future developments.

6.3.2 Building models into cryo-EM density maps

While density maps at near-atomic resolutions were relatively uncommon even a decade ago, it is now common to achieve better than 4 Å resolution with single-particle cryo-EM (Frank 2017, Glaeser 2019). State-of-the-art structures are even pushing toward 1 Å resolution and can resolve densities for individual atoms (Nakane *et al* 2020, Yip *et al* 2020). As such, computational modeling tools for generating full atomistic models in cryo-EM density maps have also become standard fare in analyzing a reconstruction; more is presented about this in section 6.5.

Briefly, today's modeling tools can be divided into two general classes; template-based and template-free/*de novo*. These two classes can also be further categorized based on their level of automation, with some tools requiring extensive user input while others are nearly completely automated and only require one to supply inputs and basic information. Regardless of which type of modeling tool and their level of automation, each of the tools have some inherent limitations and/or caveats for their use. As an example, the secondary-structure-element based modeling tools found in Gorgon require the accurate identification of secondary structure in both the structure and sequence. Errors in predicting the location and/or length can lead to poor topological assignments and incorrect models. Likewise, poorly resolved regions in a density map may introduce errors in placing side chains in all-atom models.

6.3.2.1 A simple, practical guide to modeling

While there have been a number of different modeling, refinement, and validation utilities developed over the last decade, they all follow the same basic procedures: map preparation, modeling, and assessment. A number of these modeling utilities are 'integrated', incorporating multiple steps into their modeling pipeline. It should be noted that these modeling procedures are highly dependent on map resolution. A general 'pipeline' for building atomistic models from a density map is shown in figure 6.4.

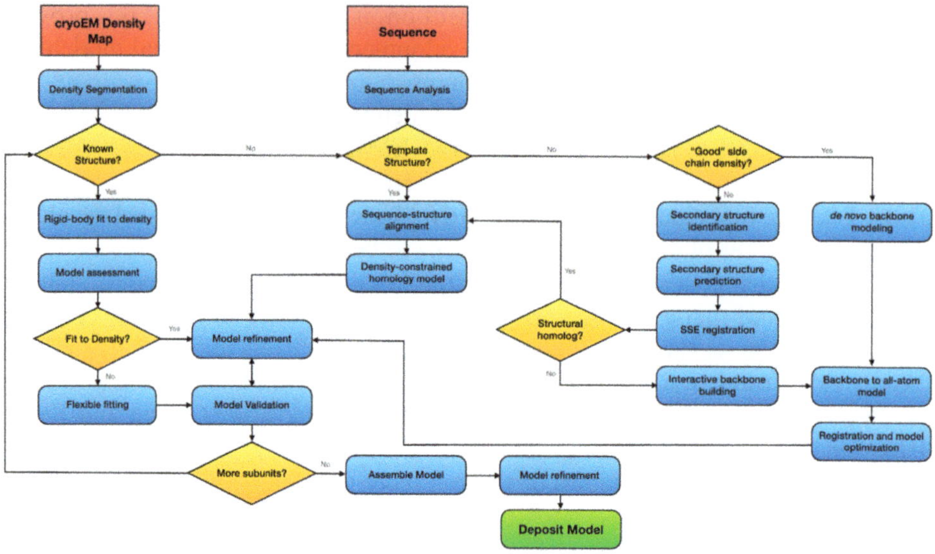

Figure 6.4. Modeling pathway. A general flowchart for model generation from a cryo-EM density map is shown, assuming that only the map and sequences of the components of the map are known (red boxes). The pathway ends in a refined macromolecular model (green box) that is deposited, along with the map, in their respective databases.

The first phase of modeling involves map preparation, which itself is divided into two additional phases. Various types of map filtration techniques can be used to improve map resolvability, eliminate noise, and enhance structural detail (Jiang *et al* 2003, Terwilliger *et al* 2020a, 2020b). In the case of automated *de novo* model building, filtering should result in a clearly distinguishable path through the density. It is important to visually inspect the map after filtering, examining the map for resolution-appropriate features. For example, one would expect to see the separation of β-strands at better than 4.7 Å resolution or holes in aromatic side chains at better than 2.0 Å resolution. Once filtered, density maps are then segmented into individual subunits; symmetry searches are often used at this step to identify related subunits in an assembly. Regardless if the subsequent modeling utilities operate on a single subunit or an entire complex at once, it is good practice to perform an initial segmentation to obtain a better understanding of the individual components and how they may relate to each other.

The next phase of the modeling pipeline is the construction of the actual protein/lipid/nucleic acids within the structure. As with filtering and segmentation, the individual programs utilize a variety of different inputs, as well as produce a variety of different outputs (Malhotra *et al* 2019, Baker *et al* 2010a, 2010b). Logically, the selection of the appropriate model and any transformations (i.e. Cα backbone to all-atom) are necessary before a model is refined using either/both interactive and automated refinement steps. In some instances, modeling and refinement may actually be iterated until a final solution is obtained (Terwilliger *et al* 2020c).

In the final phase of modeling, models are validated on a number of criteria, ranging from model stereochemistry to fit to density (Kryshtafovych *et al* 2018, Lawson *et al* 2021). As mentioned, different modeling utilities fit and measure with different model characteristics, however, all of the validation utilities require a model and/or a map(s). The resulting model statistics are a good general measure of quality, but do not necessarily reflect all shortcomings of the model and its agreement to the map. Once validated, the corresponding map and model can be deposited in their respective database.

Of course, each of the aforementioned steps comes with certain conditions and caveats that are crucial in generating accurate atomistic models in near-atomic-resolution cryo-EM density maps (Terwilliger *et al* 2020c, Kim *et al* 2020). While there are numerous points of emphasis in modeling, one of the most common considerations is that the map origin and coordinate system may be different among various software programs. It is often best to 'reset' the map origin before beginning any map manipulations and modeling, in order to ensure proper map/model registration. Another issue that may arise is that the map scaling may not be correct and the map must be rescaled in order to build accurate models. A number of model-building programs, such as Phenix (Terwilliger *et al* 2018b), now incorporate a map-scaling routine in their pipeline. Also a part of many modeling utilities is an in-built mechanism for segmenting and modeling symmetrical copies of individual subunits (Terwilliger *et al* 2020b, 2020c). In the case where the user must segment individual subunits manually or semi-automatically (Pintilie *et al* 2009, 2010) it is preferential to isolate regions containing slightly more than the suspected monomeric subunit, as mistakenly truncating a map could result in a poor model. Additionally, unresolved regions in the density map, changes in local resolvability or anisotropic density distribution may affect modeling results. While gaps are not necessarily resolved by map filtration, filtering a density map based on local resolution can help to avoid some issues.

In the following sections we will discuss some of the more common model-building, refinement, and validation tools used in cryo-EM today. The following software list is not meant to be exhaustive, but rather a survey of popular tools, potential advantages/disadvantages, limitations, and options.

6.3.2.2 Model building: template based

As mentioned, the original cryo-EM density modeling tools were restricted to feature recognition and rigid-body fitting of known structures. With the advent of higher-resolution structures and improved computing resources, it has become possible to explore flexibly fitting known structures or homology models into cryo-EM density maps. These types of utilities, collectively termed template-based modeling, span a wide range of algorithms, from complete molecular-dynamics simulations to simple geometric morphing.

The following is a list of common interactive and automated template-based modeling tools.

CryoFit: CryoFit is an automated cryo-EM density-fitting method that uses a reduced model molecular-dynamics potential for fitting and is available as part of the Phenix software package (Kim *et al* 2019). It has two key advantages: speed and the preservation of stereochemistry. CryoFit does not require a multi-node workstation or cluster, and it can produce accurate, flexibly-fit models in just a few hours. Secondary and tertiary contact potentials can be specified to keep contacts intact during the fitting process.

DireX: A standalone utility, DireX performs geometry-based conformational sampling of protein structures under restraints (Wang and Schröder 2012). The restraints can be a density map or distance restraints making DireX useful for both cryo-EM and other structural modalities. DireX combines structural information with the experimental restraints using the Deformable Elastic Networks.

Flex-EM: Flex-EM divides fitting of models into density maps into two steps, a rigid-body fitting followed by a flexible, refinement step (Topf *et al* 2008). Rigid-body fitting can be carried out using any fitting methods. Flex-EM flexible fitting of the model to the map begins by fitting the components to approximate positions in the map and refining the model with conjugate gradients minimization and simulated annealing molecular-dynamics Flex-EM is part of the comparative modeling suite, Modeller (Webb and Sali 2016).

Gorgon: Gorgon takes a rather unique approach to flexible fitting of a known structure to a density map (Baker *et al* 2011, Dou *et al* 2017). Using feature recognition, the secondary-structure elements in a map are first identified. The known structure or model is then fitted to the density map by matching the secondary-structure elements in the map with those in the model for an initial rigid-body registration. Local geometric deformations are then computed based on secondary-structure element positions, from which the model is then iteratively deformed until it is optimally fitted to the density map. While this method is extremely quick, contains its own initial rigid-body fitting routine and can operate on an entire complex, it does not inherently 'know' about protein structure. As a result, a quick, automated refinement step with a program like Phenix is required to complete the flexible fitting.

MDFF: Part of NAMD (Phillips *et al* 2020), a molecular-dynamics package, with a graphical interface in the visualization software VMD (Humphrey *et al* 1996), molecular-dynamics flexible fitting (MDFF) uses external forces, which are proportional to the gradient of the density map, in a molecular-dynamics simulation to flexibly fit models into cryo-EM density maps (Trabuco *et al* 2008). To prevent overfitting, MDFF uses harmonic restraints to preserve the secondary structures. While MDFF is computationally intensive, NAMD is available for single-node individual workstations to Linux clusters and supports GPU acceleration.

NMFF: NMFF (Tama *et al* 2004b) is a package of programs that performs multi-resolution flexible fitting in single-particle cryo-EM, tomography, and other lower resolution density maps. NMFF flexible fitting is based on identifying a set of low-frequency normal-mode vectors and constructing a multi-resolution elastic-network representation to maximize the correlation between the map and model.

Norma: Norma (Suhre *et al* 2006) is a freely available, standalone software suite that can model large conformational changes using cryo-EM density maps as a constraint. It is based on UROX (Siebert and Navaza 2009) and elNemo (Suhre and

Sanejouand 2004). In Norma flexible fitting a set of amplitudes are determined for a number of low-frequency normal modes that minimize a misfit parameter.

6.3.2.3 Model building: de novo

In contrast to template-based modeling, *de novo* modeling attempts to generate models directly from a density map without an explicit structural template. Like template-based modeling however, there are a number of different algorithms for model building, ranging from simple geometric optimizations to full-atom molecular dynamics. Accordingly, the complexity of these model-building processes also spans an equally large range. Some of the geometric-based methods take only seconds, on a standard desktop computer, to generate a backbone model for an entire complex. Force-field based methods, on the other hand, can take days on a supercomputer to construct a single subunit. Interestingly enough, most of today's model-building software can produce structures of very similar quality and reliability irrespective of compute time.

The following is a list of common interactive and automated *de novo* model-building tools.

ARP/wARP: Originally developed for use in x-ray crystallography, ARP/wARP is a software suite used to iteratively build models in density maps (Langer *et al* 2008). It combines pattern recognition in density maps, hybrid modeling and maximum likelihood-based refinement. Modeling in ARP/wARP is available through a graphical interface or via a web service.

Buccaneer: Buccaneer is a program for automatically tracing protein chains in experimental density maps that uses an oriented electron-density likelihood target function to identify potential Cα positions (Hoh *et al* 2020). Promising seed positions (Cαs) in the map are grown to extended chain fragments. Chain fragments are then assembled into an initial chain trace. Buccaneer is particularly well suited for low-resolution (<4 Å) density maps.

ISOLDE: Implemented as a plug-in to ChimeraX (Goddard *et al* 2018), ISOLDE is an immersive environment for model building in low- to medium-resolution density maps (Croll 2018). At its core, ISOLDE utilizes molecular-dynamics engines to allow users to help guide the fitting and refinement of a model to the density map. Additionally, ISOLDE contains real-time validation tools that can be marked directly on the model.

MAINMAST: Main-chain model tracing using spanning tree from an EM map (MAINMAST) is a fully automated *de novo* modeling method that directly traces a protein's main chain or builds missing fragments of a model in the density map, where Cαs are represented as a tree-graph (Terashi and Kihara 2018). MAINMAST creates a set of potential models and computes a confidence level to indicate the accuracy of regions in the model. Similar to Pathwalking, described next, Mainmast only computes a Cα backbone trace; conversion of the backbone model to a full-atom model and subsequent all-atom refinement is performed with other utilities. MAINMAST is available as a standalone program.

Pathwalking: Pathwalking is a suite of utilities that treat cryo-EM model building as a geometric optimization problem framed around the traveling salesman problem (TSP) (Baker *et al* 2012a, Chen and Baker 2018, Chen *et al* 2016). Essentially, a set of pseudoatoms populate a density map—in TSP parlance pseudoatoms are cities—from which a minimal protein path distance is calculated. The advantages of this approach are that it is extremely quick (<5 s for even the largest problems), does not require large-scale computing resources, produces relatively accurate first-approach $C\alpha$-backbone models even at relatively modest resolutions (~5 Å), and can generate models for multiple subunits simultaneously. However, user interaction is required to check and tweak the backbone path before assigning a sequence to the path and generating an all-atom model. Pathwalking is available as part of EMAN2 (Tang *et al* 2007) and PHENIX (Liebschner *et al* 2019), as well as having a graphical plug-in for UCSF Chimera.

Phenix map_to_model: The map_to_model tool in Phenix, which is also described in section 6.5.3, is a fully integrated and automated procedure that requires only a map, nominal resolution, sequence, and symmetry as inputs (Terwilliger *et al* 2018a, 2020c). Differentiating it from most *de novo* tools, map_to_model begins with a map sharpening procedure to improve connectivity and structural details, followed by subunit segmentation using the symmetry information. Secondary-structure features are then identified followed by main-chain construction using a template-based method. Real-space restraints are then used to optimize the model; a final assembly can then be generated from all of the subunits, including symmetry related positions. Phenix's map_to_model utility scales with map and model complexity but can result in highly accurate models.

6.3.3 Model refinement

Following the generation of an all-atom model, the goal of model refinement is to simultaneously improve both model fit-to-density and stereochemistry, such as bond lengths, angles, rotamers, etc (DiMaio *et al* 2009, Murshudov 2016, Nicholls *et al* 2018, Baker *et al* 2013, Wang *et al* 2014, Afonine *et al* 2018b). Typically, these routines target near-atomic-resolution cryo-EM maps, but lower resolution model optimization can be performed, albeit with additional constraints, to ensure that overfitting does not occur and the model maintains proper stereochemistry. Unlike x-ray crystallographic model refinement, cryo-EM model refinement is more of an optimization step, where only the model is adjusted. Ultimately, cryo-EM refinement will take a first-approach model and improve it so the resulting model is an accurate representation of the experimental data.

Generally, the cryo-EM model refinement process is iterative and cycles through automated model refinement routines and manual adjustments, which may include fixing missing atoms or broken bonds (if any occur), model-registration errors and adjusting large domains which may not be converging into the density (i.e. stuck in local minima). Typically, these model optimization routines require a large amount of computational time as they do both global and local fitting, in addition to performing geometry minimization. All of the refinement routines typically monitor a host of important stereochemical properties, and thus they have some built in validation measures. An example illustrating the refinement and validation of the bacteriophage P22 capsid structure (Hryc *et al* 2017) is shown in figure 6.5(A).

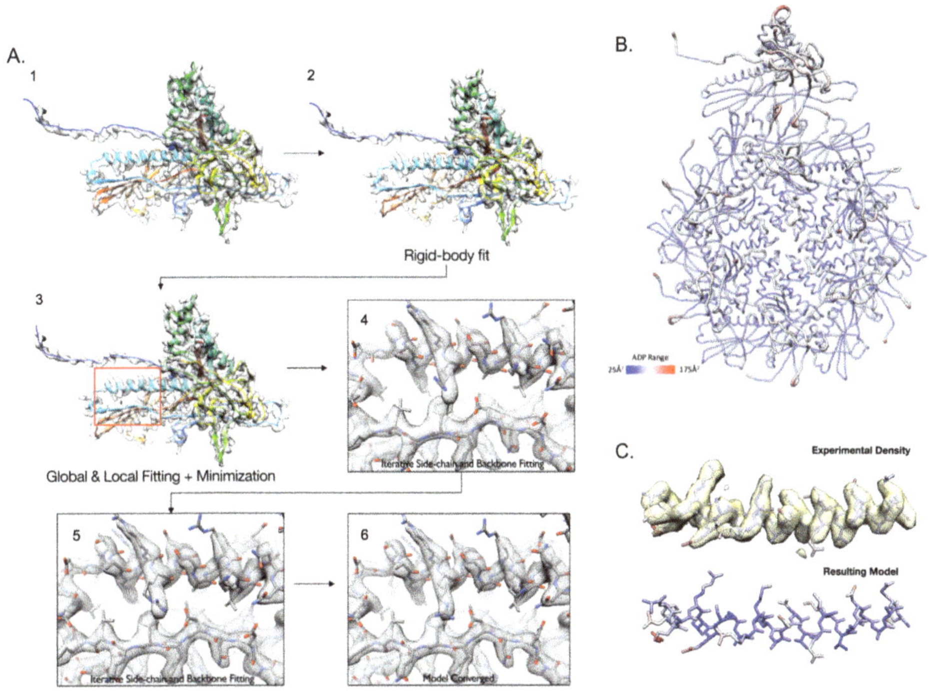

Figure 6.5. Model optimization of P22. In (A) a simplified model optimization routine is shown for the 3.3 Å cryo-EM density map of P22. (1) The initial model (shown in rainbow representation) has not been optimized to the map. Additionally, the model has been translated with respect to the map. (2) The first step in the model optimization process is to perform a coarse-grain rigid-body fit of the model to the map. This improves the general fit-to-density. (3) The model then goes through a series of local and global fits along with energy minimization to maintain and improve stereochemistry. (4) Side-chain and backbone fitting is then iterated at a finer level. (5) Note the improved backbone fitting, but non-converged side chains. (6) Finally, the model converges to its final optimized state. In (B) an asymmetric unit is shown with the average ADP per-residue mapped onto the model. In (C) a zoomed-in view of the experimental density and the corresponding model with atoms colored based on B factors are shown.

The following is a list of common interactive and automated model refinement tools.

COOT: While not originally developed for refining models in cryo-EM density maps, COOT has become an essential tool in interactive refinement (Emsley *et al* 2010). Unlike the aforementioned models, which are mostly automated, COOT is an interactive tool that allows the user to actively monitor fit to density, model geometry, clashes, and a host of other model statistics. COOT is often used in combination with automated refinement methods. In earlier versions of COOT and other cryo-EM image processing software, map origin and model centering in the map were an issue.

Gromacs: Gromacs uses a correlation-driven molecular-dynamics (CDMD) method for refining atomistic models into near-atomic-resolution cryo-EM density maps (Igaev *et al* 2019). Force fields and thermodynamic sampling are used to automatically improve the real-space correlation between the model without user input or additional constraints. The resulting method is touted as being resolution-independent, stereochemically accurate and fully automated. Refinements can be carried out on multi-core workstations and servers, although reported examples may take several days for refinement on a server depending on the size and complexity of the system.

Phenix real-space refinement: As a module in Phenix, real_space_refine is designed to optimize not only the model's fit to density, but also improve model stereochemistry (Afonine *et al* 2018b). Phenix real-space refinement features a number of options that allow the user to control the level of refinement, as well as options for enabling different geometrical constraints. Phenix real-space refinement requires only a map and model, it is therefore easy to use in combination with any of the aforementioned modeling tools, although computational time scales with the size of the map and the number of atoms. Phenix refinement is also capable of producing B factors, indicating relative certainty of atom placement (see below) (figure 6.5(B) and (C)). While relatively 'quick', requiring minutes to several hours depending on the size, complexity, and options, Phenix real-space refinement can accurately locally refine models, although it is limited in its ability to correct large modeling errors, such as register shifts.

Rosetta: Rosetta provides an automated method for improving manually built models in near-atomic-resolution cryo-EM maps (Wang *et al* 2016). This procedure utilizes a multi-stage approach for automatic refinement, including methods for automatically detecting and correcting problematic residues, selecting models with good density agreement and geometry, correcting magnification errors, improving side-chain-optimization, and estimating uncertainty in a model. These utilities allow for the correction of backbone errors that significantly deviate from the starting model, as well as optimizing other regions of the model and producing high-certainty refinements. Rosetta refinement is considerably slower than Phenix's real-space refinement, however, it is capable of identifying and fixing longer-range model errors.

UCSF Chimera/ChimeraX: Like COOT, UCSF's Chimera, and its successor, ChimeraX, provide an interactive environment to examine both the map and model (Goddard *et al* 2018, Pettersen *et al* 2004). While Chimera does provide a number of mechanisms to monitor model quality and fit to density, it does not offer the same ease of use in whole-model refinement that COOT provides. Rather Chimera provides a convenient utility for small, directed refinements to a model. Chimera also provides a whole host of other cryo-EM utilities and offers an excellent environment for simultaneous map/model viewing and analysis.

6.3.4 Model validation

The goal of cryo-EM model validation is to ensure that the model accurately reflects the experimental density map while maintaining proper stereochemistry. Similar to model building, model validation is resolution dependent, in that if you have higher resolution features, it is easier to ensure that modeled density is properly interpreted.

Currently, model validation falls into two categories: fit-to-density and model geometry assessment; both of which are continuously optimized during model refinement. As previously discussed, model optimization/refinement is guided by validation measures to ensure that the model is continuously improving. While some model-validation routines have been well established from x-ray crystallography, newer validation routines adopt cryo-EM map-specific validation, such as using the Fourier shell correlation to assess fit to density (Ramírez-Aportela *et al* 2020a). As such, some of the validation 'tools' consist more of concepts and best practices rather than discrete software utilities. The following outlines some of the current model-validation best practices.

Fit to density: As a model should be an accurate reflection of the experimental density map, one the most obvious validation assessments is directly comparing the map and model. Computing the cross-correlation between map and model is typically the most-common measurement of fit-to-density. In order to compute correlations, a simulated map, at a declared resolution (typically the resolution of the experimental map), is generated from the model and map values are then compared to those in the experimental EM map.

This method has the advantage of being able to compare modeled regions at various levels; such as atoms, side chains, secondary structures, and even protein subunits (Hryc *et al* 2017). Because one can target specific protein levels, this becomes useful when assessing the fit at various resolutions. For near-atomic resolutions it is common for cross-correlations to be computed at the side-chain level along with an overall cross-correlation score for the complete map versus model fit (figure 6.3). This allows one to assess regions where the model fits the data well and regions where the fit is not as ideal, which can easily be visualized with programs such as UCSF's Chimera. While this is a great tool to assess fit to density, it does have the disadvantage that there is no true standard to compare values between different datasets.

It should be noted that these cross-correlation scores can be computed in most modeling packages. For instance, Phenix real-space refinement computes these correlation values during optimization and displays the results when complete.

Map–model FSC: In cryo-EM, a map-to-map correlation can be reported in terms of spatial frequency using the Fourier shell correlation (FSC); this is often used as the primary mechanism for reporting resolution. The FSC provides a general understanding of the resolution at which the two datasets are similar. When adapted for map–model assessment, it can provide another metric for assessing how well they agree. Here, a map is generated from the molecular model, similar to what is done when computing cross-correlations (Hryc *et al* 2017, Ramírez-Aportela *et al* 2020a). This calculated map replicates the experimental data in most or all of its detail at the reported resolution. Once the calculated map is generated an FSC curve is computed between the calculated map and the experimental map.

The map–model FSC curve provides a lot of useful information, such as general fit-to-density; the closer these two curves (map–model and map–map) are in shape, there is a better indication of good map-to-model agreement. When comparing the resolution values of the map–map FSC (measured at 0.143) and the map–model FSC (measured at 0.5), one can obtain an indication of how close the model represents the experimental data and if overfitting has occurred; the map–model FSC would report a higher resolution than the experimental data. Moreover, one can generate map–model FSC curves for half maps and compare results to ensure that model refinement is correctly fitting the data (figure 6.6(A)). While the FSC is helpful, it does not assure authenticity of map features, nor does it specify what regions of the map are well modeled and vice versa.

Map–model FSCs can be generated in almost all of today's image processing suites. Phenix provides a simple command-line utility to generate a map–model FSC.

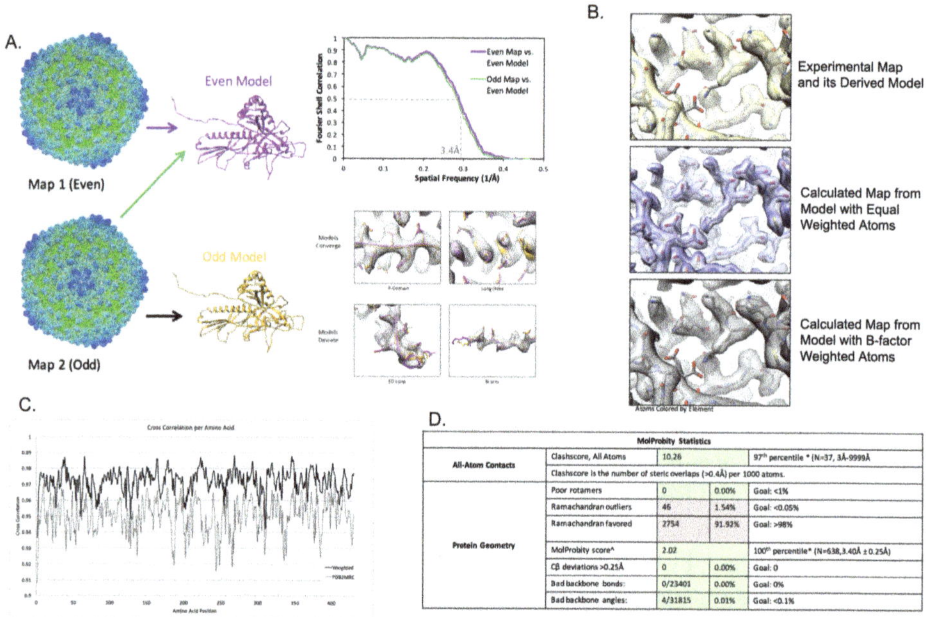

Figure 6.6. Model validation of P22. In (A) half datasets from the P22 reconstruction (even/odd) density maps were used to generate two models with Pathwalker. FSC curves were then generated comparing simulated density from one half-data-set model to the two independent split density maps. Regions with strong density reveal that the two models converge and regions of the map with greater uncertainty reveal that the two models deviate. The experimental map density and superimposed model are shown in (B). Typically, when creating a map from a model, all atoms are weighted equally, as shown (middle panel), however, this is not a proper representation of the experimental density map (top panel). The model-derived map, with proper ADPs (bottom panel) faithfully recreates the experimental density map, including uncertainty/weak density in the map. In (C) the cross-correlation values were computed per amino acid, between the experimental map and the calculated map. A 4 Å zone around the amino acid was used to isolate the density and an average was taken across the asymmetric unit. All amino acids were better represented using the B factor map as opposed to the equally weighted map. Results from MolProbity are shown in (D). The MolProbity score combines the clashscore, rotamer, and Ramachandran evaluations into a single score, normalized to be on the same scale as x-ray crystallography resolution.

6.3.5 Model uncertainty

In molecular modeling there is an amount of uncertainty that exists when placing atoms within a density map. This uncertainty is difficult to capture and display with a single EM map and model. To begin to understand the level of uncertainty within the data and the model optimization process, it is possible to turn to the maps generated during the map reconstruction/refinement phase. As part of the gold-standard reconstruction process, two independent maps are computed from the half datasets. While the resolvability in these half maps is typically not as good as that of the combined map, for near-atomic-resolution density maps they are typically sufficient to generate two independent models optimized against the two half-map datasets (figure 6.6(A)). While not strictly a validation tool, assessing the independent models provides an understanding of uncertainty within the map (Herzik *et al* 2019). If atoms in the half-dataset models vary greatly, then there is weak, or poorly resolved, density constraining the model and thus divergence occurs.

As an example, models for the P22 bacteriophage capsid (shown in figure 6.6) were generated independently using Pathwalking for each half-dataset map (~3.4 Å in resolution) (Hryc *et al* 2017). When comparing the two independent models, a 0.56 Å root mean square deviation (RMSD) between the Cα atom positions of the two independent models indicates the reproducibility of the data and a relatively small amount of uncertainty. Some regions did show divergence, where some loop regions had weak density and/or some amount of intrinsic amount of disorder. Unsurprisingly, the two independent models agreed well with the model built using the full dataset (~0.6 Å RMSD for both datasets).

Anisotropic displacement parameters (ADPs), also referred to as B factors or temperature-factors, have been traditionally used in x-ray crystallography to describe the uncertainty in atom location (Trueblood *et al* 1996). This uncertainty may be due to temperature-dependent atomic vibrations (which would impact the ability to localize atom position in a structure) (Coppens 1975) or intrinsic disorder in the structure (International Tables for Crystallography 2020). By applying ADPs per atom, the model can then describe anisotropic displacements either derived from the data or modeling errors. This provides an uncertainty per atom in the form of a Gaussian fall-off, that is to say that the atom location has strong probability at a given position and, as one moves further away, there is less certainty or probability of the atom existing in that location.

B factor values determined during real-space refinement with Phenix correlate well with the uncertainty generated from the two independent models. In the case of P22, the correlation between high B factors and model variation is due to density fall-off at loop regions, and the flexibility of the N- and C-terminus. In Phenix real-space refinement, ADPs can be iteratively improved along with the overall fit to density. Ultimately, a properly weighted model can correctly represent the experimental data that were used to derive it (figure 6.6(B)). When comparing the map and model through cross-correlation and FSC curves, using this properly weighted, calculated map is the ideal way to represent the model (figure 6.6(C)). Finally, this calculated map may also provide the ability to refine the experimental density map further (discussed in the section 'Toward the future').

EMRinger: While EMRinger does not assess all aspects of a model, it is used to analyse and validate side-chain positions (Barad *et al* 2015). From the model a normalized map value at each potential $C\gamma$ position is computed around the χ_1 dihedral angle. The distribution of map values can then be plotted by dihedral angle, providing local information about both the map and the atomic model. The most probable position of the $C\gamma$ atom of the side chain can be inferred from the peak in the distribution. Both a global and local assessment can identify poorly modeled/refined portions of the model. EMRinger is available both as standalone software and as part of Phenix.

MolProbity/CaBlam: Originally created to assess x-ray crystallography models, MolProbity (Davis *et al* 2007) and its new feature called CaBlam (Prisant *et al* 2020) focus on validating the atomistic models in terms of geometry and stereochemistry. Owing to their wide adoption in x-ray crystallography as model-validation standards, cryo-EM modeling has also adopted these tools to analyse cryo-EM-based models to the same level of scrutiny. While the aforementioned tools (fit-to-density and map/model FSC) measure map–model agreement, they do not measure model quality; models must also have proper bond distances, no unusual clashes within and outside of a subunit, rotamers should be acceptable, and torsion angles should be acceptable based on the Ramachandran plot. All of these model checks are done in MolProbity and a comparison to other models at a similar resolution helps to indicate overall model quality (figure 6.6(D)). Finally, for lower resolution models, CaBlam (the C-alpha based low-resolution annotation method), provides a complementary analysis that measures whether peptide-plane orientations are compatible with local Ca geometry. Both MolProbity and CaBlam are available through a web portal or through Phenix.

***Q*-score**: The *Q*-score is a parameter to characterize the resolvability of individual atoms in a cryo-EM density map (Pintilie *et al* 2020). It measures the fall-off of map values around an atom. By averaging map values at an increased distance from all atoms, a map profile can be computed, which is then compared to a reference Gaussian falloff. The result is similar to a cross-correlation, but it has the advantage of not being influenced by resolution or masking. Currently, it is implemented as a plug-in to UCSF Chimera.

TEMPy2: An object-oriented Python library, TEMPy2 is designed to assess the goodness-of-fit between a model and density map, or between two density maps, using a variety of different scoring functions (Farabella *et al* 2015). TEMPy2 can also generate ensembles of alternative fits in assessing model fit. Furthermore, TEMPy2 can integrate data from cross-linking and evaluate models with their respect to expected distance distributions.

6.3.6 Model deposition

Although model deposition is covered fully in section 6.6, a few comments are included here for the sake of completeness. While not necessarily an explicit step in model building; map and model deposition are critical aspects of archiving and making the experimental data available to the public. With the growing number and diversity of maps and models, the EMDataResource (https://www.emdataresource.org/) was established to provide standards for map and model deposition, as well as a unified portal for deposition and retrieval. The EMDataResource is now a global resource for three-dimensional electron microscopy data, including archiving, retrieval, news, events, software tools, data standards, validation methods, and community challenges. Today, the wwPDB OneDep system provides an efficient interface for simultaneous map and model deposition (Young *et al* 2017), while the EMDataBank at PDBe provides a feature rich site for map and model retrieval (Velankar *et al* 2016).

6.3.7 Revisiting the cryo-EM model challenge

While citing improved resolution and the number of deposited density maps and models can provide a quantitative metric for the increased adoption of cryo-EM, perhaps the most significant way to understand the progress in the field is to look at the most recent cryo-EM model challenge. While the first two model challenges took place over multiple months and explored a host of modeling topics, the 2019 Cryo-EM Map-based Model Metrics Challenge was more focused and specifically sought to (i) assess model quality using the current modeling software, (ii) evaluate reproducibility of modeling results, and (iii) compare different model evaluation metrics. Rather than giving users/developers months to build models, modelers were provided a smaller set of targets for analysis over the course of ~6 weeks, with assessments and presentation of the results approximately 2–3 weeks later. Additionally, datasets for human apoferritin were provided at three resolutions (1.8–3.1 Å) to assess modeling metrics at different resolutions. Tools established in the prior challenge for evaluation and analysis were expanded to include the new set of models (Kryshtafovych *et al* 2018).

In contrast to results from previous challenges, all reported methods produced relatively accurate and reproducible models by the 13 participating teams (figure 6.7). In addition to assessing model-building performance, the challenge led to the recommendation of ‘multiple scoring parameters to provide full and objective annotation and assessment of the model, reflective of the observed density in the cryo-EM map’ (Lawson *et al* 2021). As a whole, the 2019 Model Metrics Challenge represents a maturation of the cryo-EM modeling field, where users should now be able to use any of the published methods to achieve highly reliable models.

6.3.8 Toward the future

In x-ray crystallography the model is constantly being used to refine phases. With these current cryo-EM modeling methods it may be possible to use the model to assess map features and improve the reconstruction. Using a model-derived map as a noise-free template, particle orientations could be refined and thereby produce an improved density map. However, this comes with the risk of introducing model bias

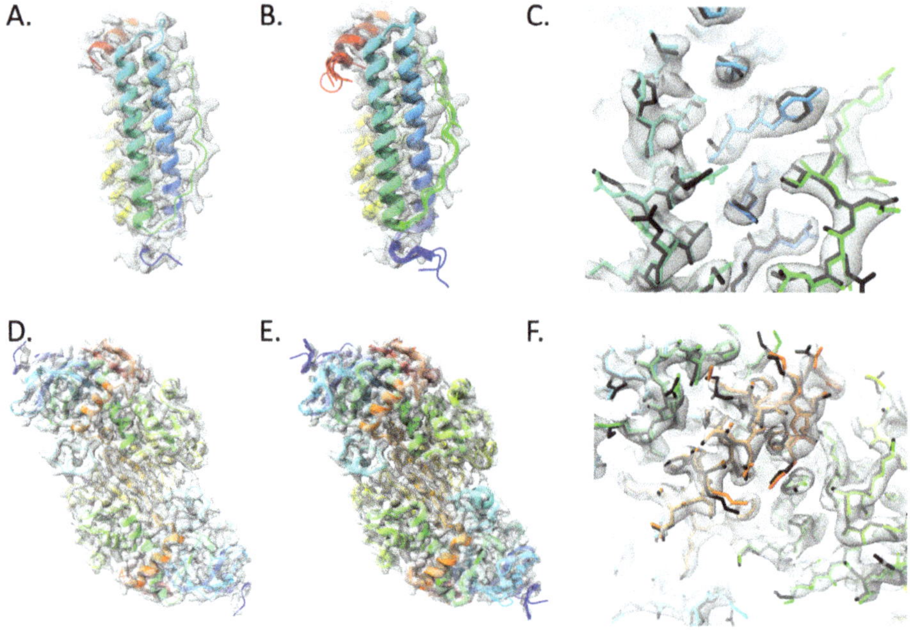

Figure 6.7. 2019 Cryo-EM Model Metrics Challenge. The deposited *de novo* models from the challenge are shown for two datasets, T0103 human apoferritin (3.1 Å resolution, panels (A)–(C)) and T0104 horse liver alcohol dehydrogenase (2.9 Å resolution, panels (D)–(F)). In (A) and (D), the known structure, rainbow colored from the N- (blue) to the C-terminus (red) is shown superimposed on the map. The submitted models, represented as ribbons, from their respective maps are shown in (B) and (E). The models are rainbow colored from the N- to C- termini. The average RMSD for T0103 was 0.78 Å (0.44–2.69 Å), while T0104 had an average RMSD of 0.90 Å (0.38–3.34 Å) when compared to their respective known structures, PDBID: 3AJO; PDBID: 6NBB. Zoomed-in views of one model compared to the known structure (black) are shown in (C) (from Pathwalking) and (F) (from Phenix map_to_model).

during refinement. Introduction of bias could be mitigated by verifying that the model-derived map has no more (or less) density at the modeled positions than is present in the experimental map. The absence of bias could be assured by removing model features, such as side chains or a set of secondary structures, or through the introduction of 'wrong' features (i.e. mutating side chains) and monitoring these features during successive rounds of map refinement. If the altered features do indeed return to their original state, then model bias should not be an issue. In the P22 example (Hryc *et al* 2017), these experiments were tested and no bias was indicated (figure 6.8(A)). As such, it was hypothesized that using a model-derived map, which contains no noise and only authentic density features, in the refinement process could not only be possible, but advantageous to the final reconstruction.

In P22 the atomistic model was used to compute a map and re-determine particle orientations. The updated map did indeed reveal a suppression of noise in both the exterior and interior capsid density and a slight resolution improvement (figure 6.8(B)). Moreover, after optimizing the molecular model against the newly generated density map, the B factors were reduced by an average of ~24 Å^2 from the

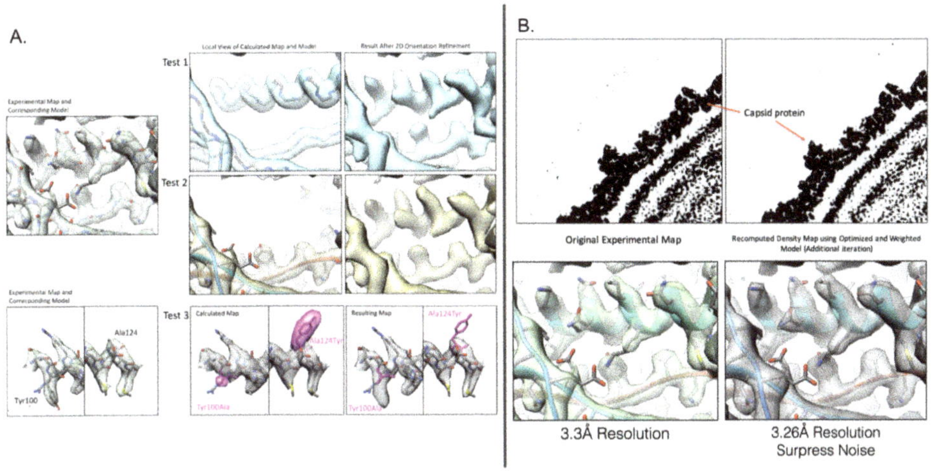

Figure 6.8. Map–model refinement of P22. Panel (A) outlines the testing of 'computational mutagenesis' of different regions to determine if these regions return after subsequent rounds of refinement and reconstruction. (Test 1) A complete capsid at 3.3 Å resolution was generated based on a molecular model, but without side chains and carbonyl oxygens. A local view of the calculated map is shown in the middle panel with the model that was used to generate the map. After multiple rounds of 2D orientation refinement a new map is generated, which now has high-resolution detail for its side chains. (Test 2) A second test where the spine helix was removed and after multiple iterations of 3D orientation refinement the helix reappears with similar features when compared to the experimental data. (Test 3) A mutation test was completed *in silico* by altering the model. Two amino acids per subunit were swapped as shown in magenta. After 3D orientation refinement using the calculated map as an initial model the resulting map did not show any artifacts from the mutation and was similar to that of the experimental map. (Left) Experimental data shown for comparison. In (B) a slice through the final 3.3 Å cryo-EM density map and the map refined against the cryo-EM model with one additional iteration is shown. Note the reduction in noise at and outside the capsid surface and the strength of the (unmodeled) disordered DNA inside the capsid shell, although the fact that DNA density is not present in the model but present in the reconstruction reinforces that it is not noise. After one round of refinement with the model, the nominal resolution is improved by ~0.1 Å; the corresponding model is shown.

original average B factor of ~97 Å^2, allowing for more precise localization of atomic positions. The re-refined models had an RMSD of 0.22 Å with respect to the original model, while the EMRinger score improved from 3.58 to 4.33. These results indicate that the resulting map and model are a more accurate representation of the structure.

Pushing resolution and being able to model higher-resolution features is not the only advancement in cryo-EM modeling. Focused classification has unlocked the ability to generate an ensemble of maps from classified particles. While this grouping of particles may result in reduced-resolution density maps, they provide the opportunity to generate an ensemble of potentially biologically relevant models. Working backwards from the density map to the particle level allows for the identification of individual subunit states within the particle. This method was used in 2017 on GroEL, where subunits were classified into three dominant states and models were subsequently built for each state (figure 6.9) (Roh *et al* 2017). Each subunit of the 14-subunits in the complex were assigned to one of the classes, thereby creating a gallery of models in various, independent states.

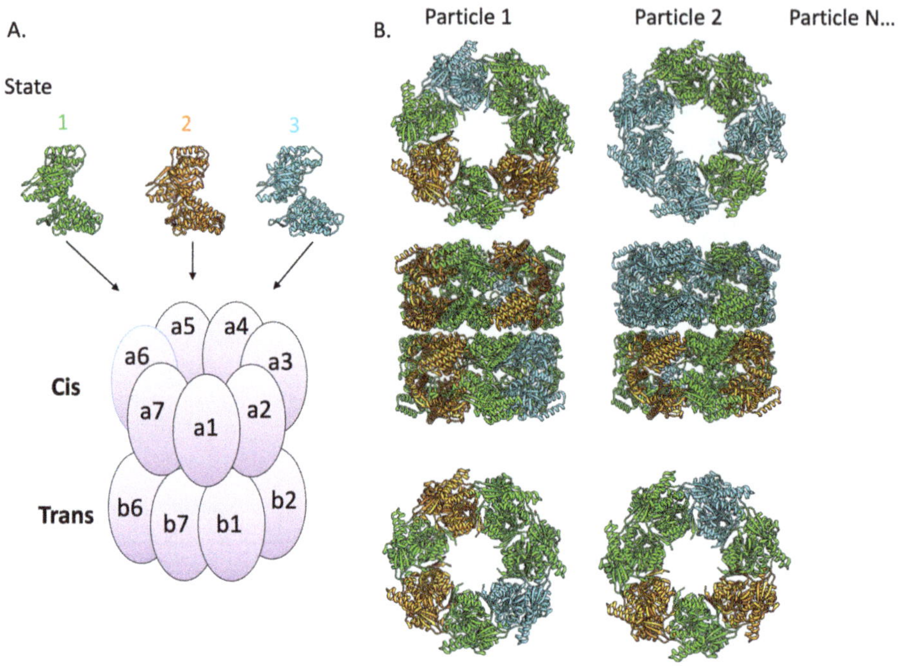

Figure 6.9. Conformational variants in a single apo-GroEL oligomer. In (A) three functional-state models for a GroEL monomer, using maps generated from the focused classification results, are shown. These subunit conformations were localized to their respective position in each GroEL complex and a full multimeric GroEL model was assembled. Shown in (B) are models for individual GroEL complexes.

6.3.9 Conclusions

In a little over a decade, cryo-EM model building has come from ad hoc model generation to nearly fully realized automated model building. Today there are dozens of freely available programs, as standalone utilities or packaged within larger software suites, to automatically generate models using individual workstations to large computing clusters. As evidenced by the increased accuracy and reproducibility in the models produced in the cryo-EM model-building challenges, model building has matured to the point where users can reliably construct models irrespective of their software of choice. While improvement in methods may indeed result in better models, cryo-EM modeling is now squarely focused on how these models can answer the underlying biological questions and how they can help to improve or better represent the underlying experimental data.

Acknowledgments

This work was supported by a grant from the National Institutes of Health (P01GM063210). We would like to thank Wah Chiu, Michael Schmid, and Donghua Chen for their work on the bacteriophage P22 structure and Mariah Baker for comments on the manuscript.

6.4 Quality evaluation of cryo-EM map-derived models

Grigore Pintilie[1], *Catherine L Lawson*[2], *and Wah Chiu*[1,3]
[1]*Department of Bioengineering, James H Clark Center, Stanford University, Stanford, CA, 94305, USA*
[2]*Institute for Quantitative Biomedicine and Research Collaboratory for Structural Bioinformatics, 16 Rutgers, The State University of New Jersey, Piscataway, NJ 08854, USA*
[3]*Division of Cryo-EM and Bioimaging, SSRL, SLAC National Accelerator Laboratory, Stanford University, Menlo Park, CA 94025, USA*

6.4.1 Introduction

Single-particle cryo-EM has moved very rapidly to reach map resolutions between 1.2 and 5 Å, predominantly clustered around 3–4 Å. Cryo-EM single-particle reconstruction is a process in which many 2D images of particles picked from micrographs are combined to form 3D maps (Glaeser *et al* 2007, Frank 2006). Since the spatial distribution of thousands to hundreds of thousands of atoms is complex and difficult to comprehend, each map needs to be annotated and represented in the form of model coordinates. Resolvability of structural features can vary throughout a 3D map and can be influenced by various factors such as the presence of different conformations across the particles used (Herzik *et al* 2019), preferential specimen orientations on embedding ice (Tan *et al* 2017), inaccuracies in electron microscope imaging parameter estimation (Zivanov *et al* 2020), radiation damage (Bammes *et al* 2010), and reconstruction artifacts such as over-refinement of noise (Henderson 2013). The overall quality and resolvability of the map influences how models are obtained and also their accuracy. Hence, the process of building or fitting a model into the map also directly reflects the quality of the map, which is the theme of this section.

We focus here on methods for validating the map-derived model. Model validation examines whether the observed density is accurately described and the model obeys the biochemical properties of the proteins, nucleic acids, and their associated and surrounding molecules. Other sections in this chapter also describe various procedures in model building including methods for fitting known models or building models *de novo* (Emsley *et al* 2010, Birmanns *et al* 2011, Baker *et al* 2012b, Pintilie and Chiu 2012, Terwilliger *et al* 2018b, DiMaio and Chiu 2016). The global estimated resolution of the map by the gold-standard method (Henderson *et al* 2012) offers one indication of what modeling tools may be applied. Local resolution estimation methods can also play an important part in analyzing such maps (Kucukelbir *et al* 2014, Vilas *et al* 2018).

This section primarily focuses on map–model metrics, which consider both the map and model together. Such metrics can inform us how well models fit into high-resolution maps. In some cases, they can also tell us to what degree various features in the model (e.g. atoms or residues) are actually resolved in the map. We demonstrate several such metrics on models built into cryo-EM maps with resolutions ranging from 1.2 to 4.0 Å. Model-only metrics are also briefly reviewed in the final section; they are important to show whether the model is biophysically realistic and/or has similar properties to other models/atomic structures (e.g. from x-ray crystallographic structures).

6.4.2 Map–model metrics

Map–model metrics are useful to indicate how well a model matches a map. Many such metrics have been proposed and applied to cryo-EM maps and models over time, including atom inclusion and average density value (Rossmann 2000, Lagerstedt *et al* 2013), cross-correlation (Joseph *et al* 2017a, Afonine *et al* 2018a), Fourier shell correlation (Harauz and Heel 1986), EMRinger (Barad *et al* 2015), *Z*-scores (Pintilie *et al* 2016, Pintilie and Chiu 2018), and *Q*-score (Pintilie *et al* 2020). They have been developed for various reasons at different stages of attainable resolutions in the cryo-EM field. Recent efforts involving a group of experts are underway to characterize and find relationships between different metrics (Lawson *et al* 2021). In this section, some of these metrics will be described and illustrated with examples. They are summarized in table 6.1.

6.4.2.1 Map–model FSC

A Fourier shell correlation (FSC) plot can be used to assess how well an atomic model matches a cryo-EM map. To calculate such a plot, a map is first generated from the model, e.g. with *phenix.fmodel* (Afonine *et al* 2018a), EMAN pdb2mrc (Bell *et al* 2018), or Chimera *molmap* (Goddard *et al* 2007). The model-derived map and the cryo-EM map are then transformed into Fourier components and the correlation is calculated at small frequency/resolution increments or 'shells' in the Fourier domain. A benefit of such an FSC map–model plot is to show at what resolutions the map and model agree. When the model matches the map well, the FSC plot is typically high at low frequencies and begins to drop at higher frequencies.

Figure 6.10 illustrates the map–model FSC for a cryo-EM map and model of apoferritin at 1.75 Å resolution (EMD:20026). Panel (A) shows the entire map and the x-ray model PDB:3ajo initially fitted to each of 24 symmetrically related positions (the complex has octahedral symmetry). The model was subsequently refined with *phenix.real_space_refine* into the map. It is common that the x-ray model does not necessarily fit to a cryo-EM map readily without refinement for various legitimate reasons. The FSC curve between the map and the model starts at ~0.9 for a frequency close to 0 and stays relatively high until a drop below 0.5 at a frequency of ~0.5 $Å^{-1}$ or 2 Å. This is close to the estimated resolution of the map (1.75 Å).

FSC curves between the cryo-EM map and atomic models of a single protein, helix, and residue can be computed as exemplified in figure 6.10(B), (C), (D), and (E), respectively. The cryo-EM map is masked (by Phenix) with the part of the model being considered. The FSC curve for the entire model, protein, and helix all fall below 0.5 at 0.50 $Å^{-1}$ or 2.0 Å resolution, even though the plots up until this point have varying shapes. The FSC curve for the single residue (Phe51) on the other

Table 6.1. List of map–model metrics with brief summary. The scores marked with * have been shown to correlate to the FSC-based resolution of the cryo-EM map for well-fitted models; hence they reflect both map–model fit and also map quality. Map–model FSC and CC scores also require a model–map to be generated; inputs for this include atom types and positions, desired resolution, and for some methods, scattering factors and B factors.

Map–model metric	Summary
*Map-model FSC	Compares cryo-EM map and map generated from model (model–map) at different frequencies/resolutions.
Atom inclusion	Percentage of atoms that are within a chosen threshold/contour level.
Average density value	Average of map values at atom positions.
Cross-correlation (CC)	Similarity of the cryo-EM map and model-map.
**Z*-score	Compares CC of model as placed into the map to CC of other placements.
*EMRinger	Compares map values at rotated Cγ atom to favorable rotameric values (not applicable to all residues).
**Q*-score	Cross-correlation of map values around each atom with a reference Gaussian.

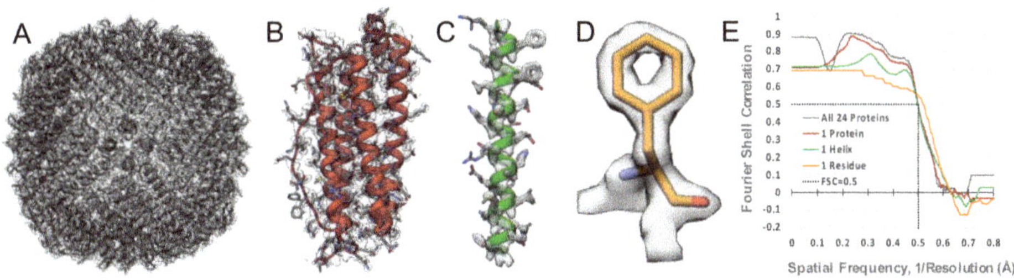

Figure 6.10. Map–model FSC plots for apoferritin at 1.75 Å resolution (EMD:20026). (A) The entire cryo-EM map and model, (B) one of the 24 proteins in the complex, (C) a single helix and (D) a single residue, Phe51 (E) FSC plots.

hand stays above 0.5 to a higher frequency of 0.52 Å^{-1} or 1.9 Å resolution. This may indicate slightly better resolution in that part of the map compared to the FSC-based resolution for the entire map, protein, or a single helix.

Map–model FSCs are very informative to use on entire maps and models, where the effects of masking are small (the mask boundary is small compared to the overall volume of the map and model). However, they become harder to apply on smaller features, as the mask boundary becomes large compared to the volume being measured. Hence, even when looking at a single residue, for example, using a soft mask means that other densities are being considered, not just that residue. Thus, such residue-level map–model FSC scores are typically not performed or reported.

6.4.2.2 Atom inclusion and average density value

The atom inclusion metric indicates how many atoms are inside a 3D contour display of the cryo-EM map at a given value or threshold (Rossmann 2000, Lagerstedt *et al* 2013). As an example, for the residue pictured in figure 6.11(B), six of the seven atoms in the residue (Asp92) are inside the map contour, hence the atom inclusion score is 6/7 or 0.86. One limitation of this metric is that it is heavily dependent on the chosen threshold. As the threshold is lowered, more and more atoms tend to be inside the contour, and the score goes up, even though atoms may be in areas of noise rather than signal. The average map density value at all atom positions can also be calculated and used as a metric, which avoids this problem. However, the latter is still limited in that it does not consider density values in the map that are around the atoms, rather than just at their positions.

6.4.2.3 Cross-correlation

The cross-correlation metric is calculated between a cryo-EM map and a map derived from the model. Unlike atom inclusion, cross-correlation does not depend on a given threshold, but map resolution is an input parameter for the model-derived map. A simple way to derive the model-derived map is to apply a Gaussian-filter to a map where all atoms contribute positive values at the nearest grid points (e.g. in UCSF Chimera, using the *molmap* command). The width of the Gaussian-filter controls the resolution of the derived map. A more physically realistic way to generate model-derived maps is to apply electron-scattering factors and temperature (B-) factors to each atom (e.g. using *fmodel* in Phenix). Example maps derived using both methods are shown in figure 6.11(C)–(F) at two resolutions of 2.0 and 3.5 Å.

Once a model-map is derived from the atom coordinates, the cross-correlation is calculated between values in the model-derived map and cryo-EM map. There are two main formulations of the CC score. The first considers the values in the two maps as they are without modification; an example of this is the SMOC score in TEMPy (Joseph *et al* 2017a). A second formulation, used in Phenix, subtracts the mean or average value from all values before performing the cross-correlation

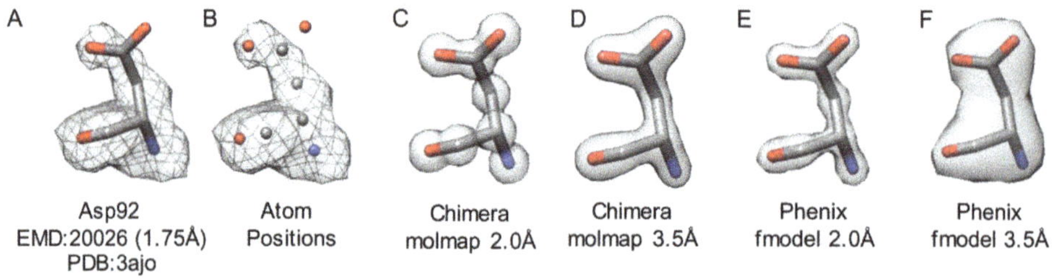

Figure 6.11. Illustrations of different aspects of the calculation of atom inclusion (AI) and cross-correlation (CC) scores for residue Asp92 in a 1.75 Å map of apoferritin (EMD:20026), using the fitted and refined x-ray model PDB:3ajo. (A) The model atoms are colored by element (gray: carbon; red: oxygen; blue: nitrogen) and bonds are indicated by tubes between atoms. Atom inclusion scores consider just atom positions, as shown in (B). CC scores are calculated from model-generated maps; examples are shown here at 2 Å and 3.5 Å using Gaussian blurring (C), (D), and electron-scattering factors (E), (F).

(Afonine *et al* 2018a). In Chimera, the former is simply called CC, while the latter is called CC about the mean (Goddard *et al* 2007). The advantage of the CC about the mean is that it removes dependence on the scale and offset of values in the cryo-EM map, and hence allows a better comparison of CC scores of models in different maps.

When using all values within the bounding box of the model (e.g. *CCbox* in Phenix, or global CC score in TEMPy), areas in the map that are not close to the model being considered may influence the score. To avoid the latter, masking can be applied to include values only close to the model, e.g. *CCmask* in Phenix (Afonine *et al* 2018a) or local CC in TEMPy (Joseph *et al* 2017a). Typically, *CCmask* or local CC scores are higher than *CCbox* or global CC scores as they avoid including areas away from the model which typically contain noise or other molecules not represented in the model being considered.

The main input parameters that affect CC scores include the resolution at which the model-map is generated and temperature/B factors set for each atom (when using *phenix.fmodel*). For the resolution, the FSC-estimated resolution of the map may be used, even though the resolution could vary locally. The CC score will also depend on which formulation is used, and since different software may use different forms they may also produce different values. Previous analyses have shown that CC scores can (perhaps unintuitively) be higher in lower resolution maps (Pintilie and Chiu 2018, Lawson *et al* 2021). Hence a higher CC may mean a better fit in a map, however, it does not necessarily mean a better quality or higher resolution map.

6.4.2.4 Z-scores

Z-scores are a way to compare CC scores obtained for the same model in the same map, in different positions and orientations (Pintilie and Chiu 2012, Pintilie *et al* 2016). For example, when performing an exhaustive rigid-body search with a model in a map, different fits are tried by sampling over all possible positions and orientations of the model in the map (with small position and orientation increments). At each fit a CC score is calculated. Typically, although not necessarily, the ‘correct’ fit of a model into a map is found to have the highest CC score amongst the other positions and orientations searched, because that position and orientation is where the model matches the map the best, according to the CC score.

Z-scores can be used to calculate how much better the top score is compared to other scores obtained in the map. When there is a ‘correct fit’, the *Z*-score is also typically higher, as the score for that fit tends to be significantly higher than the scores for the ‘incorrect fits’. On the other hand, if the model does not have a ‘correct’ fit into the map, i.e. the model being fitted does not match any of the molecules in the cryo-EM map substantially, then all fit scores tend to be lower and more similar around any one area being searched.

An example is shown in figure 6.12. A map of apoferritin at 1.75 Å (EMD-20026) is shown in figure 6.12(A) and segments corresponding to protein subunits created with Segger (Pintilie *et al* 2010) in figure 6.12(B). The model PDB:3ajo was then fitted to the central orange segment by rotational/exhaustive search using ‘Fit to Segments’ in Segger. In figure 6.12(C) cross-correlation scores are plotted for the top

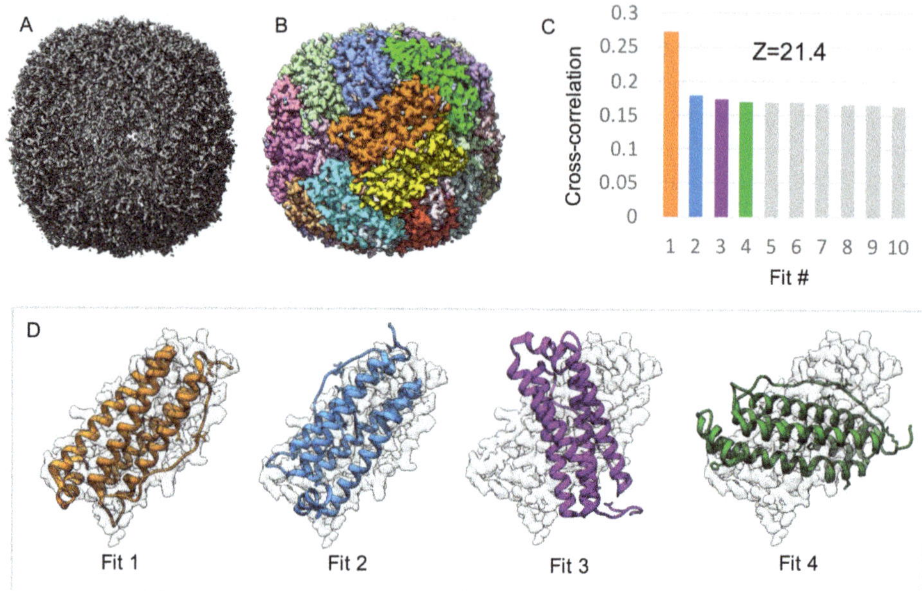

Figure 6.12. Illustration of segmenting and fitting to a map of apoferritin at 1.75 Å resolution using cross-correlation (CC) and *Z*-scores. (A) Entire map of apoferritin at 1.75 Å resolution (EMD-20026). (B) Segments obtained with Segger at a threshold of 0.015 with the 'grouping by connectivity' option. (C) CC scores for the top ten fits obtained by fitting the model PDB:3ajo to the orange segment by rotational search. (D) The top four fits shown along with the segment fitted to; the segment is shown with a transparent gray surface.

ten fits. A *Z*-score of 21.4 is obtained, meaning the CC of the top fit is 21.4 standard deviations above the mean of the next ten CC scores.

A high *Z*-scores signifies that the top fit is significantly better than other random fits and it can give higher confidence that the model is fitted correctly into the segmented map. Typically, higher *Z*-scores are obtained at higher resolution as shown here; in lower resolution maps, *Z*-scores also tend to be lower. Note that fits 3 and 4 in figure 6.12 have similar CC scores to fit 2 even though more of the model is 'outside' the segment shown in the transparent gray surface, because the CC is calculated against the entire map rather than just the segment. Also, if an exhaustive search were used across the entire map rather than the local search shown here, 24 fits with similar scores to the top fit would be expected since the apoferritin complex consists of 24 copies of the same protein.

6.4.2.5 EMRinger

The EMRinger metric is based on the idea that each amino acid side chain has certain favorable dihedral angles around the $C\alpha$–$C\beta$ bond (Barad *et al* 2015). To calculate the score, the $C\gamma$ atom (when present) is rotated in a full circle around the $C\alpha$–$C\beta$ bond, and the map density is measured at each position. If any peak(s) in this plot coincides with a favorable dihedral angle (Shapovalov and Dunbrack 2011) then the residue is considered to be rotameric (figure 6.13). The EMRinger metric

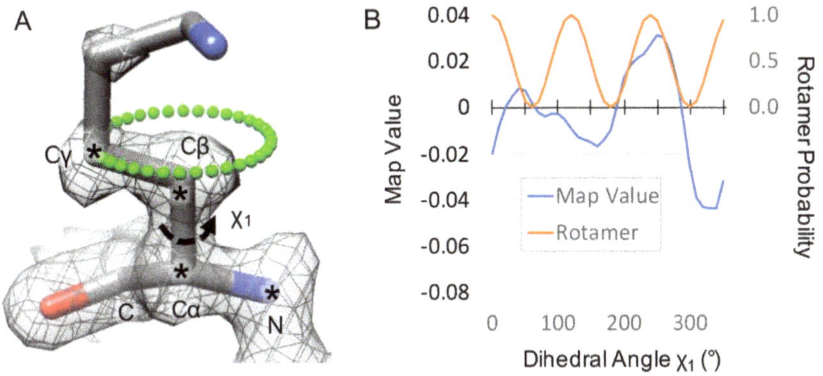

Figure 6.13. Illustration of EMRinger metric, which involves rotation of the $C\beta$ atom around the χ_1 dihedral angle (A). In (B) the plot shows that the map values coincide with a peak in the likelihood of the angle in known rotamers, hence the residue is considered rotameric. The four atoms involved in the χ_1 dihedral are marked with *.

considers how many residues are rotameric compared to the total number of residues that can be considered—the higher the proportion, the higher the score. EMRinger thus combines both map resolution and quality of the model: a well-fitted model backbone and a higher resolution map are likely to produce a higher EMRinger score. A limitation of the EMRinger metric is that it only considers map values around the rotated $C\gamma$ atom; moreover, it can only be applied to residues in proteins that have a single $C\gamma$ atom and are non-proline.

6.4.2.6 Q-*score*

The *Q*-score is another metric to evaluate how well a model matches the map, however, in a way that assesses the resolvability of individual atoms (Pintilie *et al* 2020). It does this by looking at spherical layers of points around each atom, with radii ranging from 0 to 2 Å. Only points that are closer to the atom rather than to any other atom in the model are considered, so as to avoid (although not completely eliminate) the influence from other atoms. These layers are illustrated in figure 6.14(A) for atoms OD1 and OD2 of amino acid Asp92 from apoferritin (PDB:3ajo) in a 1.75 Å map (EMD:20026).

Figure 6.14(B) shows plots of average map values versus radial distance away from the atom. For Atom OD2 this plot has a high peak at a distance of 0 Å and Gaussian-like drop away from the center of the atom. Such a density profile is typically associated with a well-resolved atom. On the other hand, atom OD1 has a low map value at a distance of 0 and the average map values at increasing distances away increase slightly before dropping again. This can be taken to mean the atom is not well-resolved.

The *Q*-score is calculated by comparing map values around each atom to Gaussian-like profiles, which we expect to see when the atom is well-resolved, via a cross-correlation score. Hence, the *Q*-score is much like the cross-correlation score, except that it is calculated at each atom and using only points that are closer to that

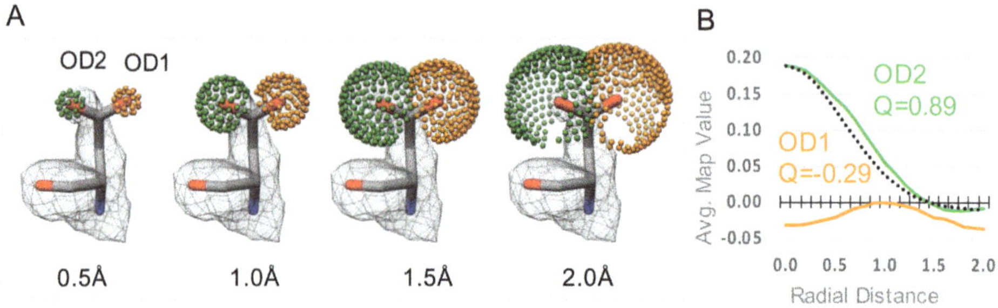

Figure 6.14. Calculation of Q-score using points around the atom at radii ranging from 0 to 2 Å. (A) Spherical layers of points are considered around two atoms (OD2 and OD1) in the residue Asp92, shown with sticks. The map is 1.75 Å apoferritin (EMD:20026), shown with transparent surface and mesh. (B) Average map values versus radial distance for each of the two atoms. For OD2, a Gaussian function is fitted and shown with black dotted line.

atom. For the two atom profiles plotted in figure 6.14(B), Q-scores are 0.89 for OD2, which is well-resolved, and −0.29 for OD1, which is not well-resolved, matching quantitatively what is seen in the map contours.

6.4.2.7 Per-amino acid residue/nucleotide-base Q*-scores*

Even though a Q-score is calculated for a single atom, Q-scores of groups of atoms can be averaged to measure the resolvability of a larger feature, such as an entire amino acid residue. Moreover, Q-scores can also be averaged for all the atoms in a model. It was shown that such Q-scores correlate very strongly to the FSC-estimated resolution of the map (Pintilie *et al* 2020). Thus, for a given estimated map resolution, we can also calculate what the expected Q-score would be. If Q-scores are lower than this expected value, it may indicate that the map is not resolved as well in those parts, although another possibility is that the model could be better fitted, refined, or re-built in those regions.

Per-residue Q-score plots are shown in figure 6.14 for protein and RNA fragments of two cryo-EM maps. figure 6.15(A) and (B) show a protein in the Csy-AcrF9 map at 2.57 Å resolution (EMD:21358, PDB:6vqv) (Zhang *et al* 2020a). Some parts of the protein are well-resolved with good side-chain densities, and the per-residue Q-scores in these parts are at or above the expected level for that resolution. Two dips can also be seen in the per-residue Q-score plot (figure 6.15(B)), corresponding to more flexible loops, where residues are not as well-resolved (marked with 1, 2, and red in the figure).

Figures 6.15(C) and (D) show the cryo-EM map of CRISPR-Csm at 3.5 Å resolution (EMD:0516 PDB:6nud) (Guo *et al* 2019). The crRNA component (yellow surface and model) is bound tightly to the rest of the complex and hence appears well-resolved, with nucleic acid bases clearly separable (figure 6.15(D)) and has Q-scores better than the expected Q-score in the 5′ end (figure 6.15(E)). On the other hand, the target ssRNA (red surface and model) is more loosely bound with fewer interactions with the complex itself, and it is held in place mostly by base-pairing with the crRNA (figure 6.15(D)). As a result, this part of the map appears to be less

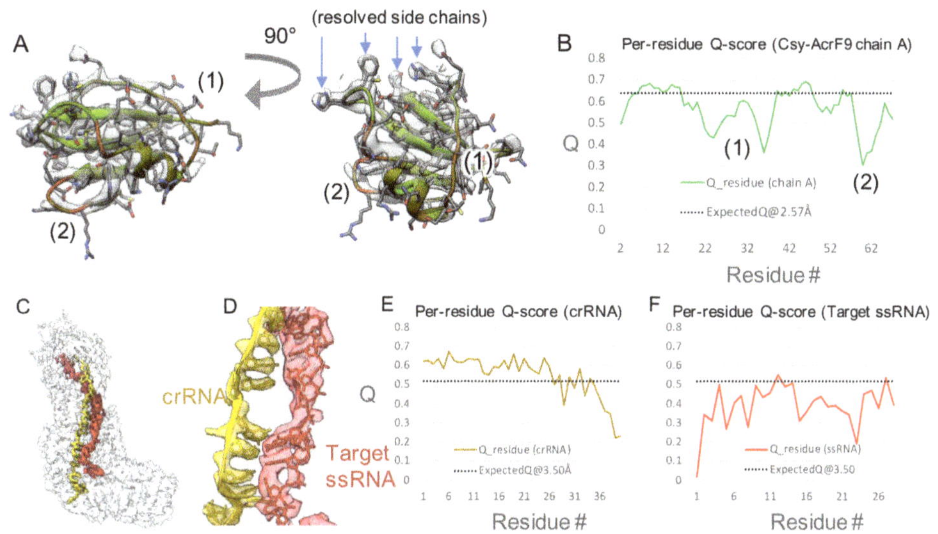

Figure 6.15. Examples of per-residue *Q*-scores for a protein (A), (B) and two RNA components (C)–(F). In (A) a protein from the Csy-Acr9 complex (EMD:21358, PDB:6vqv), chain A, colored with per-residue *Q*-scores. In (B) the per-residue *Q*-score plot shows residues with the expected *Q*-score for 2.6 Å, but some in two areas dipping below (marked with 1 and 2 in (A) and (B)). In (C), (D) is the complex CRISPR-Csm type III-A (EMD:0516: PDB:6nud), focusing on the CRISPR-RNA (crRNA) and target single-stranded RNA (ssRNA) components. The entire map is shown (the gray transparent surface) with crRNA and ssRNA segmented (yellow and red surfaces). (D) Model of crRNA and ssRNA built into the map. (E) Plot of per-residue *Q*-score of well-resolved crRNA and (F) plot of per-residue *Q*-score for less resolved target ssRNA.

resolved and, correspondingly, per-residue *Q*-scores for the ssRNA model are lower than the expected *Q*-score at this resolution (figure 6.15(F)).

6.4.2.8 Q-scores at different resolutions

Typical *Q*-scores for protein residues and nucleotides in cryo-EM maps at different resolutions are shown in figure 6.16(A) and (B), respectively. In figure 6.16(A) maps and models of apoferritin are shown. The models for all maps were based on the x-ray model of apoferritin (PDB:3ajo), first rigidly fitted to each map and then refined using Phenix. *Q*-scores are closest to the maximum value of 1 in the map at 1.25 Å and drop as the resolution of the map, or resolvability of the atom, decreases. *Q*-scores for the Asp126 seem to drop faster (a difference of 0.52) than for the polar Asn25 (0.48), hydrophobic Leu26 (0.48), and aromatic Tyr12 (0.43).

Figure 6.16(B) shows *Q*-scores for nucleotides at different resolutions for several different maps in the EMDB. In the highest resolution map shown (1.9 Å), individual-atom contours are almost visible. Toward lower resolutions, *Q*-scores drop and nucleotides become less-resolvable. Interestingly, average *Q*-scores for backbone atoms in nucleotides are higher than for the nucleotide-base atoms at 1.9 Å, i.e. backbones are better resolved at this resolution. On the other hand, at 2.2 Å and lower, *Q*-scores for the bases are higher than they are for the backbone, i.e. the backbone becomes less resolved than the bases.

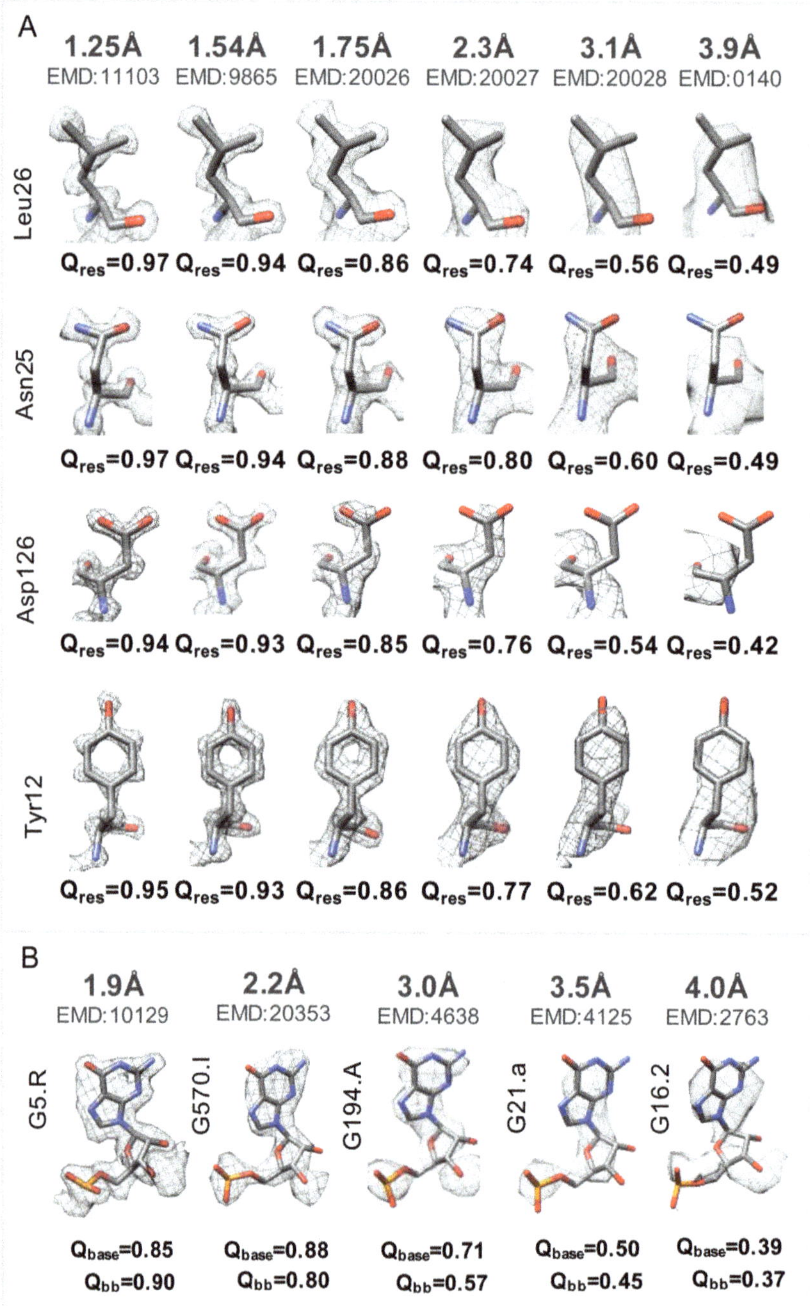

Figure 6.16. Images and Q-scores for protein amino acids (A) and nucleotides (B) at different resolutions. In (A) maps of apoferritin at different resolutions are used; the residue Q-score (Q_{res}) is shown below each residue. Each residue is indicated with vertical text on the left, e.g. Leu26 is the residue leucine at position 26 in the single chain from the model (PDB:3ajo). In (B) average Q-scores for the base (Q_{base}) and backbone atoms (Q_{bb}) are shown below each nucleotide. The nucleotide is indicated to the left of each image, e.g. G5.R is guanine, position 5, chain R, in the model associated with the corresponding EMD id.

6.4.2.9 Assessment of water molecules and ions

At resolutions of ~2.7 Å and higher it has become common to model water molecules and ions near protein molecules in cryo-EM maps (Murphy *et al* 2019, Passos *et al* 2020). An analysis using *Q*-scores confirmed that some of the water molecules and ions placed in an x-ray structure were also resolved in cryo-EM maps of the same specimen at 1.54 and 1.75 Å (Pintilie *et al* 2020). More waters were resolved in the higher resolution map and in both maps the water molecules and ions had similar distances to other nearby atoms as in the x-ray structure.

More recent maps of apoferritin at even higher resolutions (1.34 and 1.36 Å) show even more resolved water molecules (Zhang *et al* 2020b). In these two maps, waters were placed using a new automated method that considers various criteria to distinguish between water and ions (Prisant *et al* 2020). However, placing water molecules and ions may be prone to errors since they are small and a noise peak could potentially be mistaken for a water molecule or ion. Methods to validate the water placements can thus include the use of two independent maps, looking at distances between water molecules and nearby atoms, and measuring the resolvability of the water molecules.

Figure 6.17 shows an example of this validation process applied to water molecules. Water molecules were placed into two independent maps of apoferritin at resolutions of 1.34 and 1.36 Å. Figures 6.17(A) and (D) show a close up of a portion of a single protein, showing several water molecules in that region in similar positions in both maps. Of the detected water molecules, figures 6.17(B) and (E) show distances between the placed water molecules and nearby protein O or N

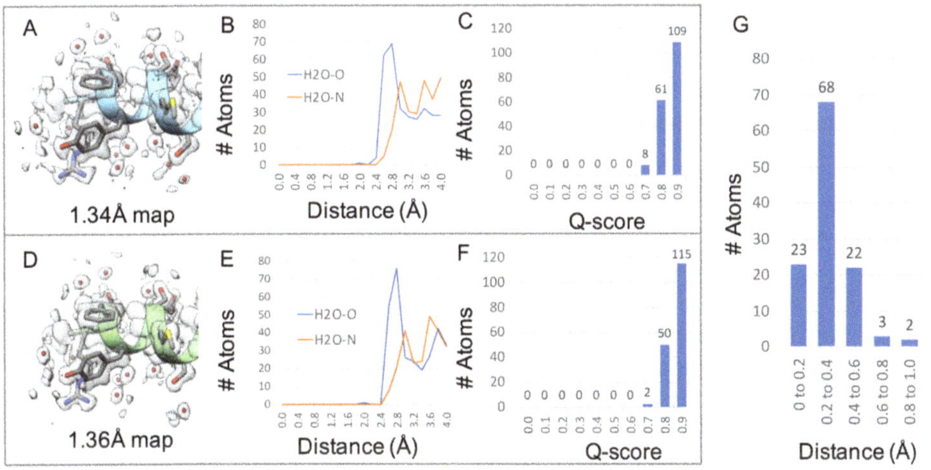

Figure 6.17. Water molecules placed in maps of apoferritin at 1.34 Å (A)–(C) and 1.36 Å (D)–(F). Water molecules are shown with red spheres in two small sections of each map (A), (D). Distances between placed water molecules and protein O atoms have peaks at 2.8 Å, similar to water atoms placed in x-ray models (B), (E). The resolvability of water molecules is high for all placed water molecules (C), (F) as shown in the histograms of *Q*-scores. When the protein chains are aligned, water molecules found in both maps are in similar positions; the plot in (G) shows the number of water atoms within ranges of distances from each other in the two maps, i.e. 23 water atoms are within 0–0.2 Å, 68 water atoms are within 0.2–0.4 Å, etc.

atoms. Even though the method considered water placements within a range of 2.3–3.3 Å, sharp peaks can be seen at a distance of 2.8 Å, matching water placements typically seen in high-resolution x-ray maps. The *Q*-scores of the placed waters within both maps are high, as shown in figures 6.17(C) and (F).

Of the 178 water molecules placed in the 1.34 Å map, a substantial number are close to one of the 167 placed in the 1.36 Å map when the protein chains are aligned: 23 are within 0.2 Å and 68 are within 0.2–0.4 Å (figure 6.17(G)). Water molecules that are in similar placements in the two maps increase confidence that these should be accurate. The other water molecules with positions that differ in the two maps raise the possibility that water positions near proteins may still vary in the same specimen across different datasets, or alternatively that water placements are still prone to some amount of 'false positive' assignment. Similar uncertainty in water detection is also seen in x-ray crystallographic structures, leading to methods that 'undowse' water molecules that do not follow certain criteria (Prisant *et al* 2020). Future research is needed to rigorously evaluate the extent of uncertainty in water molecule assignment by taking into account the dynamic properties of water around proteins.

For the placements of ions a similar analysis as described for detecting water is also possible. To determine whether a density peak is a water molecule or an ion, the distance to the nearby protein atom and whether the atom is charged or polar is considered (Prisant *et al* 2020). For example, if the distance is closer to 2.2 Å and the atom is charged, then it is more likely to be an ion. On the other hand, if the distance is closer to 2.8 Å and the atom is polar, then it is more likely to be a water molecule. When an ion is assigned it can be challenging to pin down what type of ion it is (e.g. Ca, Mg, or Zn); any information about the protein itself (what type of ion it is likely to bind) and solvent content during purification may then be considered.

6.4.3 Model-only metrics

In the previous section, map–model metrics were described along with their use in showing how well a model or parts of the model match the map, and at the same time how well the map resolves various features contained in the model (from atoms to secondary structures). Map–model metrics do not explicitly reflect the quality of the model, however, in terms of physical principles such as bond lengths, angles, and stereochemistry (dihedrals), and whether atoms in the model may be too close to each other (clashes) (figure 6.18).

Most cryo-EM maps do not resolve atoms individually and hence model-only metrics are also required to ensure that the model itself is of good quality, in particular if the model has been refined or flexibly fitted into the map. Model-only metrics are based on valence geometry (bond lengths, angles), backbone dihedral angles (Ramachandran statistics, Ramachandran *et al* 1963), side-chain dihedrals (rotameric statistics, Shapovalov and Dunbrack 2011), and the distances between nearby non-bonded atoms (clashes). Other metrics such as CaBLAM look at typical geometrical relationships between a series of consecutive CO groups relative to the local C-alpha atoms to assess the main-chain geometry of the model and to avoid

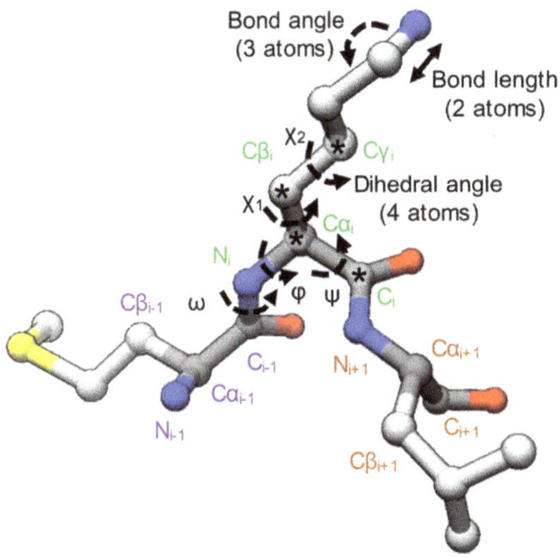

Figure 6.18. Illustration of the measurable properties of an atomic model: bond lengths (involving two atoms), angles (involving three atoms), and dihedrals (rotation around a bond, involving four atoms). Some dihedrals have commonly used symbols, e.g. the rotation around the peptide bond (C to N) is ω, around the N to Cα bond is φ, around the Cα to C bond is ψ, and around the Cα to Cβ bond is χ_1. The four atoms involved in the χ_1 dihedral are marked with *.

incorrect backbone topology (Prisant *et al* 2020). Model-only metrics are summarized in table 6.2.

Common software packages for calculating model-only scores include MolProbity (Williams *et al* 2018a), Phenix (Liebschner *et al* 2019), and Refmac (Winn *et al* 2011). Model-only metrics are very important during the process where a fitted model is refined or flexibly fitted into a cryo-EM map. Because most cryo-EM maps do not resolve individual atoms, the tendency of such a process is to move all atoms into areas of high density, or even into noise peaks, resulting in models of bad quality. Hence, refinement and flexible-fitting methods typically use restraints on atom movements to ensure such proper molecular structure, and this helps greatly to avoid overfitting the model to any noise present in the map. An analysis of overfitting using independent half maps has shown that such restraints manage to prevent overfitting when using reasonable parameters in flexible-fitting software, but also that it can happen if the parameters are highly directed toward improving map–model fit (Pintilie *et al* 2016).

The use of restraints during refinement or flexible fitting can mean that the final model may be forced to have good model-only scores. Hence, metrics that are typically not used in refinement can then become important for validation (e.g. CaBLAM). Model-only metrics also do not tell us if the model faithfully represents the cryo-EM map. For example, a model with perfect model-only score may wander 'outside' of higher-density regions, leaving them empty of model atoms, just to produce a good model-only score. In such cases, it is often the case that the model

Table 6.2. Model-only metrics with brief summaries. The scores use parameters typically derived from other high-resolution structures.

Model-only metric	Summary
Bonds	Distances between bonded atoms should be within some standard deviation from an equilibrium distance.
Angles	Angle between any three covalently bonded atoms should be within some standard deviation from an equilibrium value.
Side-chain dihedral angles	Side-chain dihedral angles should be similar to the rotamers seen in high-resolution structures (Shapovalov and Dunbrack 2011).
Backbone dihedrals	Backbone dihedral angles (ϕ, ψ) should fall within allowed areas in the Ramachandran plot (Ramachandran *et al* 1963).
Twisted peptide bond	Peptide-bond dihedral (ω) between two residues should be close to 180° for trans and 0° for cis conformations.
Cis peptide bonds	Cis peptide bonds occur very rarely, most often in prolines. Non-proline cis peptide bonds are flagged as possibly being incorrect.
Clashes	Non-bonded atoms should be more than some distance apart from each other, otherwise they are considered to clash.
CaBLAM	Pseudo-dihedrals between CO groups in adjacent residues should be compatible with the local C-alpha trace.
Undowser	Water molecules should be within a certain distance range from certain atom types (e.g. polar oxygen atoms), as previously seen in other high-resolution structures (Prisant *et al* 2020).

has to be further adjusted or refined to better fit the density (consequently to increase the model-map score), while maintaining good model-only scores.

Building a model based on a cryo-EM map often involves a trade-off between optimizing the map–model and model-only scores. The ideal model has high scores for both map–model and model-only. In practice, reaching the goal of building the ideal model may be hampered by the presence of lower resolution regions of the cryo-EM map, where atomic features may not be clearly visible, or they may be dynamic (i.e. the model may have multiple conformations). In such cases it can be difficult to obtain an un-ambiguous or single model. Map–model metrics are useful to point to such areas of ambiguity or lower resolvability, yielding lower scores compared to areas where the model fits the map better and where atomic features are better resolved.

6.4.4 Summary and conclusions

What we describe in this section includes a number of metrics that can be used to assess the models and maps in the current trend of cryo-EM structure resolution (1.2–4 Å). Some but not all of these metrics are widely used. There remains research opportunity in this area of technology development in cryo-EM to better characterize/evaluate existing metrics and/or to develop new ones. A frequently asked question (and a still-open problem) is how to annotate a cryo-EM map-derived model so that the PDB-deposited model is accompanied with metadata so that the readers of the cryo-EM model can readily relate the different resolvability in different parts of the structures (Hryc *et al* 2017). As a start, all the metrics discussed here are worth using in PDB depositions to provide such information.

Acknowledgment

The research is supported by NIH grants (R01GM079429, P41GM103832, and U24GM129541).

6.5 How algorithms from crystallography are helping electron cryo-microscopy

Paul D Adams[1,2], *Dorothee Liebschner*[1], *Thomas C Terwilliger*[3,4], *Pavel V Afonine*[1], *and Jane S Richardson*[5]

[1]*Molecular Biosciences and Integrated Bioimaging, Lawrence Berkeley National Laboratory, Berkeley, CA 94720, USA*

[2]*Department of Bioengineering, University of California at Berkeley, Berkeley, CA 94720, USA*

[3]*Bioscience Division, Los Alamos National Laboratory, Los Alamos, NM 87545, USA*

[4]*New Mexico Consortium, Los Alamos, NM 87545, USA*

[5]*Department of Biochemistry, Duke University Medical Center, Durham, NC 27710, USA*

6.5.1 Introduction

Crystallography is based on the interactions of short wavelength radiation with atoms in a crystal. In the case of x-rays this interaction involves a molecule's electron-density distribution and can thus reveal the positions of atoms in the crystal. The electron density (see figure 6.19) is obtained as the Fourier transform of the amplitude and phase of the scattered x-rays. As only the intensities of the waves can be measured, the phase information is lost and has to be inferred through careful experiments or complex computation. In contrast, electron cryo-microscopy (cryo-EM) is based on the interaction of electrons with the electrostatic field of the molecule, which is influenced by both the electrons and the nuclei of atoms.

Single-particle cryo-EM is now a commonly used method which combines 2D projection images of macromolecules into a 3D reconstruction (electrostatic potential map, or cryo-EM map; see figure 6.19). In contrast to diffraction experiments, where experimental phase information is obtained in a less direct way, cryo-EM provides experimental images of molecules. In addition, 3D reconstructions now routinely attain resolutions significantly better than 4 Å, allowing for atomic model interpretation and solving structures *de novo*.

One of the important features of the rapidly developing field of cryo-EM is the adoption of computational technologies that previously had not been available or that come from other fields. Indeed, although the underlying experiments differ

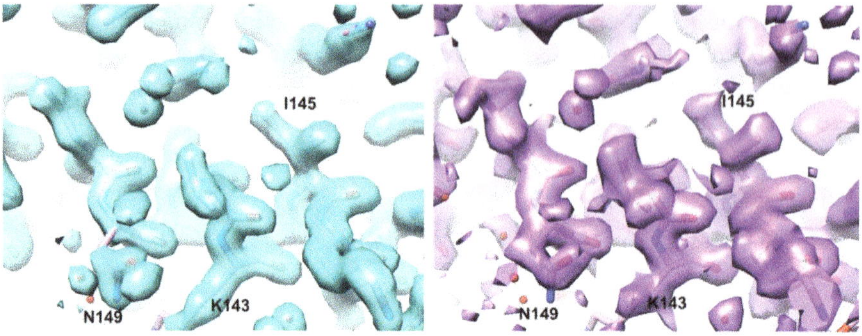

Figure 6.19. X-ray Fourier map (left) and cryo-EM reconstruction (right) for a portion of human apoferritin, both at a resolution of 1.8 Å. The model of the x-ray structure (2cei) is shown, with residues K143, I145, and N149 labeled for reference. The cryo-EM map was superimposed on the x-ray map using transformations based on refined cryo-EM and x-ray models with the Phenix tool *phenix.superpose_maps*. For purposes of clarity both maps have their voxels set to zero outside the model (i.e. at grid points further than 3 Å from the atom centers) and are contoured at twice the RMS value (i.e. 2 sigma) of each respective map.

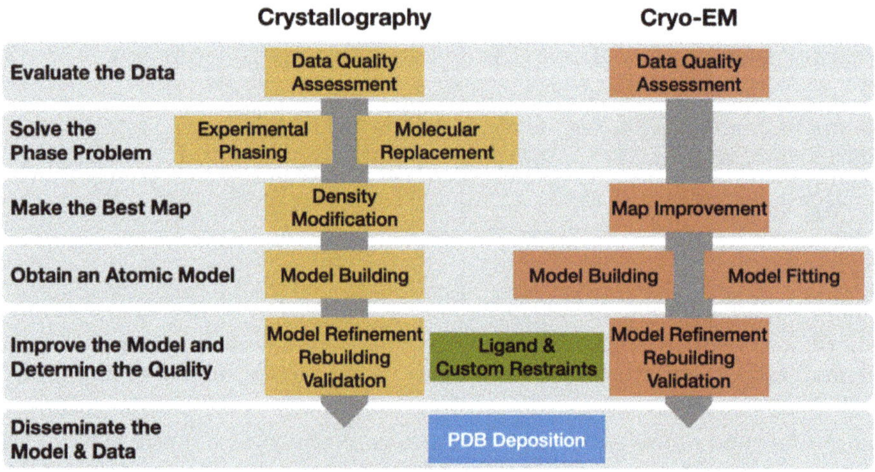

Figure 6.20. The structure-solution steps for crystallography and cryo-EM each have their nuances, but the overall workflow is the same.

between crystallography and cryo-EM, many of the computational steps involved in reaching a high-quality atomic model are similar.

Once crystallographic data have been reduced from diffraction images to Bragg intensities, or once a cryo-EM 3D reconstruction is performed yielding a three-dimensional map, there then is a common workflow (see figure 6.20). The first step consists of analyzing the derived experimental data to detect any anomalies that can complicate or even prevent structure determination. The second and third steps are to obtain the best possible map so that a model can be built. The fourth step focuses on iteratively improving the model by cycles of local rebuilding, refinement, and validation. Software tools can be used to readily analyse data from either kind of experiment.

Phenix (Adams *et al* 2002, 2010, Liebschner *et al* 2019) is a software suite that uses data from x-ray, electron, or neutron diffraction as well as from cryo-EM 3D reconstructions to determine macromolecular structures. It provides tools to address the unique properties of the experimental data, while maximally exploiting the commonalities in computational steps. New tools are regularly developed or enhanced to improve the structure-solution workflow. For example, a series of programs focusing on the analysis of cryo-EM maps and models was recently created to address the emerging needs of the cryo-EM community. Phenix has programs to tackle all of the tasks summarized in figure 6.20. In the following, we provide details of the major steps in proceeding from an initial cryo-EM 3D reconstruction to a refined and validated atomic model. Software tools for some of these steps are also provided within the CCP-EM software (Wood *et al* 2015, Burnley *et al* 2017).

6.5.2 Map improvement

The Phenix software has three tools that can be used to improve a cryo-EM reconstruction. These are automatic map sharpening using *phenix.auto_sharpen* (Terwilliger *et al* 2018c), density modification using *phenix.resolve_cryo_em* (Terwilliger *et al* 2020b), and recombination of the best parts of several maps using *phenix.combine_focused_maps* (Liebschner *et al* 2019). Map sharpening and combination of information from different maps can improve the clarity of a map, but do not change the fundamental information that is present. Density modification, in contrast, can change both the clarity of a map and the information that is present. This is possible because it can include information about a map that is known in advance and it can transfer information from one part of a map to another. We will first describe how auto-sharpening can improve a map. Then we will show how density modification can be used with two half maps, often yielding an even better map. Finally, we will describe how a composite map can be made that incorporates the best parts of several different maps, even if the different maps do not superimpose.

Map sharpening or blurring is a common approach to map improvement (Terwilliger *et al* 2018c, Jakobi *et al* 2017, Ramlaul *et al* 2019, Ramírez-Aportela *et al* 2020b). It consists of changing the emphasis in a map on low-frequency or high-frequency components. The map is represented as a sum of Fourier terms (much as in crystallography). Then the weights on Fourier terms are adjusted depending on the spatial frequency (resolution) of each term. In the Phenix *phenix.auto_sharpen* program (Terwilliger *et al* 2018c), Fourier terms are adjusted so that the mean amplitude decreases smoothly with finer resolution with an overall exponential factor B (a B-value corresponding the Wilson B factor of crystallography) up to the nominal resolution of the map, and then the mean amplitude decreases rapidly with a much larger exponential factor at higher resolution. The nominal resolution and the value of the exponential factor B determine how rapidly the amplitudes (weights) of Fourier terms decrease with resolution, and are the key parameters in this process.

In cryo-EM it is common practice to create not just a single reconstruction with all the data, but also two half maps, each obtained using half the data. A comparison of these half maps allows estimation of errors in the reconstruction. A frequency-dependent comparison of the maps (a Fourier shell correlation analysis, see section 5.2) also provides information about the accuracy of the full map as a function of spatial frequency or resolution (Rosenthal and Henderson 2003). Normally the nominal resolution of a map is taken from such an analysis of the half maps, where the resolution is the spatial frequency at which the estimated correlation between the observed map and the (unknown) true map is about 0.5. This corresponds to the resolution where the Fourier shell correlation between half maps is about 0.143.

The value of the exponential factor *B*, which determines how rapidly the amplitudes of Fourier terms fall off with resolution, is determined automatically in *phenix.auto_sharpen* by optimizing the clarity of the resulting map. In essence, the automated sharpening algorithm analyses the connected regions in the map obtained at a particular map cutoff value. It then maximizes both the connectivity of the map (fewest distinct regions) and the surface area of the map (amount of detail in each region) to yield a map that is optimal for interpretation. Note that this process is only accentuating information at some resolutions and de-emphasizing information at others; it does not introduce new information other than this resolution-dependent weighting. This auto-sharpening procedure can often dramatically improve the clarity of a map, and it is a useful and general tool. In addition to the automatic method based on connectivity and detail of a map, it can include model-based information and information obtained from comparison of half maps to optimize the clarity of a map.

Density modification is a procedure for map improvement that takes advantage of the fact that we have substantial prior knowledge about the map (Terwilliger *et al* 2020b, Scheres 2012, Kimanius *et al* 2020). For example, in a cryo-EM reconstruction it is quite certain that outside the region occupied by the macromolecule, the map should be very uniform (Terwilliger *et al* 2020b). In crystallography this corresponds to the expectation of a flat solvent. Just as important but less obvious is the expectation that in the region of the macromolecule, the distribution of map values is highly non-Gaussian. Rather, most of the map in this region has values near zero, but a small fraction (where the atoms are located) have very high values, and essentially none of the points have very low values. These expectations about map values can be combined with the experimentally obtained map values, using weights based on uncertainties in map values estimated by comparing corresponding half maps. The resulting map, incorporating both experimental information and prior knowledge, is then automatically sharpened using a resolution-dependent weighting based on the resolution-dependent estimated error in the map. Figure 6.21 illustrates the use of density modification on a high-resolution (1.8 Å) map of human apoferritin and compares it with auto-sharpening. The map shown in figure 6.21(A) is automatically sharpened with the Phenix automated sharpening algorithm, while the map shown in figure 6.21(B) has been density modified. The density-modified map shows additional details not present in the auto-sharpened map, for example there is a clear hole in the density surrounding the Y29 side chain, and the shape of the density better resembles that expected for this residue.

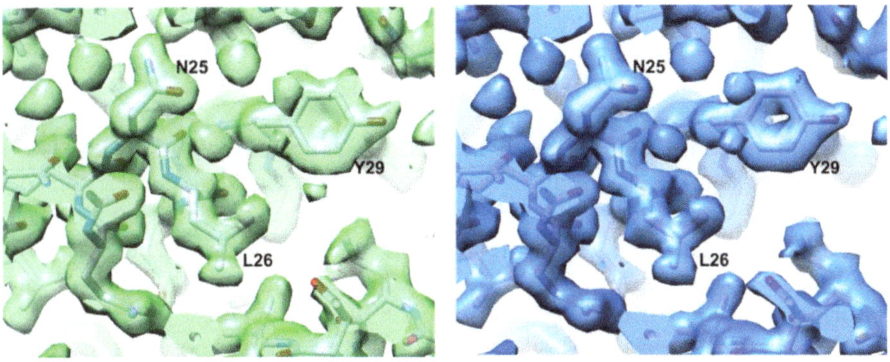

Figure 6.21. Original (left) and density-modified (right) cryo-EM maps for human apoferritin at a resolution of 1.8 Å. A model based on PDB entry 3ajo and refined against the original cryo-EM map is shown. Residues N25, L26, and Y29 are labeled for reference. The resolution-dependence of the original map is adjusted to match that of the density-modified map using *phenix.auto_sharpen* with the *external_map_sharpening* parameter. Maps are contoured to enclose equal volumes using the *phenix.map_comparison* tool. The maps in this figure can be reproduced by carrying out the apoferritin density-modification tutorial supplied with the Phenix package.

Density modification can generally be applied to a set of half maps and often leads to an improved overall map compared to simple auto-sharpening. Nevertheless, it is recommended to try both approaches; the map that shows the most clarity should be used for subsequent steps.

A third approach to map improvement is to combine the best parts of several maps. It is not uncommon for a large macromolecule to be somewhat flexible, so that even perfect images obtained from a group of molecules would not superimpose over the entire molecule. In the reconstruction step it is possible to focus attention on one part of the molecule, for example superposing just one region, yielding an accurate reconstruction of that part of the molecule but a blurry (low resolution) reconstruction for the rest, as is explained in section 4.5.3. A set of such maps, each focused on a different part of the molecule, can be combined with the Phenix tool *phenix.combine_focused_maps* to obtain a composite map that is relatively clear for all parts of the molecule. The key elements of combining different maps are: (i) the transformations needed to superimpose one map on another, (ii) the region of space where each transformation applies, and (iii) how to weight the information coming from different maps.

The approach for obtaining transformations used in combining focused maps requires that an atomic model has been built and that a single model has been refined against each map with rigid-body refinement with *phenix.real_space_refine* (see below). Rigid-body refinement allows different chains to move relative to each other but does not allow variations within a chain. Then the relationship between corresponding chains, refined based on two maps, shows how the maps are to be superimposed. Also, the region of space where each transformation applies is just the region covered by that molecule. Finally, the weights for information from each map come from the correlations between the density in that map and density expected from the model refined using that map. This weighting information can be calculated locally (with continuously changing values estimated from correlations within a small sphere around a location of interest) or globally (with correlations calculated over an entire chain).

The resulting composite map can be used in refinement or other analyses that require a map. As not every point of flexibility in the molecule has been considered, the map is imperfect, so it is still important to compare any resulting models with each original map as well. We note that the composite map will generally not superimpose directly on most of the original maps. To superpose a region of interest in an original map on the composite map, the *phenix.superpose_maps* tool can be used (once again using a rigid-body-refined model to superimpose the maps).

6.5.3 Map interpretation and model building

There are two overall approaches to obtaining an initial atomic model representing a cryo-EM map. One is to dock an existing model (or pieces of existing models) into the map, and the other is to construct an entirely new model based on the map itself. Phenix has tools to carry out both of these approaches. The *phenix.dock_in_map* tool (Liebschner *et al* 2019) can take any number of atomic models that are already known and find the location within a map where each model best fits the density. In this way an atomic model can be built up from existing models. This approach works well if the map can be represented by existing models, and if the existing models have conformations similar to those of the molecules in the reconstruction. In cases where no existing models are available the Phenix tool *phenix.map_to_model* (Terwilliger *et al* 2020c) can be used to interpret a map and create an atomic model representing it. The *phenix.map_to_model* procedure is fully automated and only requires a map, the sequence of the molecule to be built, and the nominal resolution of the map.

The automated model-building procedure starts with identifying the unique part of the map. If the cryo-EM map reconstruction process was carried out with some symmetry enforced, then the unique part of the map is a compact region that, given the reconstruction symmetry, represents the whole map. In *phenix.map_to_model*, the reconstruction symmetry can be either supplied or automatically identified. Then the region is chosen that is the most compact and most isolated (fewest connections to neighbors), and that can represent the entire map. All the model-building is then performed in this unique part of the map. At the end of the procedure the map symmetry is applied, overlapping chains are removed, and a model with symmetry is obtained.

The model-building process is carried out with several different algorithms. If the map contains more than one type of chain (e.g. both protein and RNA), the map is interpreted with each type and then the best-fitting model in each part of the map is kept. A simple model-building approach for protein chains consists of searching for helices and strands, two common and relatively regular secondary-structure types found in proteins that can be represented by standard template structures (Terwilliger 2010). In effect, this procedure is similar to the *phenix.dock_in_map* approach, which uses existing structures to represent the molecule, except that the templates are much smaller than typical protein chains. For nucleic acids the same approach can be used to find A-form RNA and B-form DNA chains. This model-building algorithm is generally used first, as it can often interpret much of the regular structure in a map.

To complete the fragments built in the previous step, the *phenix.trace_and_build* algorithm (Terwilliger *et al* 2020c) is applied. In this algorithm the connection between the end of an already-placed fragment and the next fragment that is part of the same chain is identified by examining the map at various contour levels. At very high contour levels, the ends of an already-placed fragment and any other fragments are not connected. At very low levels everything becomes connected. By starting at a high contour level and gradually lowering the level until the already-placed fragment just connects with one other fragment, a likely path of the chain can be identified. Iteration of this procedure, always requiring that the chain only have a single path and no branching, can lead to interpretation of much of a protein structure. Once the path of the chain is determined, likely positions of side chains can be identified from short branches coming off the chain. In this way the main-chain and then the side chains can be found.

A second approach for completing protein chains is the RESOLVE model-building method (Terwilliger 2003). In this method, helices and strands are extended in both the N-terminal and C-terminal directions using a library of common tripeptide fragments obtained from existing structures and choosing fragments based on their fit to the density in the map. The RESOLVE approach is also applied for building nucleic acids, using trinucleotide fragments instead of tripeptide fragments.

All of these approaches are typically used to interpret a single map. Then the Phenix tool *phenix.combine_models* is automatically run to remove overlapping fragments and choose the best-fitting model for each part of the map. Finally, symmetry (if any) is applied, overlaps produced by applying symmetry are removed again with the model combination tool, and a model with symmetry is obtained.

Figure 6.22 illustrates the application of the *phenix.map_to_model* procedure to a high-resolution (1.8 Å) map of human apoferritin. Figures 6.22(A) and (C) show two views of the map and the model that is automatically built into this map. In some regions the model fits the map quite well and in others the model is missing or the fit is not optimal. This is typical for an initial application of the *map_to_model* procedure. In many cases, particularly if the map is obtained at high resolution as in figure 6.22, some simple additional steps can yield a more complete model. In the case shown in figure 6.22(A) and (C), it was clear from examination of the map and model that residue range 6–9 was pulled out of place by the aromatic side-chain density, and 154–162 jumped across an empty space rather than following the map. The Phenix tool *phenix.sequence_from_map* (Ho *et al* 2020) can often fix this sort of problem. First, residues 154–162 were removed from the model. Then *phenix.sequence_from_map* was run with the option of *run_assign_sequence*. This automatically filled in residues 154–162 with better-fitting ones. Then the same procedure was

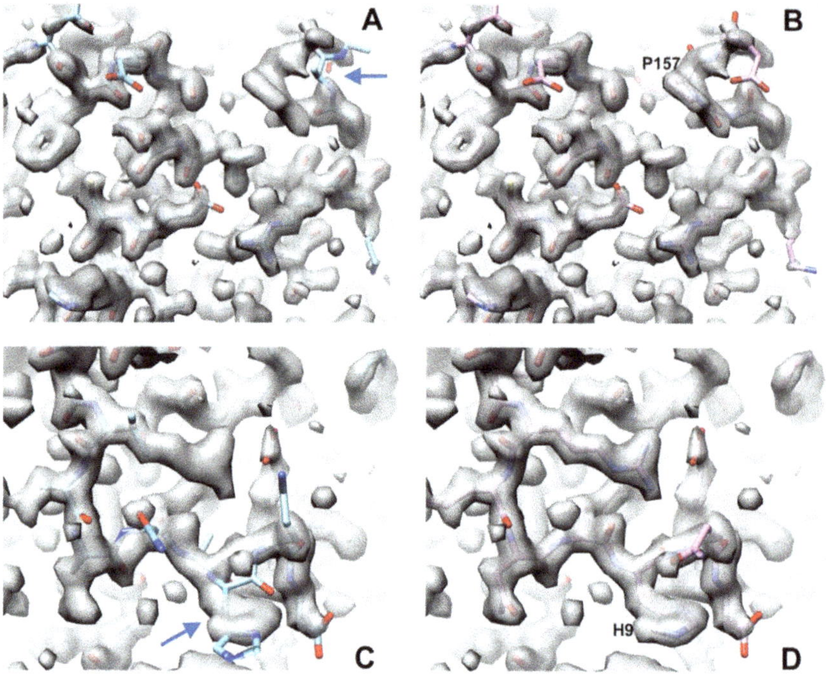

Figure 6.22. Automatic model-building (A) and (C). Portions of the model built into the density-modified cryo-EM map with *phenix.map_to_model*. (B) and (D) Model from panels (A) and (C) after two cycles of automatic sequence assignment and gap filling with *phenix.sequence_from_map* (using the option *run_assign_sequence = True*) followed by refinement with the *phenix.real_space_refine*. For further explanation see text. The maps and models in this figure can be reproduced by carrying out the apoferritin density modification and model-building tutorial supplied with the Phenix package.

carried out after deleting residues 6–9. The resulting model is shown in figure 6.22(B) and (D); it fits the map much better than the initial model.

In many cases some parts of the map are much clearer than others. In such situations automatic procedures such as *phenix.map_to_model* may build some parts of the structure and not others. There are several approaches to try and complete the model when this happens. One is to try and find an existing model that has enough similarity to be docked and give at least an indication of where the chain is located. This can be used as a starting point for either manual or automatic model building. Another possibility, which can be combined with the first, is to cut out the density in the region where the model was not built using the Phenix tool *phenix.map_box*. Then *phenix.map_to_model* can be applied to just this region. This is much faster than working with the entire map and also it may yield improved results for this region because the model-building algorithm will set working contour levels based on just the relevant part of the map.

The *phenix.map_to_model* procedure works best for maps with nominal resolutions in the range of about 3.5 Å or better, but can be used with some success for maps with resolutions up to about 4.5 Å. As the procedure is fully automated, it is reasonable to apply it as soon as a high-resolution, density-modified or auto-sharpened map is available. Once automatic model building has been carried out, normally it is necessary to complete the model with a combination of interactive graphics and automatic loop-fitting procedures, as described above for the case shown in figure 6.22.

6.5.4 Model optimization

The refinement of an atomic model with respect to the experimental data is an example of an optimization problem. The goal is to obtain an atomic model that simultaneously best fits the experimental data and obeys prior stereochemical knowledge. To achieve this, it is necessary to formulate a model with adjustable parameters. In the case of cryo-EM this is a model with atoms whose positions and mean square displacements can be varied to improve the fit to the map. To direct the refinement, a target function links the model parameters to the experimental data. Optimization algorithms are used to change the values of refinable model parameters to improve the agreement of the model with the experimental data. Refinement of atomic models against crystallographic data has been an area of research and development for several decades, leading to multiple algorithms for changing atomic positions, modeling atomic displacement, and imposing chemical and internal symmetry restraints. Fortunately, these algorithms can be reused in the context of cryo-EM. However, one of the major decisions for the refinement target function is how the model-to-data fit is calculated. In the case of crystallography this is typically formulated in reciprocal (Fourier) space. Alternatively, a target function may be formulated in terms of a map, which is then referred to as real-space refinement. In cryo-EM real-space refinement is a natural choice because the outcome of a single-particle imaging experiment is a 3D map.

The *phenix.real_space_refine* program (see figure 6.23) was specifically designed to perform refinements in real space (Afonine *et al* 2018b). The algorithm uses a simplified refinement target function that makes calculations faster, so that optimal data-restraint weights can be identified with little runtime cost. In addition to standard restraints on covalent geometry, *real_space_refine* can also make use of secondary-structure (Headd *et al* 2012), Ramachandran-plot (Oldfield 2001, Emsley *et al* 2010), and rotamer-specific restraints (Headd *et al* 2012) as well as internal molecular-symmetry constraints. The default mode performs gradient-driven minimization of the entire model. Since pure minimization often cannot correct a group modeled into the distant, incorrect local energy minimum, optimization can also be performed using simulated annealing (Brünger *et al* 1987, Adams *et al* 1997), morphing (Terwilliger *et al* 2013), rigid-body refinement (Afonine *et al* 2009), and systematic side-chain improvement (Oldfield 2001). As is the case for reciprocal-space refinement, real-space refinement should be alternated with validation and manual corrections.

Inputs to the *phenix.real_space_refine* program are an atomic model and either a real-space map (3D reconstruction) or its Fourier space equivalent. The program then iterates over a number of steps referred to as macrocycles. Each macrocycle

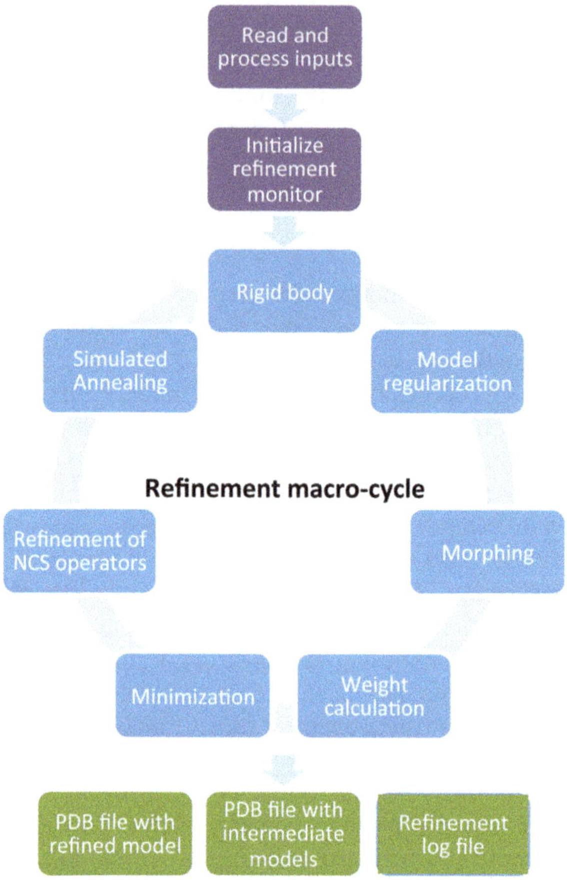

Figure 6.23. The real space refinement protocol implemented in *phenix.real_space_refine*. Reproduced from Afonine *et al* 2018b, CC BY 2.0.

begins with assessing residue side-chain fit to the map. Residues that fit the map poorly or that are rotamer outliers are subject to thorough local re-fitting, which aims to reach the best map fit while also maintaining meaningful geometry (e.g. a valid rotamer with no steric clashes). Because this re-fitting takes between 0.01 and 1 s per residue (depending on the side-chain size), this step can take a considerable amount of time for large models. Luckily this task is trivially parallelizable and therefore can be significantly accelerated if multiple processors are available. Once all residues are locally optimized, the program proceeds to optimization of all-atom coordinates against the map using gradient-driven minimization. Since the resolution of the map is finite and the map contains errors, this step requires additional information to compensate for the lack of experimental data. This information is formulated as either restraints or constraints and includes stereochemistry knowledge about local moieties, molecular packing, internal symmetry, and more. Some of this information is from tabulated libraries (Engh and Huber 1991, 2012, Vagin *et al* 2004), while in other cases it is inferred from the current model (e.g. secondary structure, internal symmetry). If symmetry was imposed on the map during the reconstruction process, symmetry constraints can be applied to the symmetry related molecules.

Other optimization strategies, such as simulated annealing or morphing, can achieve larger-scale atomic movements (~1–5 Å) in order to escape local minima and better fit the map. To save runtime, these strategies are not used by default but can be enabled when the model to map fit is poor and can be improved by substantial concerted shifts of atoms. Similarly, rigid-body optimization is not always performed but can be invoked at the initial stages if needed. Atomic displacement parameters (ADP or B factors) are refined after the positions of the atoms have been optimized, as their refinement requires atoms that fit the map accurately. At the end of the refinement process the final optimized model is written out in both PDB (Berman *et al* 2000, Burley *et al* 2017) and mmCIF (Bourne *et al* 1997) formats for further manual inspection with interactive graphics such as Coot (Emsley *et al* 2010) and ChimeraX (Goddard *et al* 2018).

6.5.5 Validation

In general, when some experimental data are described by a model, validation needs to address the quality of three different aspects: the model, the data, and how well the model explains the data. In structural biology, model and model-to-map criteria can be expressed either as global metrics or as local per-residue values.

In principle, metrics for model validation are the same between cryo-EM and crystallography, and much knowledge from past decades can be applied to cryo-EM structures. However, in practice new quality indicators are needed because typically poorer data resolution requires the tightening of geometric restraints in refinement. As a consequence, some traditional validation metrics almost always result in good scores which means they cannot be used to flag problems. As experimental data are fundamentally different between the two techniques, they require dedicated metrics to describe their quality. *Phenix.xtriage* is the Phenix tool to validate diffraction data and *phenix.mtriage* (Afonine *et al* 2018a) is the tool to assess cryo-EM reconstructions, primarily using the cross-correlation between half maps as a function of resolution, called the Fourier shell correlation (FSC) curve.

Assessing model-to-data fit has commonalities between the techniques, but also differences. The primary model-to-data metrics for crystallography are residuals between model-calculated and experimental diffraction amplitudes, measured in general by R_{work} and cross-validated by R_{free} (provided by withholding typically 5%–10% of reflections from refinement). For cryo-EM there are no experimentally measured structure factors and no equivalent to R_{free}. Instead, a real-space map is calculated from the atomic model and matched against the real-space 3D reconstruction, using various correlation coefficients (CC). The *phenix.comprehensive_validation* tool and GUI for cryo-EM report both traditional and newly developed model quality, e.g. CaBLAM (Prisant *et al* 2020) and model-to-map quality indicators (Afonine *et al* 2018a), such as the CC using the entire map, the CC within a mask, and the model-to-map FSC curve. Plots of average CCs per chain and per residue help identify problematic local regions.

6.5.5.1 Model validation

Validation of atomic models uses metrics that can be calculated from the model parameters, such as coordinates, B factors, and occupancies. Model properties are tested against basic physics and chemistry and against prior knowledge from high-resolution reference data. These reference data are generated by quality filtering experimentally determined crystallographic models at both the chain and residue level, while also taking into consideration the electron density around each residue. An extensive and still reasonably up-to-date guide to interpretation and use of model validation is available in Richardson *et al* (2018a), but there are now additional considerations for cryo-EM and for any structure solved using data worse than 2.5 Å resolution.

The Phenix *comprehensive_validation* tool calculates estimated standard deviations (RMSD) from the libraries used for restraints in refinement. The overall RMSDs (bonds, angles, dihedrals, planes, chirals) primarily reflect refinement strategy and weights. Bond angle outliers, when they do occur, almost always flag a serious problem of some kind, although the source of the problem is often not the bond angle itself but a result of other problems in the residue or in the immediate surroundings. However, the absence of angle outliers is not by itself a good measure of overall quality. The C-beta deviation is a combined measure of side-chain–backbone conformational incompatibility around the C-alpha atom (Lovell *et al* 2003). While this metric can be more informative than individual bond angles, it will not flag outliers if bond angles are tightly restrained, as is typically done at poor resolutions.

Both conformational and steric validation metrics are performed in Phenix by the same algorithms developed for the MolProbity web service at molprobity.biochem.duke.edu (Williams *et al* 2018a). Validation of local conformation uses coupled combinations of dihedral angles. Scores for Ramachandran or side-chain rotamericity (Hintze *et al* 2016) are obtained from MolProbity's smooth multi-dimensional distributions, in six categories for Ramachandran that show very different outlier contours: general, Ile/Val, Gly, pre-Pro, trans-Pro, and cis-Pro (Read *et al* 2011). RNA is validated for ribose-pucker outliers and known backbone conformers (Richardson *et al* 2008), which unfortunately do not hold for DNA. Pucker assignment as C3′-endo or C2′-endo uses only the observables most reliable to quite low (3.5–4 Å) resolution—the phosphate position and the glycosidic bond direction. It is also used to set pucker-dependent dihedral targets in refinement.

Steric validation (H-bonds, van der Waals, clashes) is accomplished by adding and optimizing hydrogen atoms with *phenix.reduce* and calculating the contacts with *phenix.probe* (Word *et al* 1999a, 1999b). The overall measure, called the clashscore, is the number of serious clashes $\geqslant$0.4 Å overlap per 1000 atoms. Individual clashes mean that at least one of the clashing atom groups is misplaced.

Several kinds of rare but serious problems are now flagged in validation when they occur. Genuine cis-nonPro peptides are usually functional but only occur for 1:3000 residues and have been grossly overused, in particular at poor resolution or in poor local density (Croll 2015, Williams *et al* 2018b). Cis-nonPro peptides cannot be justified by the data at resolutions worse than about 2.5 Å and should be modeled only if known from other data. Twisted peptides (omega off planar >30°) are nearly always wrong.

In the Phenix GUI, results from residue-specific validations of each type are listed as a table and clicking on an outlier will recenter the graphics window on that atom or residue in COOT or PyMol (Schrödinger 2015). Maps are displayed if available, and most types of outlier flags are shown, not just the one linked. These direct links reduce the effort to fix problems. A 3D overview of all local outliers on the model lets the user locate the major problem clusters. This is available in the KiNG Java-based viewer by running *phenix.kinemage*.

X-ray crystallography and its validation and correction methods were primarily developed and are quite mature for the mid-resolution range of about 1.4–2.7 Å. This is still the range for a great majority of PDB-deposited structures. The Phenix or MolProbity local validations described above are highly diagnostic in this range. Some things can be corrected automatically, manual corrections are fairly straightforward, and structural biologists have used these tools to improve statistics such as clashscores, NQH flips, and ribose-pucker outliers by 3–4-fold in the past 20 years (Williams *et al* 2018a). Cryo-EM structures in this resolution range are also handled well, although correcting local misfittings automatically is a development priority for large structures.

6.5.5.2 Lower resolution and cryo-EM

At resolutions between 2.5 and 3 Å, the main-chain density loses the features that make it possible to accurately position backbone CO groups. As a consequence, there can be significant deviations in peptide orientation, and in turn this makes Ramachandran, H-bond, secondary-structure assignments unreliable and moves many of the backbone donor-acceptor pairs far out of range. Toward 4 Å or in mobile regions, the map can be so weak or featureless that restraints yield 'perfect' geometries that show no outliers. However, this does not mean that the resulting model conformations correspond to those in the sample. The CaBLAM analysis was developed to help circumvent those limitations and give meaningful validation of backbone conformation at poor resolutions (Williams *et al* 2018a). It uses virtual C-alpha atom dihedrals to determine the local course of the chain, and a virtual CO dihedral to diagnose where the peptide orientations are incompatible with that C-alpha trace. In particular for cryo-EM, CaBLAM seems to be the most effective tool for protein backbone model validation and correction at 3–4 Å resolution (Richardson *et al* 2018b, Lawson *et al* 2021).

6.5.6 Validation-guided corrections

The stage between an initial model and deposition is an iterative cycle of validation, rebuilding, and refinement. The rebuilding process is one of gradual improvement, and even at deposition the goal is not zero outliers, but as few outliers as feasible (Richardson *et al* 2018a). Manual rebuilding is still a necessary step at some point in most structure solutions, because downhill refinement may not be able to cross barriers into a more correct local-minimum conformation. Manual corrections are usually performed in COOT, as linked from the Phenix validation GUIs (under development for ChimeraX as well). At poor resolutions, interactive guidance in Isolde (Croll 2018) can try out larger changes, even for local sequence realignment. Validation metrics such as all-atom clashes, poor real-space correlations, and extreme outliers of any kind are suggestive of where and what to try. Very sparse backbone H-bonds are also an indicator of probably incorrect backbone conformation.

Any strong validation outliers are diagnostic if they occur, but their absence at low resolution can be a symptom of overfitting (such as restraining Ramachandran values) rather than of a correct model. At least currently, CaBLAM outliers are not subject to such overfitting, so they provide less biased overall and local quality indicators. Most but not all 1% CaBLAM outliers flag a genuine problem, and a high fraction of them represent large needed changes in peptide orientation rather than small tweaks across a contour boundary. When in a helix or sheet they can often be corrected manually by making the secondary structure more regular. Figure 6.24(a) shows two adjacent CaBLAM outliers at 3.2 Å and figure 6.24(b) shows the correct conformation at 1.53 Å, achieved by a large rotation of the central peptide (O in red). That could be fixed without reference to the high-resolution structure, by extending the regular helix one residue. Such problems in cryo-EM may be caused by Ramachandran restraints that move residues into the edge of the nearest local minimum. When CO groups cannot be seen in the map, they may be modeled in a very incorrect orientation, which risks placing them in the wrong part of the Ramachandran plot. When Ramachandran restraints move these residues to the nearest minimum, they might be more 'correct' in the sense of the Ramachandran distribution, but their orientation does not correspond to the 'true' conformation in the sample. Figure 6.24(c) shows this effect for examples from the structures in figure 6.24(a) and (b).

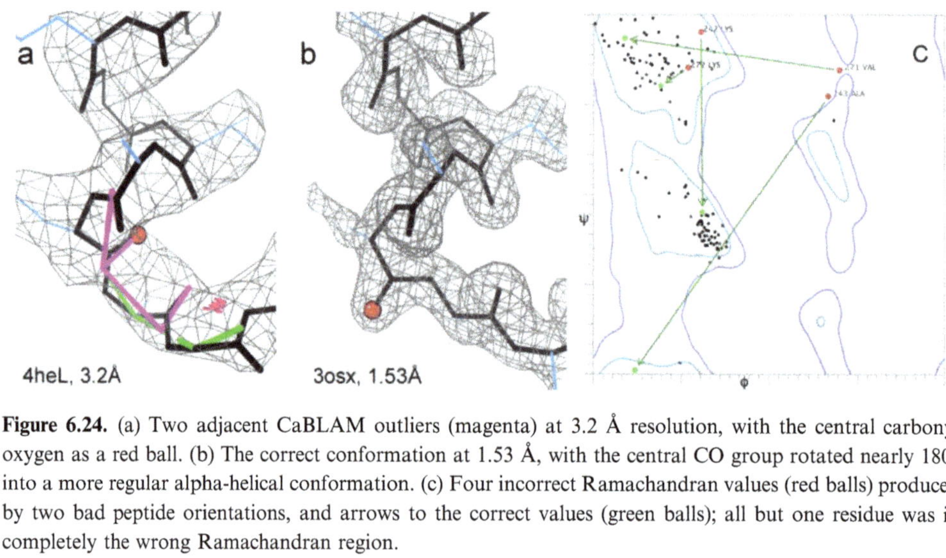

Figure 6.24. (a) Two adjacent CaBLAM outliers (magenta) at 3.2 Å resolution, with the central carbonyl oxygen as a red ball. (b) The correct conformation at 1.53 Å, with the central CO group rotated nearly 180° into a more regular alpha-helical conformation. (c) Four incorrect Ramachandran values (red balls) produced by two bad peptide orientations, and arrows to the correct values (green balls); all but one residue was in completely the wrong Ramachandran region.

Automated rebuilding steps are gradually increasing in scope and effectiveness, and there are many of them in Phenix. Global or large-scale tools include the initial cycles in Autobuild, plus morphing, CryoFit (Kim *et al* 2019), simulated annealing, RNA corrections in ERRASER (Chou *et al* 2016), and the conformational sampling in Isolde or *phenix.mr_rosetta* (DiMaio *et al* 2011). An important mid-scale tool is systematic side-chain improvement, which tries alternative rotamer choices to repack an interacting set of side chains such as a local hydrophobic core. It is performed in *phenix.real_space_refine* but is useful for x-ray crystallography as well as cryo-EM. Another is the *rebuild_in_place* option in RESOLVE, which will try out other conformational possibilities across a loop or other stretch that has shown problems in the context of the whole model.

6.5.7 Conclusions

Single-particle cryo-EM has advanced at an amazing pace over the last five years as a result of major advances in microscope and detector technology. Had there not been existing software tools already developed for crystallography it is likely that the process of building, refining, and validating an atomic model would have become a major bottleneck in cryo-EM. Fortunately, it has been possible to repurpose many of the crystallographic tools and keep pace with the other technical advances in the cryo-EM field. However, there are many challenges that are more specific to cryo-EM that require continued development. For example, the technique is well suited to studying very large molecular complexes. The size of these structures and their complexity will require further methods development to increase the speed of analysis, and approaches that facilitate manual intervention, which is currently challenging for large structures.

Cryo-EM data, although dramatically better than before the revolution, is still often at low enough resolution to present challenges for model building, refinement, and validation. Improvements in all of these methods are required to increase model accuracy, in particular as the rate of structure determination increases with better sample preparation, and microscope and detector technology. There are also indications that dramatically better resolution may be achieved in cryo-EM (Nakane *et al* 2020, Yip *et al* 2020), albeit maybe only for a subset of samples. This very high-resolution data is a godsend for model building and validation, but presents some challenges for model optimization, which will require further development to efficiently and accurately handle the data resolution and model detail—including features such as experimentally resolved hydrogen atoms. All in all, cryo-EM has a very bright future and there is still much to be done to take it to the limit of what is possible.

Acknowledgments

This work was supported by the NIH (grant GM063210 to PDA, JSR, and TCT) and the Phenix Industrial Consortium. This work was supported in part by the US Department of Energy under Contract No. DE-AC02–05CH11231.

6.6 Archiving structures and data

Catherine L Lawson and Brian P Hudson
Institute for Quantitative Biomedicine, Research Collaboratory for Structural Bioinformatics, Rutgers, The State University of New Jersey, Piscataway, NJ 08854, USA

6.6.1 Introduction

Structures that have been determined using cryo-EM single-particle analysis (SPA) and their associated experimental data can be accessed by the wider scientific community through public data archives.

There are three major data archives for cryo-EM-derived data relevant to SPA practitioners (figure 6.25). The Electron Microscopy Data Bank (EMDB) distributes density maps (Lawson *et al* 2016). EMDB currently holds nearly 12,000 maps, the majority of which (approximately 80%) are derived from SPA. The Protein Data Bank (PDB) holds atomic-coordinate models derived from all major structural biology methods (wwPDB Consortium 2019) and it currently has over 5000 EM-derived entries with ~90% derived from SPA. The Electron Microscopy Public Image Archive (EMPIAR), established in 2012, distributes raw image data and image stacks (Iudin *et al* 2016).

The organizations that manage and provide access to EMDB and PDB archives strive to follow FAIR best practice methods to ensure that data are findable, accessible, interoperable, and reusable (Wilkinson *et al* 2016).

The EMDataResource (EMDR) project aims to ensure that data archiving and validation standards for cryo-EM structures are coordinated internationally.

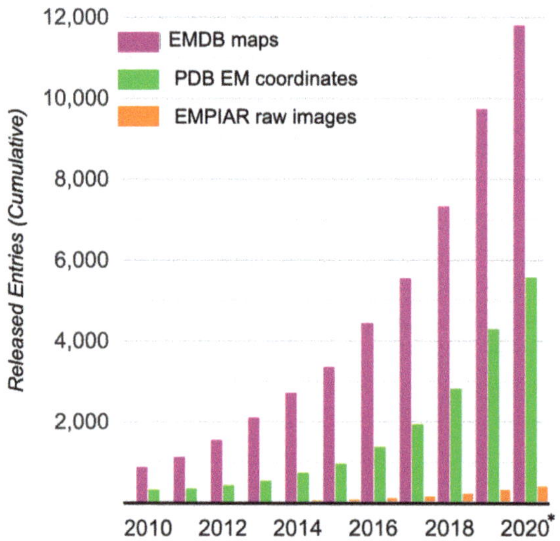

Figure 6.25. Public archives for cryo-EM structural data. Growth and cumulative holdings of EM Data Bank (EMDB), Protein Data Bank (PDB) EM-based coordinates, and the EM Public Image Archive (EMPIAR) over the past decade. Plot sources: EMDR and EMDB-EMPIAR statistics pages (*2020 data are through mid-August).

The project website (emdataresource.org) maintains up-to-date information about archiving and retrieving cryo-EM structure data, news, events, software tools, data standards, validation methods, and community challenges (Lawson *et al* 2016, 2021). Since 2007, EMDB has been operated jointly by EMDR, a collaborative project between the National Center for Molecular Imaging (NCMI) and the Research Collaboratory for Bioinformatics (RCSB) in the USA and the European Molecular Biology Laboratory-European Bioinformatics Institute (EMBL-EBI) in the UK.

The Worldwide PDB organization (wwPDB.org) manages the PDB archive and ensures that the PDB is freely and publicly available to the global community (wwPDB Consortium 2019). The current wwPDB membership consists of the RCSB Protein Data Bank (RCSB PDB) and the Biological Magnetic Resonance Data Bank (BMRB) in the USA, EMBL-EBI in the UK, and Protein Data Bank Japan (PDBj). It is anticipated that EMDB will join wwPDB in 2021.

Many major journals that publish results of cryo-EM research now require experimental density maps and their fitted coordinate models to be deposited, as do many funding agencies, and there are many good reasons to do so. Public availability enables independent validation of map interpretation and modeling, comparative analyses, and large-scale bioinformatics studies. Deposition of raw image data, although voluntary, provides essential benchmarks for cryo-EM software development and algorithm improvements.

6.6.2 Single-particle cryo-EM structure deposition

Each EMDB entry consists of a single density map (primary map) along with associated metadata describing each aspect of the experiment that produced the map, including sample details, specimen preparation, imaging, reconstruction, and density map interpretation. An EMDB entry may also include supplementary files such as FSC curves, masks, half maps, and additional maps appended with descriptors.

In cases where a single-particle cryo-EM experiment yields multiple reconstructions, each map that is analysed/described in a publication should be deposited as a separate EMDB entry.

An atomic-coordinate model fitted to an EMDB primary map is archived in PDB.

Cryo-EM reconstructions and their derived coordinate models may be jointly deposited to EMDB and PDB via the worldwide Protein Data Bank's OneDep system (Young *et al* 2017) (deposit.wwpdb.org). The system ensures uniform data collection worldwide, with review/biocuration assigned according to the geographic location of the depositor, and official validation reports are issued for each deposited primary map and/or model following biocuration.

Deposition for all maps and coordinate models in a publication is ideally completed prior to submitting a manuscript for review. The EMDB and PDB accession IDs, which are assigned post-submission, can then be included in the manuscript text. Official wwPDB validation report PDFs can be uploaded as part of the manuscript submission, providing key validation statistics about the structure to journal editors and manuscript reviewers (Gore *et al* 2017). Validation statistics are

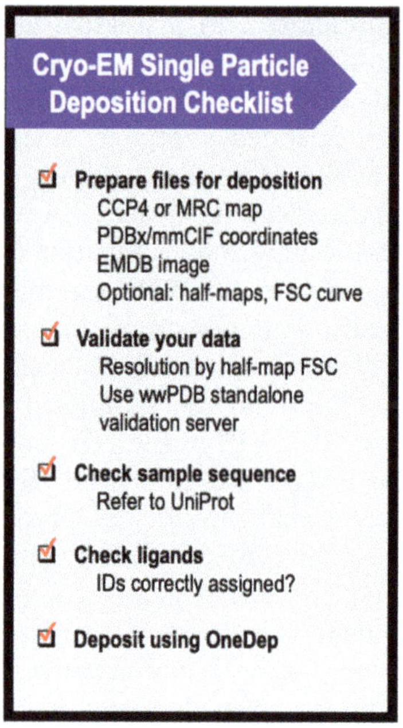

Figure 6.26. Checklist for deposition of structures (maps and coordinates) determined using SPA. Additional guidance can be found at wwpdb.org/deposition/faq.

generated according to recommendations of community task forces. For electron microscopy, initial recommendations were developed in 2010 (Henderson *et al* 2012); updated recommendations are anticipated soon based on a meeting held in 2020.

We present here a checklist for deposition of a structure determined using SPA (figure 6.26), and describe below, step-by-step, how to prepare for joint deposition of a density map to EMDB and a coordinate model to PDB, using the OneDep system.

6.6.3 Preparing files for deposition

Density map. The primary map of the EMDB entry should be the map that was interpreted by the researcher team and described in their publication. This could either be a map directly produced from a particle stack via reconstruction software ('raw map') or a map that has been computationally manipulated to facilitate interpretation (e.g. by masking, filtering, and/or sharpening). Allowed file formats for map-file upload are mrc and ccp4.

Two key map-related data items are manually entered during deposition: the primary map voxel size in the X-, Y-, and Z-directions, and a recommended contour level (density threshold). It is critical to enter the voxel dimensions correctly since they determine the scale at which the map will be displayed and/or analysed. The depositor-provided recommended contour level should yield an isosurface view that is representative of the imaged specimen.

Atomic-coordinate model. Fitted atomic coordinates are valuable both for interpreting and validating the results of a single-particle cryo-EM experiment.

The model preparation strategy that is most appropriate to optimize fit to the map will be strongly dependent on the final resolution of the reconstruction. At resolutions worse than ~4–5 Å, modeling is typically limited to rigid-body and/or domain-wise flexible fitting of source coordinates obtained from previous experimental research efforts. At higher resolutions, residue positions and conformations can be directly recognized from map features, and atom positions can be computationally optimized with appropriate stereochemical and geometric restraints.

At any resolution deposited models should incorporate known polymer sequences (protein or nucleic acid) and conform to basic principles of stereochemistry and geometry in terms of bond lengths, angles, chirality, and avoidance of clashes. Modeling and optimization software packages often provide basic assessments of model quality (geometry, stereochemistry, clashes).

Model refinement packages typically also provide one or more measures of the fit of the model coordinates to the density map, such as masked or unmasked correlation coefficient or map–model FSC curve. Most of the measures that have been devised for assessing model fit-to-map over the past decade are suitable for monitoring the progress of model optimization against a single map (Lawson *et al* 2021).

During computational optimization the relative weighting of the model's fit to the map versus stereochemical restraints should be carefully monitored. Overfitting of the map can result in poor model quality, while overly strong stereochemical restraints can result in poor map fit.

At resolutions of ~2.5 Å or better, a density map may appear to provide evidence for ordered solvent or ions, or better order in some parts of the structure relative to others. It is advisable to be cautious about adding solvent/ions or including B factors. Do putative water or ion peaks have strong spherical density and appropriate geometry relative to chelating atoms, and is the density convincingly strong in both half maps? Can relative order parameters (B factors) be satisfactorily modeled at a domain or residue level, rather than atom by atom? The answers to these questions will affect how one approaches modeling atoms into density.

The official archival file format of the PDB is the PDB exchange/macromolecular crystallographic information file (PDBx/mmCIF; mmcif.wwpdb.org). It is strongly recommended to upload atomic-coordinate models derived from single-particle cryo-EM to the OneDep deposition system in this format. The original PDB format, although still widely used, cannot properly represent large assemblies, such as ribosomes, owing to character/digit limits on chain and atom IDs. In contrast, PDBx/mmCIF is able to accommodate assemblies of unlimited size and complexity. Most major refinement packages offer the option to output coordinates in PDBx/mmCIF. Alternately, coordinates can be converted using PDB_Extract (pdbextract.wwpdb.org).

If a model follows strict point-group or helical symmetry, symmetry matrices need to either be included within the coordinate file or uploaded in a separate file, entered via the OneDep interface, or provided to biocuration staff via the OneDep communication page.

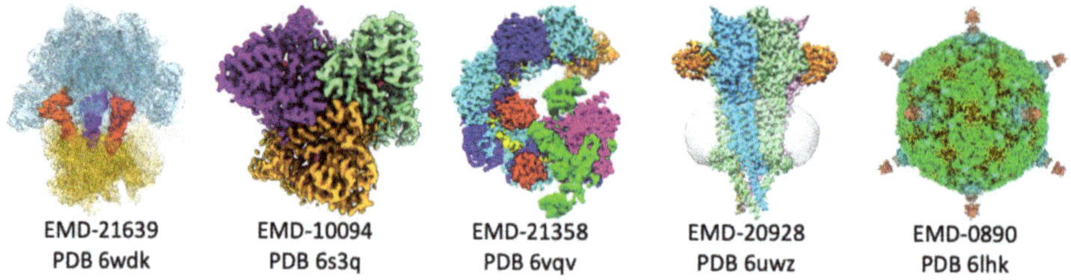

Figure 6.27. Representative EMDB images of recently released EM structures. Color is used effectively in these images to highlight important features.

EMDB map image. It is mandatory for the depositor to upload a representative image of the EMDB primary map. The uploaded image must be free from copyright restrictions with a size of approximately 500 × 500 pixels. The image is used by EMDB data delivery services (see 'Accessing Archived Cryo-EM Data' below); it provides an important first impression of entry content (figure 6.27).

Additional file uploads for the EMDB entry. Additional files related to the primary map may be included as part of the EMDB entry. These may include the raw map produced by the reconstruction program (if not deposited as the primary map), the independent half maps, masks, and/or FSC curve data. Deposition of these additional files is strongly encouraged as it allows others to independently interpret the experimental results and verify the reported resolution. As with the primary map, additional maps must be uploaded to OneDep in either mrc or ccp4 format. FSC curve data are uploaded as an xml file. Many major software packages will produce the curve data in the required format. Curve data in the accepted file format can also be obtained by uploading half maps to the FSC validation server hosted at EMBL-EBI.

6.6.4 Data validation

Map resolution. The independent half-map ('gold-standard') reconstruction procedure has become a standard practice to guard against overfitting of particle orientations; it is also used to estimate map resolution (Henderson *et al* 2012). The reported resolution of the primary map is a single number entered by the depositor. It is normally based on the Fourier shell correlation (FSC) curve of final independent half maps read at the threshold value of FSC = 0.143. Currently, EMDB only records and reports the overall resolution of the primary map. Local resolutions are not recorded or reported, although some depositors will upload a local resolution map as an additional file.

Validation services. The wwPDB provides a validation service that enables depositors to upload map and coordinate files to generate a preliminary wwPDB validation report that provides information about structure quality (figure 6.28). Following guidance from the 2010 EM Validation Task Force (Henderson *et al* 2012), current validation reports for cryo-EM structures provide basic statistics

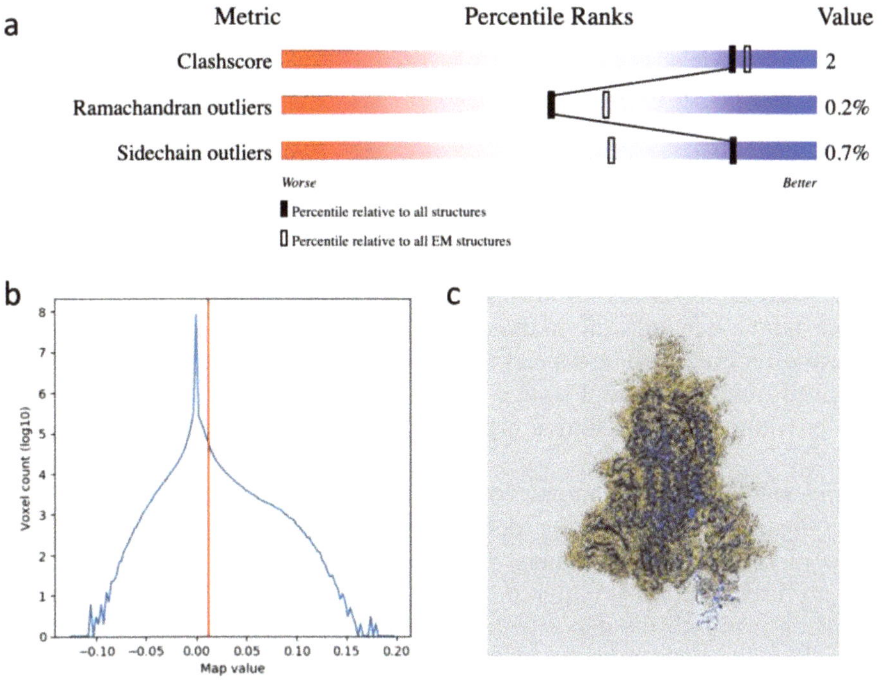

Figure 6.28. Example wwPDB validation report elements for cryo-EM structures. The structure shown is a SARS-CoV-2 spike (EMD-22253/PDB 6xm0). (a) Key metrics sliders. Validation score values for model coordinates are listed on the right and their rankings are displayed within red-to-blue gradient bars both with respect to all structures in the PDB (solid rectangles) and with respect to all EM structures in PDB (unfilled rectangles). (b) Map density-value distribution, with recommended contour level indicated by a red vertical line. A large peak at 0 is typical; it indicates that masking has been applied. (c) Overlay of coordinate model and map contoured at recommended contour level. The enclosed volume in the image is supplied in the validation report in the context of a volume versus contour level plot (not shown). In this example, the enclosed volume is 168 nm^3, or ~1/3 of the expected volume based on reported molecular weight (0.43 MDa). The tight contour level selected by the depositors excludes a small map region with lower density at the lower right.

about the quality of the model both in absolute terms and relative to other structures in the PDB archive. Since 2019 validation reports also incorporate a series of orthogonal 2D images and statistics plots of the primary map (density-value distribution, volume versus density value, rotationally averaged power spectrum), as well as atom inclusion statistics that provide a simple measure for model fit-to-map (Lagerstedt *et al* 2013, Abbott *et al* 2018). The report format and contents for SPA-derived structures were recently reviewed by a large panel of cryo-EM experts at a wwPDB-sponsored workshop held at EMBL-EBI in January 2020. A white paper that will enumerate a number of recommendations for improvements to validation is currently in preparation.

EMDR maintains lists of available services and software for validating cryo-EM structure data at emdataresource.org/validation.html.

6.6.5 Sample sequence and ligands

Each unique modeled-polymer entity within a structure needs to be described separately as part of a OneDep deposition. The depositor-provided sequence for each entity must match the coordinate sequence and it must also include any unmodeled portions. The depositor may also optionally select the most appropriate sequence database reference for each polymer entity. As a matter of policy, the PDB uses only UniProt references for proteins and Genbank references for nucleic acids. If a sequence database reference is not selected by the depositor or if a depositor-selected reference does not comply with the PDB's referencing policy, the closest sequence database match will (in most cases) be selected by PDB biocurators during the biocuration process. The depositor must also identify whether the polymer entity was derived from the natural source or was produced by recombinant or synthetic means, providing relevant source organism and expression host information when they do so.

Ligand validation is an important checkpoint during deposition wherein non-polymer ligands present in the deposited coordinates are checked against the contents of the wwPDB Chemical Component Dictionary. In cases of an exact match, no further action is required on the part of the depositor. In cases where no exact match is found, the depositor is given the opportunity to visually review the ligand(s) in question. In most cases, in particular when the ligand is novel and is not yet incorporated into the Chemical Component Dictionary, the depositor needs only to confirm the identity of the ligand or provide an image or SMILES string for use by biocurators during processing. In some cases, however, ligand review will reveal issues that require the depositor to correct and re-upload their coordinate file.

6.6.6 Deposition using OneDep

The OneDep system is designed to take the depositor step-by-step through the deposition process. The first step is to register and identify the type(s) of experimental data that will be deposited. Following registration, an email is sent with instructions on how to begin the deposition session.

Once logged into the session a communication module is available to address any questions or issues that may arise with your deposition to wwPDB biocuration staff. Also, options are provided that allow one to copy over data saved from a previous completed deposition. If you have multiple maps/models to deposit this can be a huge timesaver. However, all saved data items should then be carefully reviewed within each deposition session and any that do not apply to the current deposition should be updated.

You are next directed to upload your density map, coordinates, and associated files. As mentioned above, the preferred upload format for coordinates is PDBx/mmCIF since this is the master file format used by the OneDep system for processing, validation, and distribution (although at the time of writing, the legacy PDB format is still accepted for cryo-EM coordinates).

Completion of file uploads triggers a file-checking process and generation of an upload summary page. The upload summary page should be carefully checked and any flagged issues should be addressed by file replacements, if needed.

Completion of file upload (or replacement upload) also triggers generation of a preliminary validation report. This step takes some time, so an email is sent to the depositor when it is ready for review.

Once the file upload process is completed, depositors are directed to fill in data items that were not provided in file uploads. A series of fillable form pages (appropriate to the type of experiment indicated during the initial registration step) will become accessible and can be navigated via the left-hand menu. On each page, all mandatory fields required for submission that are empty are highlighted in red. When all mandatory information for a page is filled in, a green icon appears for that page in the navigation menu. Rather than speeding through each page and entering only mandatory fields, it is strongly recommended to fill all fields relevant to the experiment, as described below.

Experimental metadata. An archived structure has highest value to the larger scientific community when it is accompanied by complete metadata describing all aspects of the experiment that gave rise to it. EMDB and PDB share a core data dictionary that enables description of the experimental methods used to determine the cryo-EM map and its fitted model (table 6.3). For each of the major sections of the dictionary, the major data items that a depositor should be prepared to address in a OneDep deposition are indicated below.

Sample description. An imaged macromolecular assembly is described in two different ways: as a hierarchical list of subassemblies and as a list of distinct macromolecular entities. For example, the hierarchical description of a virus/antibody-Fab complex identifies the full complex as the parent plus two subassemblies: virus and Fab, while the list of macromolecular entities (compulsory for PDB entries) defines each unique polymer entity in the assembly, e.g. VP1, VP2, VP3, VP4 of the virus plus the heavy chain and light chain of the Fab, associating each with a sequence accession ID.

Sample molecular weight. Assembly and assembly component molecular weights are provided in units of megadaltons (MDa), unless the structure is helical, in which case molecular weight is specified per unit length in units of kilodaltons per nanometer.

Specimen. Specimen preparation includes description of the buffer composition, the specimen grid type, specimen support films, vitrification procedure, and vitrification instrument.

Imaging. The microscope model, detector, energy filter (if used), and operating parameters are described.

Particle selection. The number of particles used to generate the reconstruction are recorded. The depositor is encouraged to enter a textual description of the workflow/process that led to the final particle selection, in particular when one or more focus classification steps were used.

Reconstruction. The half-map FSC-measured resolution and symmetry imposed are recorded.

Table 6.3. Data dictionary categories for EM structure depositions to EMDB and PDB. The table links directly to the corresponding category descriptions at mmcif.wwpdb.org.

Top level	Specimen preparation	Image processing and reconstruction	Experimental data
em_experiment	em_buffer	em_3d_reconstruction	em_map[a]
em_software	em_buffer_component	em_image_processing	em_structure_factors[a]
Sample description	em_crystal_formation	em_particle_selection	em_layer_lines[a]
em_entity_assembly	em_embedding	em_volume_selection	
em_entity_assembly_molwt	em_sample_support	em_ctf_correction	
em_entity_assembly_naturalsource	em_specimen	em_2d_crystal_entity	
em_entity_assembly_recombinant	em_staining	em_3d_crystal_entity	
em_virus_entity	em_vitrification	em_helical_entity	
em_virus_natural_host	em_fiducial_markers[a]	em_single_particle_entity	
em_virus_shell	em_focused_ion_beam[a]	em_euler_angle_assignment[a]	
Data collection	em_grid_pretreatment[a]	em_final_classification[a]	
em_diffraction	em_high_pressure_freezing[a]	em_start_model[a]	
em_diffraction_shell	em_shadowing[a]	Structure analysis	
em_diffraction_stats	em_support_film[a]	em_3d_fitting	
em_image_recording	em_tomography[a]	em_3d_fitting_list	
em_image_scans	em_tomography_specimen[a]	em_fsc_curve[a]	
em_imaging	em_ultramicrotomy[a]		
em_imaging_optics			

[a] Archived only by EMDB.

Structure analysis. Information is collected about the methods used for fitting the map, which may be either based on a previously determined structure or the result of a *de novo* structure determination.

Software. Software package information is collected for all steps in the experiment, including data collection, image processing, reconstruction, and model fitting. Identify and review output logs from each of these steps prior to beginning the deposition.

Once all of the final files have been uploaded, the data fields have been filled, and the preliminary validation report generated from the uploaded files has been reviewed and approved, the deposition is complete and can be submitted.

6.6.7 Post-deposition: what happens next?

Upon submission, the PDB coordinate deposition and/or EMDB map deposition will be assigned unique PDB and/or EMDB identifiers, respectively. The deposition will then be reviewed and processed by wwPDB biocuration staff. The geographical location specified during initial setup of the deposition session will determine the biocuration site (USA, UK, or Japan).

When biocuration is complete, the depositor is sent a notice with the processed data files, copies of the wwPDB validation report, and a letter detailing issues that the depositor may wish to review and address. The original deposition session remains available for review of specific details, or to send correspondence to wwPDB biocurators. The deposition session is locked post-submission, but a depositor can request that it be unlocked should they need to make changes directly or upload new files.

When the depositor approves the entry, or (if there are no major issues) three weeks have passed, the entry is placed on hold. The hold period is one year from the date that the deposition was originally submitted, i.e. the date that PDB and/or EMDB IDs were issued. Data will be released, i.e. made public in the PDB and/or EMDB archives, under any of these conditions:

- The depositor requests release.
- The primary citation describing the determination of the structure has been published electronically.
- The editors of the publishing journal request structure release.
- One year has passed from the date of initial submission without either publication of the primary citation or a release request from either depositor or journal.

For an entry that has not yet been released, a depositor can request its withdrawal so long as neither primary citation publication has occurred nor a release request from the publishing journal has been received by the wwPDB. During this period, it is also possible to make any necessary changes or corrections to coordinates, maps, or metadata by requesting that wwPDB biocuration staff unlock their original deposition session so that the depositor can make changes or upload revised files. It is assumed that any changes made pertain to the original data collection.

Depositors wishing to provide coordinates, maps, and metadata based on a new data collection must initiate a new deposition for that purpose.

Minor revisions may continue to be made to a PDB entry or EMDB entry after release. Changes to metadata are considered minor revisions and can be made by biocurators at the request of depositors. It is also possible for a depositor to make changes to the coordinates of a PDB entry post-release. This is considered a major revision. When requested, the deposition session is unlocked so the depositor can provide new coordinates. The entry is subsequently re-processed and re-released by wwPDB biocurators. PDB policy requires that the experimental data (i.e. the map) remain unchanged and the number of allowed revisions is limited to one per PDB entry per year and three per research group per year. Changes to experimental data require a completely new deposition for both map (EMDB) and associated coordinates (PDB). These new entries receive new PDB and EMDB IDs and supersede the entries they are intended to replace. The replaced entries are then labeled as obsolete in the archives.

6.6.8 Accessing cryo-EM structure data

Cryo-EM structure data can be accessed in multiple ways: via web-based search, programmatic retrieval, and direct downloads.

Web-based search services enable users to identify sets of structures based on common keywords (e.g. ribosome or CRISPR) or selected data values (e.g. method type = single particle, or map symmetry = icosahedral). Collected data for individual structures can be reviewed via summary pages, and primary maps/coordinate models can be displayed in 3D viewers (figure 6.29). Search services based on EMDB map

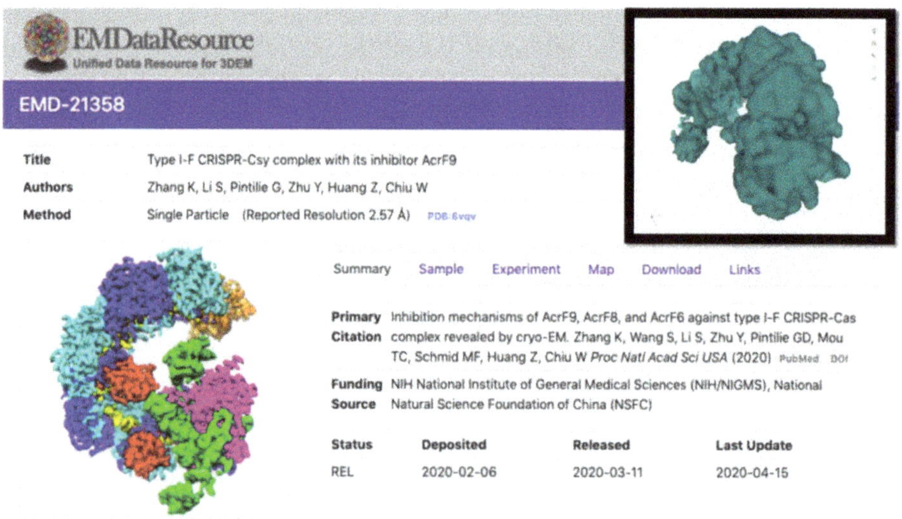

Figure 6.29. Example atlas page for an EMDB entry. Tabs provide information about the sample, experiment, and map parameters, access to file downloads, and links to additional resources. Inset at top right: viewer enables users to examine the primary map of the EMDB entry in 3D.

data and/or PDB EM model data are offered worldwide by EMDR, PDBj, RCSB PDB, EMBL-EBI, and the Spanish National Research Council (see EMDR's search page for links). Statistical trends of the archived data may also be viewed (e.g. as shown in figure 6.25).

Users who prefer to access cryo-EM structure metadata within their own software and/or by building their own databases can query EMDB and PDB data via application programming interfaces (APIs), or archive files can be directly downloaded. Links to API services and download instructions are available on the EMDR website.

6.6.9 Conclusions

In this section we have provided an overview of the data archiving process for Cryo-EM SPA-derived structures. Joint deposition of a map and model in the wwPDB OneDep system results in an EMDB map entry and associated PDB model entry with shared experimental metadata. The resulting collection of archived data can be accessed in multiple ways for independent validation, comparative analyses, and large-scale bioinformatics studies. We encourage depositors to carefully review and validate their data files prior to deposition according to these guidelines, and to populate all relevant experimental metadata fields.

Acknowledgment

EMDataResource is supported by NIH grant R01GM079429.

References

Abbott S, Iudin A, Korir P K, Somasundharam S and Patwardhan A 2018 EMDB web resources *Curr. Protoc. Bioinformatics* **61** 5.10.1–12

Abeysinghe S, Ju T, Baker M L and Chiu W 2008a Shape modeling and matching in identifying 3D protein structures *Comput. Aided Des.* **40** 708–20

Abeysinghe S S, Baker M, Chiu W and Ju T 2008b Segmentation-free skeletonization of grayscale volumes for shape understanding *2008 IEEE Int. Conf. on Shape Modeling and Applications* (Piscataway, NJ: IEEE) pp 63–71

Abriata L A, Tamò G E, Monastyrskyy B, Kryshtafovych A and Dal Peraro M 2018 Assessment of hard target modeling in CASP12 reveals an emerging role of alignment-based contact prediction methods *Proteins* **86** 97–112

Adams P D *et al* 2010 PHENIX: a comprehensive Python-based system for macromolecular structure solution *Acta Crystallogr.* D **66** 213–21

Adams P D, Grosse-Kunstleve R W, Hung L W, Ioerger T R, McCoy A J, Moriarty N W, Read R J, Sacchettini J C, Sauter N K and Terwilliger T C 2002 PHENIX: building new software for automated crystallographic structure determination *Acta Crystallogr.* D **58** 1948–54

Adams P D, Pannu N S, Read R J and Brünger A T 1997 Cross-validated maximum likelihood enhances crystallographic simulated annealing refinement *Proc. Natl Acad. Sci. USA* **94** 5018–23

Afonine P V, Grosse-Kunstleve R W, Urzhumtsev A and Adams P D 2009 Automatic multiple-zone rigid-body refinement with a large convergence radius *J. Appl. Crystallogr.* **42** 607–15

Afonine P V, Klaholz B P, Moriarty N W, Poon B K, Sobolev O V, Terwilliger T C, Adams P D and Urzhumtsev A 2018a New tools for the analysis and validation of cryo-EM maps and atomic models *Acta Crystallogr.* D **74** 814–40

Afonine P V, Poon B K, Read R J, Sobolev O V, Terwilliger T C, Urzhumtsev A and Adams P D 2018b Real-space refinement in PHENIX for cryo-EM and crystallography *Acta Crystallogr.* D **74** 531–44

Ahmed A and Tama F 2013 Consensus among multiple approaches as a reliability measure for flexible fitting into cryo-EM data *J. Struct. Biol.* **182** 67–77

Bai X, McMullan G and Scheres S H W 2015 How cryo-EM is revolutionizing structural biology *Trends Biochem. Sci.* **40** 49–57

Baker M L *et al* 2013 Validated near-atomic resolution structure of bacteriophage Epsilon15 derived from cryo-EM and modeling *Proc. Natl Acad. Sci. USA* **110** 12301–6

Baker M L, Abeysinghe S S, Schuh S, Coleman R A, Abrams A, Marsh M P, Hryc C F, Ruths T, Chiu W and Ju T 2011 Modeling protein structure at near atomic resolutions with Gorgon *J. Struct. Biol.* **174** 360–73

Baker M L, Baker M R, Hryc C F and Dimaio F 2010a Analyses of subnanometer resolution cryo-EM density maps *Methods Enzymol.* **483** 1–29

Baker M L, Baker M R, Hryc C F, Ju T and Chiu W 2012b Gorgon and Pathwalking: macromolecular modeling tools for subnanometer resolution density maps *Biopolymers* **97** 655–68

Baker M L, Jiang W, Rixon F J and Chiu W 2005 Common ancestry of herpesviruses and tailed DNA bacteriophages *J. Virol.* **79** 14967–70

Baker M L, Jiang W, Wedemeyer W J, Rixon F J, Baker D and Chiu W 2006 *Ab initio* modeling of the herpesvirus VP26 core domain assessed by cryoEM density *PLoS Comput. Biol.* **2** e146

Baker M L, Ju T and Chiu W 2007 Identification of secondary structure elements in intermediate-resolution density maps *Structure* **15** 7–19

Baker M L, Zhang J, Ludtke S J and Chiu W 2010b Cryo-EM of macromolecular assemblies at near-atomic resolution *Nat. Protoc.* **5** 1697–708

Baker M R, Rees I, Ludtke S J, Chiu W and Baker M L 2012a Constructing and validating initial Cα models from subnanometer resolution density maps with Pathwalking *Structure* **20** 450–63

Bammes B E, Jakana J, Schmid M F and Chiu W 2010 Radiation damage effects at four specimen temperatures from 4 to 100K *J. Struct. Biol.* **169** 331–41

Bammes B E, Rochat R H, Jakana J and Chiu W 2011 Practical performance evaluation of a 10k × 10k CCD for electron cryo-microscopy *J. Struct. Biol.* **175** 384–93

Barad B A, Echols N, Wang R Y-R, Cheng Y, DiMaio F, Adams P D and Fraser J S 2015 EMRinger: side chain-directed model and map validation for 3D cryo-electron microscopy *Nat. Methods* **12** 943–6

Bell J M, Chen M, Durmaz T, Fluty A C and Ludtke S J 2018 New software tools in EMAN2 inspired by EMDatabank map challenge *J. Struct. Biol.* **204** 283–90

Benkert P, Tosatto S C E and Schomburg D 2008 QMEAN: a comprehensive scoring function for model quality assessment *Proteins* **71** 261–77

Berman H M, Westbrook J, Feng Z, Gilliland G, Bhat T N, Weissig H, Shindyalov I N and Bourne P E 2000 The Protein Data Bank *Nucleic Acids Res.* **28** 235–42

Birmanns S, Rusu M and Wriggers W 2011 Using Sculptor and Situs for simultaneous assembly of atomic components into low-resolution shapes *J. Struct. Biol.* **173** 428–35

Bourne P E, Berman H M, McMahon B, Watenpaugh K D, Westbrook J D and Fitzgerald P M D 1997 Macromolecular crystallographic information file *Macromol. Crystallogr.* B **277** 571–90
Bowman B R, Baker M L, Rixon F J, Chiu W and Quiocho F A 2003 Structure of the herpesvirus major capsid protein *EMBO J.* **22** 757–65
Brünger A T, Kuriyan J and Karplus M 1987 Crystallographic R factor refinement by molecular dynamics *Science* **235** 458–60
Burley S K, Berman H M, Kleywegt G J, Markley J L, Nakamura H and Velankar S 2017 Protein data bank (PDB): the single global macromolecular structure archive *Methods Mol. Biol.* **1607** 627–41
Burnley T, Palmer C M and Winn M 2017 Recent developments in the CCP-EM software suite *Acta Crystallogr.* D **73** 469–77
Chen D-H *et al* 2011 Structural basis for scaffolding-mediated assembly and maturation of a DsDNA virus *Proc. Natl Acad. Sci. USA* **108** 1355–60
Chen M and Baker M L 2018 Automation and assessment of de novo modeling with Pathwalking in near atomic resolution cryoEM density maps *J. Struct. Biol.* **204** 555–63
Chen M, Baldwin P R, Ludtke S J and Baker M L 2016 *De novo* modeling in cryo-EM density maps with Pathwalking *J. Struct. Biol.* **196** 289–98
Chen V B, Arendall W B, Headd J J, Keedy D A, Immormino R M, Kapral G J, Murray L W, Richardson J S and Richardson D C 2010 MolProbity: all-atom structure validation for macromolecular crystallography *Acta Crystallogr.* D **66** 12–21
Cheng A, Tan Y Z, Dandey V P, Potter C S and Carragher B 2016 Strategies for automated cryoEM data collection using direct detectors *Methods Enzymol.* **579** 87–102
Chiu W, Baker M L, Jiang W, Dougherty M and Schmid M F 2005 Electron cryomicroscopy of biological machines at subnanometer resolution *Structure* **13** 363–72
Chou F-C, Echols N, Terwilliger T C and Das R 2016 RNA structure refinement using the ERRASER-Phenix pipeline *Methods Mol. Biol.* **1320** 269–82
Coppens P 1975 Thermal vibrations in crystallography by B T M Willis and A W Pryor *Acta Crystallogr.* A **31** 879–9
Croll T I 2015 The rate of cis-trans conformation errors is increasing in low-resolution crystal structures *Acta Crystallogr.* D **71** 706–9
Croll T I 2018 ISOLDE: a physically realistic environment for model building into low-resolution electron-density maps *Acta Crystallogr.* D **74** 519–30
Davis I W, Leaver-Fay A, Chen V B, Block J N, Kapral G J, Wang X, Murray L W, Arendall W B III, Snoeyink J and Richardson J S 2007 MolProbity: all-atom contacts and structure validation for proteins and nucleic acids *Nucleic Acids Res.* **35** W375–83
Degiacomi M T and Dal Peraro M 2013 Macromolecular symmetric assembly prediction using swarm intelligence dynamic modeling *Structure* **21** 1097–106
DiMaio F *et al* 2011 Improved molecular replacement by density- and energy-guided protein structure optimization *Nature* **473** 540–3
DiMaio F and Chiu W 2016 Tools for model building and optimization into near-atomic resolution electron cryo-microscopy density maps *Methods Enzymol.* **579** 255–76
DiMaio F, Song Y, Li X, Brunner M J, Xu C, Conticello V, Egelman E, Marlovits T C, Cheng Y and Baker D 2015 Atomic-accuracy models from 4.5-angstrom cryo-electron microscopy data with density-guided iterative local refinement *Nat. Methods* **12** 361–5
DiMaio F, Tyka M D, Baker M L, Chiu W and Baker D 2009 Refinement of protein structures into low-resolution density maps using Rosetta *J. Mol. Biol.* **392** 181–90

Dou H, Burrows D W, Baker M L and Ju T 2017 Flexible fitting of atomic models into cryo-EM density maps guided by helix correspondences *Biophys. J.* **112** 2479–93

Emsley P, Lohkamp B, Scott W G and Cowtan K 2010 Features and development of Coot *Acta Crystallogr.* D **66** 486–501

Engh R A and Huber R 1991 Accurate bond and angle parameters for x-ray protein structure refinement *Acta Crystallogr.* A **47** 392–400

Engh R A and Huber R 2012 Structure quality and target parameters *International Tables for Crystallography: Crystallography of Biological Macromolecules* ed E Arnold, D M Himmel and M G Rossmann (Chester: International Union of Crystallography) pp 474–84

Farabella I, Vasishtan D, Joseph A P, Pandurangan A P, Sahota H and Topf M 2015 TEMPy: a Python library for assessment of three-dimensional electron microscopy density fits *J. Appl. Crystallogr.* **48** 1314–23

Faruqi A R and Henderson R 2007 Electronic detectors for electron microscopy *Curr. Opin. Struct. Biol.* **17** 549–55

Frank J 2006 *Three-Dimensional Electron Microscopy of Macromolecular Assemblies: Visualization of Biological Molecules in Their Native State* 2nd edn (Oxford: Oxford University Press)

Frank J 2009 Single-particle reconstruction of biological macromolecules in electron microscopy —30 years *Q. Rev. Biophys.* **42** 139–58

Frank J 2017 Advances in the field of single-particle cryo-electron microscopy over the last decade *Nat. Protoc.* **12** 209–12

Garzón J I, Kovacs J, Abagyan R and Chacón P 2007 ADP_EM: fast exhaustive multi-resolution docking for high-throughput coverage *Bioinformatics* **23** 427–33

Glaeser R M 2019 How good can single-particle cryo-EM become? What remains before it approaches its physical limits? *Annu. Rev. Biophys.* **48** 45–61

Glaeser R M, Downing K, DeRosier D, Chiu W and Frank J 2007 *Electron Crystallography of Biological Macromolecules* (Oxford: Oxford University Press)

Goddard T D, Huang C C and Ferrin T E 2007 Visualizing density maps with UCSF Chimera *J. Struct. Biol.* **157** 281–7

Goddard T D, Huang C C, Meng E C, Pettersen E F, Couch G S, Morris J H and Ferrin T E 2018 UCSF ChimeraX: meeting modern challenges in visualization and analysis *Protein Sci.* **27** 14–25

Gore S *et al* 2017 Validation of structures in the Protein Data Bank *Structure* **25** 1916–27

Guo M *et al* 2019 Coupling of ssRNA cleavage with DNase activity in type III-A CRISPR-Csm revealed by cryo-EM and biochemistry *Cell Res.* **29** 305

Harauz G and Heel M V 1986 Exact filters for general geometry three dimensional reconstruction *Optik (Stuttg.)* **73** 146–56

Headd J J, Echols N, Afonine P V, Grosse-Kunstleve R W, Chen V B, Moriarty N W, Richardson D C, Richardson J S and Adams P D 2012 Use of knowledge-based restraints in phenix.refine to improve macromolecular refinement at low resolution *Acta Crystallogr.* D **68** 381–90

Henderson R *et al* 2012 Outcome of the first electron microscopy validation task force meeting *Structure* **20** 205–14

Henderson R 2013 Avoiding the pitfalls of single particle cryo-electron microscopy: Einstein from noise *Proc. Natl Acad. Sci. USA* **110** 18037–41

Herzik M A, Fraser J S and Lander G C 2019 A multi-model approach to assessing local and global cryo-EM map quality *Structure* **27** 344–58.e3

Heymann J B *et al* 2018 The first single particle analysis map challenge: a summary of the assessments *J. Struct. Biol.* **204** 291–300

Heymann J B 2018 Map challenge assessment: fair comparison of single particle cryoEM reconstructions *J. Struct. Biol.* **204** 360–7

Hintze B J, Lewis S M, Richardson J S and Richardson D C 2016 MolProbity's ultimate rotamer-library distributions for model validation *Proteins* **84** 1177–89

Ho C-M, Li X, Lai M, Terwilliger T C, Beck J R, Wohlschlegel J, Goldberg D E, Fitzpatrick A W P and Zhou Z H 2020 Bottom-up structural proteomics: cryoEM of protein complexes enriched from the cellular milieu *Nat. Methods* **17** 79–85

Hoh S W, Burnley T and Cowtan K 2020 Current approaches for automated model building into cryo-EM maps using Buccaneer with CCP-EM *Acta Crystallogr.* D **76** 531–41

Hryc C F *et al* 2017 Accurate model annotation of a near-atomic resolution cryo-EM map *Proc. Natl Acad. Sci. USA* **114** 3103–8

Humphrey W, Dalke A and Schulten K 1996 VMD: visual molecular dynamics *J. Mol. Graph* **14** 33–8

Igaev M, Kutzner C, Bock L V, Vaiana A C and Grubmüller H 2019 Automated cryo-EM structure refinement using correlation-driven molecular dynamics *eLife* **8** e43542

International Tables for Crystallography 2020 *International Tables for Crystallography: Reciprocal Space* vol B 3rd edn (New York: Wiley)

Iudin A, Korir P K, Salavert-Torres J, Kleywegt G J and Patwardhan A 2016 EMPIAR: a public archive for raw electron microscopy image data *Nat. Methods* **13** 387–8

Jakobi A J, Wilmanns M and Sachse C 2017 Model-based local density sharpening of cryo-EM maps *eLife* **6** e27131

Jiang W, Baker M L, Jakana J, Weigele P R, King J and Chiu W 2008 Backbone structure of the infectious Epsilon15 virus capsid revealed by electron cryomicroscopy *Nature* **451** 1130–4

Jiang W, Baker M L, Ludtke S J and Chiu W 2001 Bridging the information gap: computational tools for intermediate resolution structure interpretation *J. Mol. Biol.* **308** 1033–44

Jiang W, Baker M L, Wu Q, Bajaj C and Chiu W 2003 Applications of a bilateral denoising filter in biological electron microscopy *J. Struct. Biol.* **144** 114–22

Joseph A P, Lagerstedt I, Patwardhan A, Topf M and Winn M 2017a Improved metrics for comparing structures of macromolecular assemblies determined by 3D electron-microscopy *J. Struct. Biol.* **199** 12–26

Joseph A P, Malhotra S, Burnley T, Wood C, Clare D K, Winn M and Topf M 2016 Refinement of atomic models in high resolution EM reconstructions using Flex-EM and local assessment *Methods* **100** 42–9

Joseph A P, Polles G, Alber F and Topf M 2017b Integrative modelling of cellular assemblies *Curr. Opin. Struct. Biol.* **46** 102–9

Kawabata T 2008 Multiple subunit fitting into a low-resolution density map of a macromolecular complex using a Gaussian mixture model *Biophys. J.* **95** 4643–58

Kc D B 2017 Recent advances in sequence-based protein structure prediction *Brief Bioinform.* **18** 1021–32

Kim D N, Gront D and Sanbonmatsu K Y 2020 Practical considerations for atomistic structure modeling with cryo-EM maps *J. Chem. Inf. Model* **60** 2436–42

Kim D N, Moriarty N W, Kirmizialtin S, Afonine P V, Poon B, Sobolev O V, Adams P D and Sanbonmatsu K 2019 Cryo_fit: democratization of flexible fitting for cryo-EM *J. Struct. Biol.* **208** 1–6

Kim D N and Sanbonmatsu K Y 2017 Tools for the cryo-EM gold rush: going from the cryo-EM map to the atomistic model *Biosci. Rep.* **37** 6

Kim S J *et al* 2018 Integrative structure and functional anatomy of a nuclear pore complex *Nature* **555** 475–82

Kimanius D, Zickert G, Nakane T, Adler J, Lunz S, Schonlieb C-B, Oktem O and Scheres S 2020 Exploiting prior knowledge about biological macromolecules in cryo-EM structure determination *BioRxiv* 2020.03.25.007914

Kong Y and Ma J 2003 A structural-informatics approach for mining β-sheets: locating sheets in intermediate-resolution density maps *J. Mol. Biol.* **332** 399–413

Kong Y, Zhang X, Baker T S and Ma J 2004 A structural-informatics approach for tracing β-sheets: building pseudo-Cα traces for β-strands in intermediate-resolution density maps *J. Mol. Biol.* **339** 117–30

Kryshtafovych A, Adams P D, Lawson C L and Chiu W 2018 Evaluation system and web infrastructure for the second cryo-EM model challenge *J. Struct. Biol.* **204** 96–108

Kryshtafovych A, Schwede T, Topf M, Fidelis K and Moult J 2019 Critical assessment of methods of protein structure prediction (CASP)—round XIII *Proteins Struct. Funct. Bioinf.* **87** 1011–20

Kucukelbir A, Sigworth F J and Tagare H D 2014 Quantifying the local resolution of cryo-EM density maps *Nat. Methods* **11** 63–5

Kuhlman B and Bradley P 2019 Advances in protein structure prediction and design *Nat. Rev. Mol. Cell Biol.* **20** 681–97

Kühlbrandt W 2014 Cryo-EM enters a new era *eLife* **3** e03678

Lagerstedt I, Moore W J, Patwardhan A, Sanz-Garcia E, Best C, Swedlow J R and Kleywegt G J 2013 Web-based visualisation and analysis of 3D electron-microscopy data from EMDB and PDB *J. Struct. Biol.* **184** 173–81

Lander G C *et al* 2009 Appion: an integrated, database-driven pipeline to facilitate EM image processing *J. Struct. Biol.* **166** 95–102

Langer G, Cohen S X, Lamzin V S and Perrakis A 2008 Automated macromolecular model building for x-ray crystallography using ARP/WARP version 7 *Nat. Protoc.* **3** 1171–9

Lasker K, Topf M, Sali A and Wolfson H J 2009 Inferential optimization for simultaneous fitting of multiple components into a CryoEM map of their assembly *J. Mol. Biol.* **388** 180–94

Lawson C L *et al* 2011 EMDataBank.Org: unified data resource for cryoEM *Nucleic Acids Res.* **39** D456–64

Lawson C L *et al* 2016 EMDataBank unified data resource for 3DEM *Nucleic Acids Res.* **44** D396–403

Lawson C L *et al* 2021 Cryo-EM model validation recommendations based on outcomes of the 2019 EMDataResource challenge *Nat. Methods* **18** 156–64

Lawson C L, Berman H M and Chiu W 2020b Evolving data standards for cryo-EM structures *Struct. Dyn.* **7** 014701

Lawson C L and Chiu W 2018 Comparing cryo-EM structures *J. Struct. Biol.* **204** 523–6

Liebschner D *et al* 2019 Macromolecular structure determination using x-rays neutrons and electrons: recent developments in Phenix *Acta Crystallogr.* D **75** 861–77

Lindert S, Stewart P L and Meiler J 2009 Hybrid approaches: applying computational methods in cryo-electron microscopy *Curr. Opin. Struct. Biol.* **19** 218–25

Lopéz-Blanco J R and Chacón P 2013 iMODFIT: efficient and robust flexible fitting based on vibrational analysis in internal coordinates *J. Struct. Biol.* **184** 261–70

Lovell S C, Davis I W, Arendall W B, de Bakker P I W, Word J M, Prisant M G, Richardson J S and Richardson D C 2003 Structure validation by Cα geometry: ϕ, ψ and Cβ deviation *Proteins* **50** 437–50

Ludtke S J, Baker M L, Chen D-H, Song J-L, Chuang D T and Chiu W 2008 *De novo* backbone trace of GroEL from single particle electron cryomicroscopy *Structure* **16** 441–8

Ludtke S J, Lawson C L, Kleywegt G J, Berman H M and Chiu W 2011 Workshop on the validation and modeling of electron cryo-microscopy structures of biological nanomachines *Pac. Symp. Biocomput. Pac. Symp. Biocomput.* 369–73

Malhotra S, Träger S, Dal Peraro M and Topf M 2019 Modelling structures in cryo-EM maps *Curr. Opin. Struct. Biol.* **58** 105–14

McGuffin L J 2008 The ModFOLD server for the quality assessment of protein structural models *Bioinformatics* **24** 586–7

Mendez J H, Mehrani A, Randolph P and Stagg S 2019 Throughput and resolution with a next-generation direct electron detector *IUCrJ* **6** 1007–13

Milazzo A-C, Cheng A, Moeller A, Lyumkis D, Jacovetty E, Polukas J, Ellisman M H, Xuong N-H, Carragher B and Potter C S 2011 Initial evaluation of a direct detection device detector for single particle cryo-electron microscopy *J. Struct. Biol.* **176** 404–8

Moult J 2005 A decade of CASP: progress, bottlenecks and prognosis in protein structure prediction *Curr. Opin. Struct. Biol.* **15** 285–9

Murphy B J, Klusch N, Langer J, Mills D J, Yildiz Ö and Kühlbrandt W 2019 Rotary substates of mitochondrial ATP synthase reveal the basis of flexible F_1–F_o coupling *Science* **364** eaaw9128

Murshudov G N 2016 Refinement of atomic structures against cryo-EM maps *Methods Enzymol.* **579** 277–305

Murshudov G N, Skubák P, Lebedev A A, Pannu N S, Steiner R A, Nicholls R A, Winn M D, Long F and Vagin A A 2011 REFMAC5 for the refinement of macromolecular crystal structures *Acta Crystallogr.* D **67** 355–67

Nakane T *et al* 2020 Single-particle cryo-EM at atomic resolution *Nature* **587** 152–6

Nicholls R A, Tykac M, Kovalevskiy O and Murshudov G N 2018 Current approaches for the fitting and refinement of atomic models into cryo-EM maps using CCP-EM *Acta Crystallogr.* D **74** 492–505

Oldfield T J 2001 A number of real-space torsion-angle refinement techniques for proteins, nucleic acids, ligands and solvent *Acta Crystallogr.* D **57** 82–94

Orzechowski M and Tama F 2008 Flexible fitting of high-resolution x-ray structures into cryoelectron microscopy maps using biased molecular dynamics simulations *Biophys. J.* **95** 5692–705

Pandurangan A P, Shakeel S, Butcher S J and Topf M 2014 Combined approaches to flexible fitting and assessment in virus capsids undergoing conformational change *J. Struct. Biol.* **185** 427–39

Pandurangan A P and Topf M 2012 Finding rigid bodies in protein structures: application to flexible fitting into cryoEM maps *J. Struct. Biol.* **177** 520–31

Pandurangan A P, Vasishtan D, Alber F and Topf M 2015 γ-TEMPy: simultaneous fitting of components in 3D-EM maps of their assembly using a genetic algorithm *Structure* **23** 2365–76

Passos D O *et al* 2020 Structural basis for strand-transfer inhibitor binding to HIV intasomes *Science* **367** 810–4

Peng J and Xu J 2011 RaptorX: exploiting structure information for protein alignment by statistical inference *Proteins* **79** 161–71

Pettersen E F, Goddard T D, Huang C C, Couch G S, Greenblatt D M, Meng E C and Ferrin T E 2004 UCSF Chimera—a visualization system for exploratory research and analysis *J. Comput. Chem.* **25** 1605–12

Phillips J C *et al* 2020 Scalable molecular dynamics on CPU and GPU architectures with NAMD *J. Chem. Phys.* **153** 044130

Pintilie G, Chen D-H, Haase-Pettingell C A, King J A and Chiu W 2016 Resolution and probabilistic models of components in CryoEM maps of mature P22 bacteriophage *Biophys. J.* **110** 827–39

Pintilie G and Chiu W 2012 Comparison of Segger and other methods for segmentation and rigid-body docking of molecular components in cryo-EM density maps *Biopolymers* **97** 742–60

Pintilie G and Chiu W 2018 Assessment of structural features in cryo-EM density maps using SSE and side chain *Z*-scores *J. Struct. Biol.* **204** 564–71

Pintilie G, Zhang J, Chiu W and Gossard D 2009 Identifying components in 3D density maps of protein nanomachines by multi-scale segmentation *IEEENIH Life Sci. Syst. Appl. Workshop* **2009** 44–7

Pintilie G, Zhang K, Su Z, Li S, Schmid M F and Chiu W 2020 Measurement of atom resolvability in cryo-EM maps with *Q*-scores *Nat. Methods* **17** 328–34

Pintilie G D, Zhang J, Goddard T D, Chiu W and Gossard D C 2010 Quantitative analysis of cryo-EM density map segmentation by watershed and scale-space filtering, and fitting of structures by alignment to regions *J. Struct. Biol.* **170** 427–38

Prisant M G, Williams C J, Chen V B, Richardson J S and Richardson D C 2020 New tools in MolProbity validation: CaBLAM for CryoEM backbone, UnDowser to rethink 'waters,' and NGL Viewer to recapture online 3D graphics *Protein Sci.* **29** 315–29

Ramachandran G N, Ramakrishnan C and Sasisekharan V 1963 Stereochemistry of polypeptide chain configurations *J. Mol. Biol.* **7** 95–9

Ramírez-Aportela E *et al* 2020a FSC-Q: a cryoEM map-to-atomic model quality validation based on the local Fourier shell correlation *Nat. Commun.* **12** 42

Ramírez-Aportela E *et al* 2020b Automatic local resolution-based sharpening of cryo-EM maps *Bioinformatics* **36** 765–72

Ramlaul K, Palmer C M and Aylett C H S 2019 A local agreement filtering algorithm for transmission EM reconstructions *J. Struct. Biol.* **205** 30–40

Read R J *et al* 2011 A new generation of crystallographic validation tools for the protein data bank *Structure* **19** 1395–412

Richardson J S *et al* 2008 RNA backbone: consensus all-angle conformers and modular string nomenclature (an RNA Ontology Consortium contribution) *RNA* **14** 465–81

Richardson J S, Williams C J, Hintze B J, Chen V B, Prisant M G, Videau L L and Richardson D C 2018a Model validation: local diagnosis, correction and when to quit *Acta Crystallogr.* D **74** 132–42

Richardson J S, Williams C J, Videau L L, Chen V B and Richardson D C 2018b Assessment of detailed conformations suggests strategies for improving cryoEM models: helix at lower resolution, ensembles, pre-refinement fixups, and validation at multi-residue length scale *J. Struct. Biol.* **204** 301–12

Roh S-H, Hryc C F, Jeong H-H, Fei X, Jakana J, Lorimer G H and Chiu W 2017 Subunit conformational variation within individual GroEL oligomers resolved by cryo-EM *Proc. Natl Acad. Sci. USA* **114** 8259–64

Roseman A M 2000 Docking structures of domains into maps from cryo-electron microscopy using local correlation *Acta Crystallogr.* D **56** 1332–40
Rosenthal P B and Henderson R 2003 Optimal determination of particle orientation, absolute hand, and contrast loss in single-particle electron cryomicroscopy *J. Mol. Biol.* **333** 721–45
Rossmann M G 2000 Fitting atomic models into electron-microscopy maps *Acta Crystallogr.* D **56** 1341–9
Roy A, Kucukural A and Zhang Y 2010 I-TASSER: a unified platform for automated protein structure and function prediction *Nat. Protoc.* **5** 725–38
Sachse C, Chen J Z, Coureux P-D, Stroupe M E, Fändrich M and Grigorieff N 2007 High-resolution electron microscopy of helical specimens: a fresh look at tobacco mosaic virus *J. Mol. Biol.* **371** 812–35
Sali A and Blundell T L 1993 Comparative protein modelling by satisfaction of spatial restraints *J. Mol. Biol.* **234** 779–815
Saltzberg D, Greenberg C H, Viswanath S, Chemmama I, Webb B, Pellarin R, Echeverria I and Sali A 2019 Modeling biological complexes using integrative modeling platform *Methods Mol. Biol.* **2022** 353–77
Scheres S H W 2012 A Bayesian view on cryo-EM structure determination *J. Mol. Biol.* **415** 406–18
Schrödinger L L C 2015 The PyMOL Molecular Graphics System, Version 1.8 https://www.schrodinger.com/products/pymol
Senior A W *et al* 2020 Improved protein structure prediction using potentials from deep learning *Nature* **577** 706–10
Shapovalov M V and Dunbrack R L 2011 A smoothed backbone-dependent rotamer library for proteins derived from adaptive kernel density estimates and regressions *Structure* **19** 844–58
Shen M and Sali A 2006 Statistical potential for assessment and prediction of protein structures *Protein Sci.* **15** 2507–24
Siebert X and Navaza J 2009 UROX 2.0: an interactive tool for fitting atomic models into electron-microscopy reconstructions *Acta Crystallogr.* D **65** 651–8
Stagg S M, Lander G C, Pulokas J, Fellmann D, Cheng A, Quispe J D, Mallick S P, Avila R M, Carragher B and Potter C S 2006 Automated cryoEM data acquisition and analysis of 284742 particles of GroEL *J. Struct. Biol.* **155** 470–81
Suhre K, Navaza J and Sanejouand Y H 2006 NORMA: a tool for flexible fitting of high-resolution protein structures into low-resolution electron-microscopy-derived density maps *Acta Crystallogr.* D **62** 1098–100
Suhre K and Sanejouand Y-H 2004 ElNémo: a normal mode web server for protein movement analysis and the generation of templates for molecular replacement *Nucleic Acids Res.* **32** W610–4
Tama F, Miyashita O and Brooks C L 2004a Normal mode based flexible fitting of high-resolution structure into low-resolution experimental data from cryo-EM *J. Struct. Biol.* **147** 315–26
Tama F, Miyashita O and Brooks C L 2004b Flexible multi-scale fitting of atomic structures into low-resolution electron density maps with elastic network normal mode analysis *J. Mol. Biol.* **337** 985–99
Tan Y Z *et al* 2017 Addressing preferred specimen orientation in single-particle cryo-EM through tilting *Nat. Methods* **14** 793–6
Tan Y Z, Cheng A, Potter C S and Carragher B 2016 Automated data collection in single particle electron microscopy *Microscopy* **65** 43–56

Tang G, Peng L, Baldwin P R, Mann D S, Jiang W, Rees I and Ludtke S J 2007 EMAN2: an extensible image processing suite for electron microscopy *J. Struct. Biol.* **157** 38–46

Terashi G and Kihara D 2018 *De novo* main-chain modeling with MAINMAST in 2015/2016 EM model challenge *J. Struct. Biol.* **204** 351–9

Terwilliger T C 2003 Automated main-chain model building by template matching and iterative fragment extension *Acta Crystallogr.* D **59** 38–44

Terwilliger T C 2010 Rapid model building of alpha-helices in electron-density maps *Acta Crystallogr.* D **66** 268–75

Terwilliger T C, Adams P D, Afonine P V and Sobolev O V 2018a Map segmentation automated model-building and their application to the cryo-EM model challenge *J. Struct. Biol.* **204** 338–43

Terwilliger T C, Adams P D, Afonine P V and Sobolev O V 2018b A fully automatic method yielding initial models from high-resolution cryo-electron microscopy maps *Nat. Methods* **15** 905–8

Terwilliger T C, Adams P D, Afonine P V and Sobolev O V 2020c Cryo-EM map interpretation and protein model-building using iterative map segmentation *Protein Sci.* **29** 87–99

Terwilliger T C, Ludtke S J, Read R J, Adams P D and Afonine P V 2020b Improvement of cryo-EM maps by density modification *Nat. Methods* **17** 923–7

Terwilliger T C, Read R J, Adams P D, Brunger A T, Afonine P V and Hung L-W 2013 Model morphing and sequence assignment after molecular replacement *Acta Crystallogr.* D **69** 2244–50

Terwilliger T C, Sobolev O V, Afonine P V and Adams P D 2018c Automated map sharpening by maximization of detail and connectivity *Acta Crystallogr.* D **74** 545–59

Terwilliger T C, Sobolev O V, Afonine P V, Adams P D and Read R J 2020a Density modification of cryo-EM maps *Acta Crystallogr. D* **76** 912–25

Topf M, Baker M L, John B, Chiu W and Sali A 2005 Structural characterization of components of protein assemblies by comparative modeling and electron cryo-microscopy *J. Struct. Biol.* **149** 191–203

Topf M, Baker M L, Marti-Renom M A, Chiu W and Sali A 2006 Refinement of protein structures by iterative comparative modeling and CryoEM density fitting *J. Mol. Biol.* **357** 1655–68

Topf M, Lasker K, Webb B, Wolfson H, Chiu W and Sali A 2008 Protein structure fitting and refinement guided by cryo-EM density *Structure* **16** 295–307

Topf M and Sali A 2005 Combining electron microscopy and comparative protein structure modeling *Curr. Opin. Struct. Biol.* **15** 578–85

Trabuco L G, Villa E, Mitra K, Frank J and Schulten K 2008 Flexible fitting of atomic structures into electron microscopy maps using molecular dynamics *Structure* **16** 673–83

Trueblood K N, Bürgi H-B, Burzlaff H, Dunitz J D, Gramaccioli C M, Schulz H H, Shmueli U and Abrahams S C 1996 Atomic displacement parameter nomenclature. Report of a subcommittee on atomic displacement parameter nomenclature *Acta Crystallogr.* A **52** 770–81

Uziela K, Menéndez Hurtado D, Shu N, Wallner B and Elofsson A 2017 ProQ3D: improved model quality assessments using deep learning *Bioinformatics* **33** 1578–80

Vagin A A, Steiner R A, Lebedev A A, Potterton L, McNicholas S, Long F and Murshudov G N 2004 REFMAC5 dictionary: organization of prior chemical knowledge and guidelines for its use *Acta Crystallogr.* D **60** 2184–95 van Zundert G C P, Melquiond A S J and Bonvin A M J J 2015 Integrative modeling of biomolecular complexes: HADDOCKing with cryo-electron microscopy data *Structure* **23** 949–60

Velankar S *et al* 2016 PDBe: improved accessibility of macromolecular structure data from PDB and EMDB *Nucleic Acids Res.* **44** D385–95

Velázquez-Muriel J A, Sorzano C O S, Scheres S H W and Carazo J-M 2004 SPI-EM: Towards a tool for predicting CATH superfamilies in 3D-EM maps *J. Mol. Biol.* **345** 759–71

Vilas J L *et al* 2018 MonoRes: automatic and accurate estimation of local resolution for electron microscopy maps *Structure* **26** 337–44.e4

Volkmann N 2002 A novel three-dimensional variant of the watershed transform for segmentation of electron density maps *J. Struct. Biol.* **138** 123–9

Wang R Y-R, Song Y, Barad B A, Cheng Y, Fraser J S and DiMaio F 2016 Automated structure refinement of macromolecular assemblies from cryo-EM maps using Rosetta *eLife* **5** e17219

Wang Y, Shekhar M, Thifault D, Williams C J, McGreevy R, Richardson J, Singharoy A and Tajkhorshid E 2018 Constructing atomic structural models into cryo-EM densities using molecular dynamics—pros and cons *J. Struct. Biol.* **204** 319–28

Wang Z *et al* 2014 An atomic model of brome mosaic virus using direct electron detection and real-space optimization *Nat. Commun.* **5** 4808

Wang Z and Schröder G F 2012 Real-space refinement with DireX: from global fitting to side-chain improvements *Biopolymers* **97** 687–97

Webb B and Sali A 2016 Comparative protein structure modeling using MODELLER *Curr. Protoc. Bioinforma* **54** 5–6

Weis F and Hagen W J H 2020 Combining high throughput and high quality for cryo-electron microscopy data collection *Acta Crystallogr.* D **76** 724–8

Wilkinson M D *et al* 2016 The FAIR guiding principles for scientific data management and stewardship *Sci. Data* **3** 160018

Williams C J *et al* 2018a MolProbity: more and better reference data for improved all-atom structure validation *Protein Sci.* **27** 293–315

Williams C J, Videau L L, Hintze B J, Richardson D C and Richardson J S 2018b Cis-nonPro peptides: genuine occurrences and their functional roles *BioRxiv* 324517

Winn M D *et al* 2011 Overview of the CCP4 suite and current developments *Acta Crystallogr.* D **67** 235–42

Wood C, Burnley T, Patwardhan A, Scheres S, Topf M, Roseman A and Winn M 2015 Collaborative computational project for electron cryo-microscopy *Acta Crystallogr.* D **71** 123–6

Word J M, Lovell S C, LaBean T H, Taylor H C, Zalis M E, Presley B K, Richardson J S and Richardson D C 1999b Visualizing and quantifying molecular goodness-of-fit: small-probe contact dots with explicit hydrogen atoms *J. Mol. Biol.* **285** 1711–33

Word J M, Lovell S C, Richardson J S and Richardson D C 1999a Asparagine and glutamine: using hydrogen atom contacts in the choice of side-chain amide orientation *J. Mol. Biol.* **285** 1735–47

Wriggers W 2010 Using Situs for the integration of multi-resolution structures *Biophys. Rev.* **2** 21–7

wwPDB Consortium 2019 Protein Data Bank: the single global archive for 3D macromolecular structure data *Nucleic Acids Res.* **47** D520–8

Yi H *et al* 2015 Native immunogold labeling of cell surface proteins and viral glycoproteins for cryo-electron microscopy and cryo-electron tomography applications *J. Histochem. Cytochem.* **63** 780–92

Yip K M, Fischer N, Paknia E, Chari A and Stark H 2020 Breaking the next cryo-EM resolution barrier—atomic resolution determination of proteins! *bioRxiv* 2020.05.21.106740

Young J Y *et al* 2017 OneDep: unified WwPDB system for deposition, biocuration, and validation of macromolecular structures in the PDB archive *Structure* **25** 536–45

Yu I, Nguyen L, Avaylon J, Wang K, Lai M and Zhou Z H 2018 Building atomic models based on near atomic resolution cryoEM maps with existing tools *J. Struct. Biol.* **204** 313–8

Yu X, Jin L and Zhou Z H 2008 3.88 Å structure of cytoplasmic polyhedrosis virus by cryo-electron microscopy *Nature* **453** 415–9

Zhang J *et al* 2010a Mechanism of folding chamber closure in a group II chaperonin *Nature* **463** 379–83

Zhang K *et al* 2020a Inhibition mechanisms of AcrF9, AcrF8, and AcrF6 against type I-F CRISPR–Cas complex revealed by cryo-EM *Proc. Natl Acad. Sci. USA* **117** 7176–82

Zhang K, Pintilie G D, Li S, Schmid M F and Chiu W 2020b Resolving individual-atom of protein complex using commonly available 300-kV cryo-electron microscopes *Cell Res.* **30** 1136–9

Zhang X, Settembre E, Xu C, Dormitzer P R, Bellamy R, Harrison S C and Grigorieff N 2008 Near-atomic resolution using electron cryomicroscopy and single-particle reconstruction *Proc. Natl Acad. Sci. USA* **105** 1867–72

Zhang X, Zhang X and Zhou Z H 2010b Low cost, high performance GPU computing solution for atomic resolution cryoEM single-particle reconstruction *J. Struct. Biol.* **172** 400–6

Zhou Z H 2008 Towards atomic resolution structural determination by single-particle cryo-electron microscopy *Curr. Opin. Struct. Biol.* **18** 218–28

Zhou Z H, Baker M L, Jiang W, Dougherty M, Jakana J, Dong G, Lu G and Chiu W 2001 Electron cryomicroscopy and bioinformatics suggest protein fold models for rice dwarf virus *Nat. Struct. Biol.* **8** 868–73

Zhou Z H, Dougherty M, Jakana J, He J, Rixon F J and Chiu W 2000 Seeing the herpesvirus capsid at 8.5 Å *Science* **288** 877–80

Zivanov J, Nakane T and Scheres S H W 2020 Estimation of high-order aberrations and anisotropic magnification from cryo-EM data sets in RELION-3.1 *IUCrJ* **7** 253–67

www.ingramcontent.com/pod-product-compliance
Ingram Content Group UK Ltd.
Pitfield, Milton Keynes, MK11 3LW, UK
UKHW051902100926
8556IPUK00005B/738

9 780750 330374